principles of underwater sound

Robert J. Urick
Adjunct Professor, The Catholic University
of America, Washington, D.C.
Formerly Senior Research Physicist
U.S. Naval Surface Weapons Center Headquarters
White Oak, Silver Spring, Maryland

principles of underwater sound / 3d edition

First edition published in 1967 under the title "Principles of Underwater Sound for Engineers"

McGraw-Hill Book Company
New York St. Louis San Francisco Auckland
Bogotá Hamburg Johannesburg London
Madrid Mexico Montreal New Delhi
Panama Paris São Paulo Singapore
Sydney Tokyo Toronto

Library of Congress Cataloging in Publication Data

Urick, Robert J.
  Principles of underwater sound.

  Includes bibliographies and index.
  1. Underwater acoustics. 2. Sonar. I. Title.
QC242.2.U74   1983   534'.23   82-14059
ISBN 0-07-066087-5

Copyright © 1983 by McGraw-Hill, Inc. All rights reserved.
Printed in the United States of America. Except as permitted
under the Copyright Act of 1976, no part of this publication
may be reproduced or distributed in any form or by any means,
or stored in a data base or retrieval system, without the
prior written permission of the publisher.

  34567890   KGP/KGP   8987654

ISBN 0-07-066087-5

The editors for this book were Diane Heiberg and Janet Davis,
the designer was Naomi Auerbach, and the production supervisor
was Paul Malchow. It was set in Baskerville by Bi-Comp, Inc.
Printed and bound by The Kingsport Press.

contents

Preface to the Third Edition ix
Preface the First Edition xi
Abbreviations Used in the Book xiii

one The Nature of Sonar 1

- 1.1 *Historical Survey* 2
- 1.2 *Postwar Developments* 6
- 1.3 *Nonmilitary Uses of Underwater Sound* 7
- 1.4 *Military Uses of Underwater Sound* 7
- 1.5 *Some Basic Concepts* 11
- 1.6 *The Micropascal Reference Unit and the Decibel* 14
- *References* 16

two The Sonar Equations 17

- 2.1 *Basic Considerations* 18
- 2.2 *The Active and Passive Equations* 19
- 2.3 *Names for Various Combinations of Parameters* 22
- 2.4 *The Parameters in Metric Units* 23
- 2.5 *Echo, Noise, and Reverberation Level as Functions of Range* 23
- 2.6 *Transient Form of the Sonar Equations* 25
- 2.7 *Statement of the Equations* 28
- 2.8 *Limitations of the Sonar Equations* 29
- *References* 30

three Properties of Transducer Arrays: Directivity Index 31

- 3.1 *Array Gain* 33
- 3.2 *Measuring the Coherence of Sound Fields* 41
- 3.3 *Receiving Directivity Index* 42
- 3.4 *Limitations of Directivity Index* 42
- 3.5 *Transducer Responses* 44
- 3.6 *Calibration Methods* 44

- 3.7 Reciprocity Calibration 51
- 3.8 Calibration of Large Arrays 53
- 3.9 Beam Patterns 54
- 3.10 The Product Theorem and the Mills Cross 60
- 3.11 Shading and Superdirectivity 62
- 3.12 Adaptive Beam Forming 64
- 3.13 Multiplicative Arrays 65
  - References 68

four Generation of Underwater Sound: Projector Source Level 71

- 4.1 Relation between Source Level and Radiated Acoustic Power 73
- 4.2 Limitations on Sonar Power 76
- 4.3 Nonlinear Effects in Sonar 81
- 4.4 Explosions as Sources of Underwater Sound 86
  - References 97

five Propagation of Sound in the Sea: Transmission Loss, I 99

- 5.1 Introduction 99
- 5.2 Spreading Laws 100
- 5.3 Absorption of Sound in the Sea 102
- 5.4 Velocity of Sound of the Sea 111
- 5.5 Velocity Structure of the Sea 116
- 5.6 Propagation Theory and Ray Tracing 120
- 5.7 The Sea Surface 128
- 5.8 The Sea Bottom 136
  - References 143

six Propagation of Sound in the Sea: Transmission Loss, II 147

- 6.1 The Mixed-Layer Sound Channel 147
- 6.2 The Deep Sound Channel 159
- 6.3 Caustics and Convergence Zones 164
- 6.4 Internal Sound Channels 169
- 6.5 Artic Propagation 169
- 6.6 The Shallow-Water Channel 172
- 6.7 Fluctuation of Transmitted Sound 182
- 6.8 Horizontal Variability 190
- 6.9 Coherence of Transmitted Sound 193
- 6.10 Multipaths in the Sea 193
- 6.11 Deep-Sea Paths and Losses: A Summary 194
  - References 197

seven The Noise Background of the Sea: Ambient-Noise Level 202

- 7.1 Sources of Ambient Noise in Deep Water 203
- 7.2 Deep-Water Spectra 209
- 7.3 Shallow-Water Ambient Noise 211

7.4 Variability of Ambient Noise 215
7.5 Intermittent Sources of Ambient Noise 216
7.6 Effect of Depth 221
7.7 Amplitude Distribution 223
7.8 Noise in Ice-Covered Waters 224
7.9 Directional Characteristics of Deep-Water Ambient Noise 227
7.10 Spatial Coherence of Ambient Noise 231
7.11 Summary 233
      References 233

eight Scattering in the Sea: Reverberation Level 237

8.1 Types of Reverberation 237
8.2 The Scattering-Strength Parameter 238
8.3 Equivalent Plane-Wave Reverberation Level 240
8.4 Volume-Reverberation Theory 240
8.5 Surface-Reverberation Theory 244
8.6 Target Strength and Scattering Strength 246
8.7 Surface Scattering by a Layer of Volume Scatterers 247
8.8 Reverberation Level for Short Transients 248
8.9 Air Bubbles in Water 249
8.10 Volume Reverberation: The Deep Scattering Layer 255
8.11 Sea-Surface Reverberation 262
8.12 Theories and Causes of Sea-Surface Scattering 267
8.13 Sea-Bottom Reverberation 271
8.14 Under-Ice Reverberation 280
8.15 Reverberation in Shallow Water 281
8.16 Characteristics of Reverberation 281
8.17 Reverberation Prediction 285
      References 287

nine Reflection and Scattering by Sonar Targets: Target Strength 291

9.1 The Echo as the Sum of Backscattered Contributions 294
9.2 Geometry of Specular Reflection 295
9.3 Target Strength of a Small Sphere 298
9.4 Complications for a Large Smooth Solid Sphere 301
9.5 Target Strength of Simple Forms 302
9.6 Bistatic Target Strength 302
9.7 Target-Strength Measurement Methods 306
9.8 Target Strength of Submarines 308
9.9 Target Strength of Surface Ships 314
9.10 Target Strength of Mines 314
9.11 Target Strength of Torpedoes 315
9.12 Target Strength of Fish 315
9.13 Target Strength of Small Organisms 317
9.14 Echo Formation Processes 317
9.15 Reduction of Target Strength 320
9.16 Echo Characteristics 322
9.17 Summary of Numerical Values 325
      References 325

ten Radiated Noise of Ships, Submarines, and Torpedoes: Radiated-Noise Levels 328

- 10.1 Source Level and Noise Spectra 328
- 10.2 Methods of Measurement 329
- 10.3 Sources of Radiated Noise 332
- 10.4 Summary of the Sources of Radiated Noise 341
- 10.5 Total Radiated Acoustic Power 343
- 10.6 Radiated-Noise Levels 345
- 10.7 Cautionary Remark 351
  References 352

eleven Self-Noise of Ships, Submarines, and Torpedoes: Self-Noise Levels 354

- 11.1 Self-Noise Measurements and Reduction 356
- 11.2 Sources and Paths of Self-Noise 357
- 11.3 Flow Noise 360
- 11.4 Flow-Noise Reduction 365
- 11.5 Domes 367
- 11.6 Self-Noise of Cable-Suspended and Bottomed Hydrophones 370
- 11.7 Self-Noise of Towed Sonars 371
- 11.8 Self-Noise Levels 372
  References 376

twelve Detection of Signals in Noise and Reverberation: Detection Threshold 377

- 12.1 Definition of Detection Threshold 378
- 12.2 The Threshold Concept 380
- 12.3 Input Signal-to-Noise Ratio for Detection 383
- 12.4 Modifications to the ROC Curves 386
- 12.5 Estimating Detection Threshold 390
- 12.6 Effect of Duration and Bandwith 392
- 12.7 Example of a Computation 393
- 12.8 Detection Threshold for Reverberation 395
- 12.9 Tabular Summary 396
- 12.10 Auditory Detection 396
  References 403

thirteen Design and Prediction in Sonar Systems 406

- 13.1 Sonar Design 406
- 13.2 Sonar Prediction 407
- 13.3 The Optimum Sonar Frequency 408
- 13.4 Applications of the Sonar Equations 411
- 13.5 Concluding Remarks 416
  References 416

Index 417

preface to the third edition

The purpose of the third edition, like that of the second, is to present new ideas, facts, and concepts that have emerged in recent years. At the same time some older and useful material has been added as deemed appropriate by the use of the book in course teaching.

The question of metric units has been given much consideration. Although scientists and engineers today use metric units universally, operations analysts and the users of sonar equipments do not, preferring to cling to the yard instead of the meter as the unit of distance—not because of the natural human reluctance to change, but because the range scales and manuals of existing equipments are calibrated in yards and nautical miles. A curious benefit of the yard in practical work is that the nautical mile can be taken with little error to be an even 2 kiloyards (actually, it is 2.025), but it is an unhandy, uneven number of kilometers (1.853). While the yard has been retained as the unit of distance, a section on the metric conversion of the sonar parameters has been included.

The basic plan of the book has remained the same: to introduce the sonar equations after an introductory chapter and then, at the end, to illustrate their use in practical sonar problem solving after the many diverse and perplexing phenomena hidden within them have been explained in the intervening chapters.

Robert J. Urick

preface to the first edition

Underwater sound, as a specialized branch of science and technology, has seen service in two world wars. Although it has roots deep in the past, underwater sound, as a quantitative subject, may be said to be only a quarter of a century old. Its modern era began with the precise quantitative studies undertaken with great vigor during the days of World War II. In subsequent years, its literature has grown to sizable proportions, and its practical uses have expanded in keeping with man's continuing exploration and exploitation of the seas.

This book attempts to summarize the principles of underwater sound from the viewpoint of the engineer and the practical scientist. It lies squarely in the middle of the spectrum—between theory at one end and sonar technology at the other. Its intent is to provide a summary of the principles, effects, and phenomena of underwater sound and to give numerical quantitative data, wherever possible, for the solution of practical problems.

The framework of the book is the sonar equations—the handy set of relationships that tie together all the essential elements of underwater sound. The approach is, after an introductory chapter, to state the equations in a convenient form and then to discuss in subsequent chapters each one of the quantities occurring in the equations. The final chapter is largely devoted to problem solving, in which the use of the equations is illustrated by hypothetical problems taken from some of many practical applications of the subject.

In the desire to keep the book within sizable bounds, some aspects of underwater sound have had to be slighted. One is the subject of transducers—the conversion of electricity into sound and vice versa. Although transducer arrays are discussed, only one sound source—the underwater explosion—is dealt with at any length. It is felt that the design of electroacoustic transducers for generating and receiving sound is truly an art in itself, with a technology and theoretical background that deserves a book of its own. In addition, much of the basic theory of underwater sound is confined to references to the literature, and engineering matters of sonar hardware are omitted entirely. Although the book will therefore appeal neither to the theoretician nor to the hardware builder, it hopes to cover the vast middle ground between them and be of interest to both the design engineer and the practical physicist. It is based on a course given for several years at the Catholic University, Washington, D.C., and at Westinghouse Electric Co., and the Martin Co. in Baltimore.

The book owes a great deal to my colleagues at the Naval Ordnance Laboratory for many discussions and helpful criticism. In particular, Mr. T. F. Johnston, chief of the acoustics division, has been a constant encouragement and stimulant in the long and arduous task of writing the book. My students have been helpful too in providing me with a receptive and critical audience for whatever is original in the presentation.

Robert J. Urick

abbreviations used in the book

Abbreviation	Unit
A	ampere
dB	decibel
dyn	dyne
fm	fathom
ft	foot
ft/s	foot per second
g	gram
Hz	hertz
h	hour
kHz	kilohertz
kyd	kiloyard
kn	knots
μPa	micropascal
μs	microsecond
mi	mile
ms	millisecond
nmi	nautical mile
s	second
V	volt
W	watt
W·s/cm^2	watt-second per square centimeter
yd	yard

*principles of
underwater
sound*

one

the nature of sonar

Of all the forms of radiation known, sound travels through the sea the best. In the turbid, saline water of the sea, both light and radio waves are attenuated to a far greater degree than is that form of mechanical energy known as sound.

Because of its relative ease of propagation, people have applied underwater sound to a variety of purposes in their use and exploration of the seas. These uses of underwater sound constitute the engineering science of *sonar*, and the systems employing underwater sound in one way or another are *sonar systems*.

Sonar systems, equipments, and devices are said to be *active* when sound is purposely generated by one of the system components called the *projector*. The sound waves generated by the projector travel through the sea to the *target*, and are returned as sonar *echoes* to a *hydrophone*, which converts sound into electricity. The electric output of the hydrophone is amplified and processed in various ways and is finally applied to a control or display device to accomplish the purpose for which the sonar set was intended. Active sonar systems are said to *echo-range* on their targets.

Passive or *listening* sonar systems use sound radiated (usually unwittingly) by the target. Here only *one-way* transmission through the sea is involved, and the system centers around the hydrophone used to listen to the target sounds. Communication, telemetry, and control applications employ a hybrid form of sonar system using a projector and hydrophone at *both* ends of the acoustic communication path.

1.1 Historical Survey

Although the "modern age" of sonar may be said to date back more than a quarter of a century to the start of World War II, sonar has its origins deep in the past. One of the earliest references to the fact that sound exists beneath the surface of the sea, as well as in the air above, occurs in one of the notebooks of that versatile, archetypal engineer, Leonardo da Vinci. In 1490, two years before Columbus discovered America, he wrote (1)*: "If you cause your ship to stop, and place the head of a long tube in the water and place the outer extremity to your ear, you will hear ships at a great distance from you." Although this earliest example of a passive sonar system has the enviable merit of extreme simplicity, it does not provide any indication of direction and is insensitive as a result of the great mismatch between the acoustic properties of air and water. Yet the idea of listening to underwater sounds by means of an air-filled tube between the sea and the listener's ear had widespread use as late as World War I, when, by the addition of a second tube between the other ear and a point in the sea separated from the first point, a direction could be obtained and the bearing of the target could be determined.

Perhaps the first quantitative measurement in underwater and sound occurred in 1827, when a Swiss physicist, Daniel Colladon, and a French mathematician, Charles Sturm, collaborated to measure the velocity of sound in Lake Geneva in Switzerland. By timing the interval between a flash of light and the striking of a bell underwater, they determined the velocity of sound to a surprising degree of accuracy.

Later on in the nineteenth century, a number of famous physicists of the time indirectly associated themselves with underwater sound through their interest in the phenomenon of "transduction"—the conversion of electricity into sound and vice versa (2). Jacques and Pierre Curie are usually credited with the discovery in 1880 of piezoelectricity—the ability of certain crystals, when stressed, to develop an electric charge across certain pairs of crystal faces. Other physicists had dabbled in the subject before this. Charles Coulomb is said to have speculated on the possibility of producing electricity by pressure, and Wilhelm Röntgen wrote a paper on the electric charge appearing on the various faces of crystals under stress. The counterpart of piezoelectricity as a transduction process is magnetostriction, wherein a magnetic field produces a change in the shape of certain substances. The earliest manifestation of magnetostriction was the musical sounds that were heard when, about 1840, the current in a coil of wire was changed or interrupted near the poles of a horseshoe magnet. James Joule, in the 1840s, carried out quantitative measurements on the change of length associated with magnetostriction, and is commonly credited with being the discoverer of the effect.

---

* The parenthetical numbers throughout the text denote numbered references to the literature in a list of references at the end of each chapter.

These studies, and those of others in the 1840s and 1850s, were the foundation for the invention of the telephone, for which a long-disputed patent was issued in 1876 to A. G. Bell. Another nineteenth-century invention that was the mainstay of sonar systems before the advent of electronic amplifiers was the carbon-button microphone, a device which became the earliest, and still probably the most sensitive, hydrophone for underwater sound.

About the turn of the century there came into being the first practical application of underwater sound. This was the submarine bell, used by ships for offshore navigation. By timing the interval between the sound of the bell and the sound of a simultaneously sent blast of a foghorn, a ship could determine its distance from the lightship where both were installed. This system was the impetus for the founding of the Submarine Signal Company (now part of Raytheon Mfg. Co.), the first commercial manufacturer of sonar equipment in the United States. The method was never in widespread use and was soon replaced by navigation methods involving radio—especially radio direction finding.

Another pre–World War I achievement was the embryonic emergence of the first schemes for the detection of underwater objects by echo ranging. In 1912, five days after the "Titanic" collided with an iceberg, L. F. Richardson filed a patent applicaiton with the British Patent Office for echo ranging with airborne sound (2). A month later he applied for a patent for its underwater analog. These ideas involved the then-new features of a directional projector of kilohertz-frequency sound waves and a frequency-selective receiver detuned from the transmitting frequency to compensate for the doppler shift caused by the motion of the echo-ranging vessel. Unfortunately, Richardson did nothing at the time to implement these proposals. Meanwhile, in the United States, R. A. Fessenden had designed and built a new kind of moving-coil transducer for both submarine signaling and echo ranging and was able, by 1914, to detect an iceberg at a distance of 2 miles. Fessenden "oscillators" operating at frequencies near 500 and near 1,000 Hz are said (3) to have been installed on all United States submarines of the World War I period to enable them to signal one another while submerged. They remained in use until recently as powerful sinusoidal sound sources for research purposes.

The outbreak of World War I in 1914 was the impetus for the development of a number of military applications of sonar. In France a young Russian electrical engineer, Constantin Chilowsky, collaborated with a distinguished physicist, Paul Langevin, in experiments with a condenser (electrostatic) projector and a carbon-button microphone placed at the focus of a concave mirror. In spite of leakage and breakdown troubles caused by the high voltages needed for the projector, by 1916 they were able to obtain echoes from the bottom and from a sheet of armor plate at a distance of 200 meters. Later, in 1917, Langevin turned to the piezoelectric effect and use a quartz-steel sandwich to replace the condenser projector. He also employed one of the newly developed vacuum-tube amplifiers—probably the first application of electronics to underwater sound equipment. For the first time, in 1918, echoes were received from a submarine, occasionally at distances as great as 1,500 meters. Parallel British investigations with quartz projectors were conducted by a group under R. W. Boyle. The word "asdic" was coined at the time to refer to their then highly secret experiments.* World War I came to a close, however, before underwater echo ranging could make any contribution to meet the German U-boat threat.

In the meantime, extensive use had been made of Leonardo's air tube for passive listening, improved by the use of two tubes to take advantage of the binaural directional sense of a human observer. The MV device consisted (5) of a pair of line arrays of 12 air tubes each, mounted along the bottom of a ship on the port and starboard sides and steered with a special compensator. Surprising precision was achieved in determining the bearing of a noisy target; an untrained observer could find the bearing of a distant target to an accuracy of ½°. Another development of the time (5) was a neutrally buoyant, flexible line array of 12 hydrophones called the "eel," which could easily be fitted to any ship and could be towed astern away from the noisy vessel on which it was mounted. All in all, some three thousand escort craft were fitted with listening devices of various kinds in World War I. By operating in groups of two or three and using cross bearings, they could obtain a "fix" on a suspected submarine contact.

In 1919, soon after the end of World War I, the Germans published the first scientific paper on underwater sound (6). This paper described theoretically the bending of sound rays produced by slight temperature and salinity gradients in the sea, and recognized their importance in determining sound ranges. Longer ranges were predicted in winter than in summer, and the prediction was verified by measurements of transmission ranges in all seasons of the year in six shallow-water areas, including two off the east coast of the United States made presumably before the entry of this country into the war in 1917. This paper was far ahead of its time and remained unrecognized for over 60 years. It is testimony to the excellence of German physics in the early years of the present century.

The years of peace following World War I saw a steady, though extremely slow, advance in applying underwater sound to practical needs. Depth sounding by ships under way was soon developed, and by 1925, fathometers, a word coined by the Submarine Signal Company for their own equipment, were available commercially in both the United States and Great Britain. The search for a practical means of echo ranging on submarine targets was conducted in the United States by a handful of men under H. C. Hayes at the Naval Research Laboratory. The problem of finding a suitable sound projector in echo ranging was solved by resorting to magnetostrictive projectors for generating the required amount of acoustic power. Also, synthetic crystals of

---

* According to A. B. Wood (4), the word "asdic" was originally an acronym for "*anti-submarine division—ics*" from the name of the group which did the work. The suffix had the same significance as it does in the words "physics," "acoustics," etc. For many years thereafter, the word "asdic" was used by the British to refer to echo ranging and echo-ranging sonar systems generally.

Rochelle salt began to replace scarce natural quartz as the basic piezoelectric material for piezoelectric transducers. During the interwar period sonar received a great practical impetus from advances in electronics, which made possible vast new domains of amplifying, processing, and displaying sonar information to an observer.

Ultrasonic frequencies, that is, frequencies beyond the region of sensitivity of the unaided human ear, came to be used for both listening and echo ranging and enabled an increased directionality to be obtained with projectors and hydrophones of modest size. A number of small, but vital, components of sonar systems were added during this period, notably the development by the British of the range recorder for echo-ranging sonars to provide a "memory" of past events and the streamlined dome to protect the transducer on a moving ship from the noisy, turbulent environment of water flow past a moving vessel. By 1935, several fairly adequate sonar systems had been developed, and by 1938, with the imminence of World War II, quantity production of sonar sets started in the United States. By the time the war began, a large number of American ships were equipped for both underwater listening and echo ranging. The standard echo-ranging sonar set for surface ships was the QC equipment. The operator searched in bearing with it by turning a handwheel and listening for an echo with headphones or loudspeaker. If an echo was obtained, its range was noted by the flash of a rotating light or from the range recorder. Submarines were fitted with JP listening sets, consisting of a rotatable horizontal line hydrophone, an amplifier, a selectable bandpass filter, and a pair of headphones. The cost of this equipment with spares was $5,000! With such primitive sonar sets, the Battle of the Atlantic against the German U-boats was engaged and, eventually, won.

But from a scientific standpoint, perhaps the most notable accomplishment of the years between World War I and World War II was the obtaining of an understanding of the vagaries of sound propagation in the sea. Early shipboard echo-ranging sets installed in the late twenties and early thirties were mysteriously unreliable in performance. Good echoes were often obtained in the morning, but poor echoes, or none at all, were obtained in the afternoon. When it became clear that the sonar operators themselves were not to blame and that the echoes were actually weaker in the afternoon, the cause began to be sought in the transmission characteristics of the seawater medium. Only with the use of a special temperature-measuring equipment did it become evident that slight thermal gradients, hitherto unsuspected, were capable of refracting sound deep into the depths of the sea and could cause the target to lie in what is now known as a "shadow zone." The effect was called by E. B. Stephenson the "afternoon effect." As a means to indicate temperature gradients in the upper few hundred feet of the sea, A. F. Spilhaus built the first bathythermograph in 1937; by the start of World War II, every naval vessel engaged in antisubmarine work was equipped with the device. During this period also, a clear understanding was gained of absorption of sound in the sea, and remarkably accurate values of absorption coefficients were determined at the ultrasonic frequencies 20 to 30 kHz then of interest. These and other achievements of the interwar period are described in a paper by Klein (7).

On both sides of the Atlantic, as in World War I, the World War II period was marked by feverish activity in underwater sound. In the United States, a large group of scientists organized by the National Defense Research Committee (NDRC) began investigations of all phases of the subject.* Most of our present concepts as well as practical applications had their origins in this period. The acoustic homing torpedo, the modern acoustic mine, and scanning sonar sets were wartime developments. Methods for quick calibration of projectors and hydrophones began to be used, and an understanding of the many factors affecting sonar performance that are now summarized in the sonar equations was gained. Factors such as target strength, the noise output of various classes of ships at different speeds and frequencies, reverberation in the sea, and the recognition of underwater sound by the human ear were all first understood in a quantitative way during the years of World War II. Indeed, in retrospect, there is little of our fund of underwater acoustic knowledge that cannot be traced, in its rudiments, to the discoveries of the wartime period.

The Germans must be given credit for a number of unique accomplishments. One was the development of Alberich, a nonreflecting coating for submarines. It consisted of a perforated sheet of rubber cemented to the hull of a submarine and covered by a solid, thin sheet of rubber to keep water out of the air-filled perforations. This coating was effective only over a limited range of depth and frequency, and could not be kept bonded to the hull for long periods under operating conditions. Another innovation was the use of flush-mounted arrays for listening aboard surface ships. An array of this kind—given the designation GHG for *grüppen-hört-gerät* or "array listening equipment"—was installed on the cruiser "Prinz Eugen" and used with some success (8).

According to Batchelder (9), the word "sonar" was coined late in the war as a counterpart of the then-glamorous word "radar" and came into use later only after having been dignified as an acronym for *s*ound *n*avigation *a*nd *r*anging!

Underwater acoustics is now a mature branch of science and engineering, with a vast literature and a history of achievement that has only briefly been touched on in the foregoing. The historically minded reader may be referred to a number of papers on the history of sonar through the close of World War II (10–14).

1.2 Postwar Developments

The years since World War II have seen some remarkable advances in the exploitation of underwater sound, for both military and nonmilitary pur-

---

* At the end of the war the findings of that part of NDRC engaged in underwater sound were summarized in an admirable series of some 22 reports called the *NDRC Division 6 Summary Technical Reports.*

poses. On the military side, *active sonars* have grown larger and more powerful and operate at frequencies several octaves lower than in World War II. As a result, active sonar ranges are today far greater than they were during the hectic days of the war. Similarly, *passive sonars* have tended toward lower frequencies in order to take advantage of the tonal or line components in the low-frequency submarine noise spectrum.

Passive sonar arrays containing many hydrophones have been placed on the deep ocean floor to take advantage of the quiet environment and good propagation conditions existing at such low frequencies. At the same time, however, the submarine targets of passive sonars have become quieter, and have become far more difficult targets for detection than they once were. The developments of complex signal processing, in both time and space, that the emergence of digital computers and digital signal processors has made possible, have enabled much more information to be used for whatever function the sonar is called upon to perform. Finally, research in sound propagation in the sea has led to the exploitation of propagation paths not even dreamed of in earlier years; an example is the discovery of convergence zones in deep-sea propagation and their exploitation in recent years by sonars of sufficient figure of merit.

1.3 Nonmilitary Uses of Underwater Sound

A striking development of the postwar period, and one that is still taking place, is the expansion of the applicaitons of underwater sound to peaceful purposes. Originally employed only for depth sounding, underwater sound is now being used for the wide variety of purposes shown in Table 1.1. Sonar equipments for most of these applications are available commercially, and often a variety of models can be procured from different manufacturers. Indeed, spectrum crowding is beginning to occur, and some attention has been given (15) to frequency standardization, if not regulation in the manner of the Federal Communications Commission, in order to avoid interference problems in crowded coastal waters! One would expect that in future years more and more sonar sets of various kinds will accompany our expanding exploitation of the resources of the sea.

1.4 Military Uses of Underwater Sound

The useful spectrum of underwater sound for military applications covers some five or six decades of frequency. Beginning at the ultra-low-frequency end of the application spectrum, one finds the *pressure mine,* a device actuated by the reduction in pressure caused by the motion of a ship in its vicinity. This reduction of pressure is a hydrodynamic (Bernoulli) effect rather than a sound effect; it does not require an elastic medium for its existence and is not propagated to a distance. Yet in the mine it is likely to be sensed by a pressure transducer, to which it appears no different from an acoustic wave. The pressure disturbances that are produced by a ship under way contain

table 1.1 Nonmilitary Uses of Sonar

Function	Description
Depth sounding	
Conventional depth sounders	Send short pulses downward and time the bottom return
Subbottom profilers	Use lower frequencies and a high-power impulsive source for bottom penetration
Side-scan sonars	Sidewise-looking sonars for mapping the sea bed at right angles to a ship's track
Acoustic speedometers	Use pairs of transducers pointing obliquely downward to obtain speed over the bottom from the doppler shift of the bottom returns. Another method uses the time-delay correlogram of the bottom return between halves of a small split transducer (16)
Fish finding	Forward-looking active sonars for spotting fish schools
Fisheries aids	For counting, luring, or tagging individual fish
Divers' aids	Small hand-held sonar sets for underwater object location by divers
Position marking	
Beacons	Transmit a sound signal continuously
Transponders	Transmit only when suitably interrogated
Communication and telemetry	Use a sound beam instead of a wire link for transmitting information
Control	Sound-activated release mechanisms; well-head flow control devices for underwater oil wells
Miscellaneous uses	Acoustic flow meters and wave-height sensors

frequencies generally below 1 Hz. *Acoustic mines* sense the true acoustic radiation of ships and explode when the acoustic level in their passband reaches a certain value. In *minesweeping,* such mines are swept, or purposely exploded, by the sound of a powerful source towed behind the mine-sweeping vessel. Mines and mine sweeping commonly utilize frequencies of the spectrum where the acoustic output of vessels that are the targets for acoustic mines is at a maximum. The acoustic radiation of ships and submarines is also employed for *passive detection* by a hydrophone array on another vessel or on the bottom, a long distance away. Because the receiving array must be directional, so as to be able to determine the direction to the target, somewhat higher frequencies are employed in passive detection.

An example of a modern echo-ranging sonar is the large A/N-SQS-26 sonar for surface ships illustrated in Fig. 1.1. This sonar sends long, high-power pings in selectable directions both vertically and horizontally, and employs modern signal processing techniques for presenting the return to the observer. The system is remarkably complex, as its many components indicate, and provides great versatility in modes of operation. Another sonar for surface ships is the towed sonar A/N-SQS-35, a photograph of which is given in Fig. 1.2. In this sonar, the echo-ranging transducer is located inside the streamlined "fish" and is towed at a depth behind the surface vessel. By towing the transducer at depths as great as several hundred feet, shallow thermal gradients can be penetrated and, at the same time, the sonar is able to search in stern directions—something that a hull-mounted sonar is unable to do.

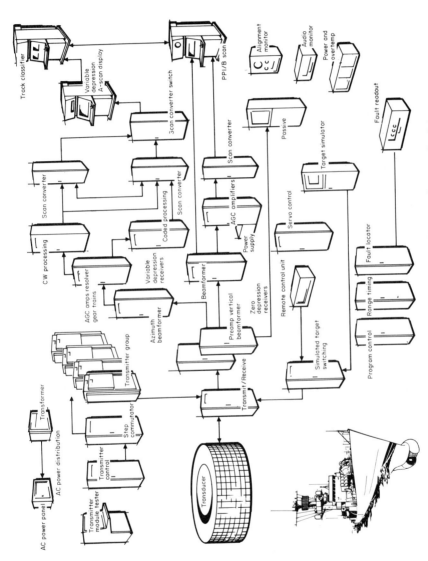

fig. 1.1 *Components of the A/N-SQS-26 sonar for surface ships. A photograph of the dome of this sonar may be seen in Fig. 11.11. (Courtesy, D. Gilchrest, General Electric Company, Syracuse, N.Y.)*

fig. 1.2 *Photograph of the towed sonar A/N-SQS-35 fitted to the stern of a destroyer.*

Homing torpedoes utilize moderately high frequencies because of the necessarily small size of their transducers and the consequently higher frequencies needed to form an adequately directional beam and to reduce noise. Homing torpedoes detect and steer toward their targets and may be either *active homers* or *passive homers,* depending on whether they echo-range or home on the radiated noise of the target. *Mine-hunting sonar systems* and *small-object locators* operate at high kilohertz frequencies by echo ranging with a highly directional transducer and a short pulse length so as to be able to detect the mine or the small object in the background of clutter (reverberation) in which it appears.

A number of special-purpose sonar equipments may be mentioned. The *underwater telephone* is a device for communicating between a surface ship and a submarine or between two submarines. The UQC-1 equipment uses the single-sideband voice modulation of an 8-kHz carrier and is the underwater analog of a radio transmitter-receiver. *Sonobuoys* are small sonar sets dropped by an aircraft for underwater listening or echo ranging. This compact, expendable device contains a miniature radio transmitter for relaying signals picked up by its hydrophone. A pictorial view of an aircraft, its sonobuoy, and a submarine target is seen in Fig. 1.3. By means of a specially packaged explosive charge also dropped by the aircraft, *explosive echo ranging,* in which the sonobuoy is used to receive and transmit echoes, can be performed. Figure 1.4 is an expanded view of a sonobuoy showing its principal components.

A relatively recent development is the neutrally buoyant flexible *towed-line*

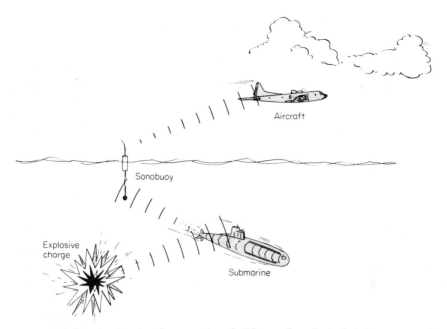

fig. 1.3 *Submarine hunting from an aircraft. The sonobuoy is the link between sound underwater and the aircraft above. The explosive charge is a sound source for echo ranging.*

array, shown in Fig. 1.5, for providing surface ships with a low-frequency passive capability for long-range detection. Such arrays may be several thousand feet long in order to obtain narrow receiving beams, and are towed a great distance astern so as to reduce the noise of the towing vessel.

1.5 Some Basic Concepts

Sound consists of a regular motion of the molecules of an elastic substance. Because the material is elastic, a motion of the particles of the material, such as the motion initiated by a sound projector, communicates to adjacent particles. A sound wave is thereby propagated outward from the source at a velocity equal to the velocity of sound. In a fluid, the particle motion is to and fro, parallel to the direction of propagation; because the fluid is compressible, this to-and-fro motion causes changes in pressure which can be detected by a pressure-sensitive hydrophone. In a plane wave of sound of pressure p, the instantaneous intensity, is related to the velocity of the fluid particles u by

$$p = \rho c u$$

where ρ = fluid density
c = propagation velocity of wave

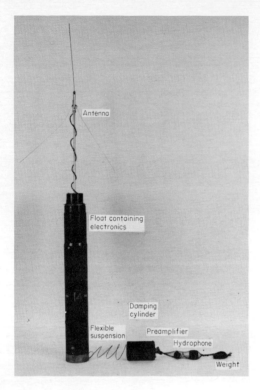

fig. 1.4 *Cutaway view of a sonobuoy.*

The proportionality factor ρc is called the *specific acoustic resistance* of the fluid. For seawater,

$$\rho c = 1.5 \times 10^5 \text{ g/(cm}^2)(\text{s})$$

For air,

$$\rho c = 42 \text{ g/(cm}^2)(\text{s})$$

Under some conditions, the proportionality factor between the velocity and the pressure is complex and is then called the *specific acoustic impedance* of the medium containing the sound wave. This is Ohm's law for acoustics; the particle velocity u may be viewed as the acoustic analog of an electric current, and the pressure p as the analog of an electric voltage.

A propagating sound wave carries mechanical energy with it in the form of the kinetic energy of the particles in motion plus the potential energy of the stresses set up in the elastic medium. Because the wave is propagating, a certain amount of energy per second will flow across a unit area oriented normal to the direction of propagation. This amount of energy per second (power) crossing a unit area is called the *intensity* of the wave. If the unit area

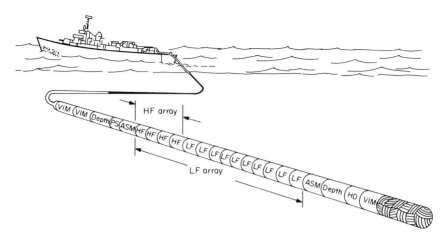

fig. 1.5 *A towed-line array. The array is neutrally bouyant and is towed at a depth determined by the amount of tow cable let out. The array has a high-frequency (HF) and a low-frequency (LF) subarray of hydrophones, together with other modules for depth sensing and vibration isolation (VIM). The array tows in a straight line behind a surface vessel on a straight course, not in the bent configuration shown in this artist's conception.*

is taken in an arbitrary direction, the intensity becomes a vector quantity analogous to the *Poynting vector* of electromagnetic propagation. In a plane wave, the instantaneous intensity is related to the instantaneous acoustic pressure by

$$I = \frac{p^2}{\rho c}$$

When the squared pressure is averaged over an interval of time, the average intensity for that interval is

$$I = \frac{\overline{p^2}}{\rho c}$$

where the bar indicates a time average. In practice, there is always some integration time inherent in the system being used, and the average intensity, rather than the instantaneous intensity, will be of practical interest. If p is in dynes per square centimeter, and ρ and c are in units of grams per cubic centimeter and centimeters per second, then the intensity I will have the units of ergs per square centimeter per second. Since 1 watt equals 10^7 ergs/s, the intensity in units of watts per square centimeter is

$$I = \frac{\overline{p^2}}{\rho c} \times 10^{-7}$$

With $\rho c = 1.5 \times 10^5$, a plane wave in water of rms pressure 1 dyn/cm² has an intensity of 0.67×10^{-12} W/cm².

For transient signals, and in problems where considerable signal distortion takes place in propagation or in encounters with the target, it is more meaningful to refer to the *energy flux density* of the acoustic wave. The energy flux density is the time integral of the instantaneous intensity, or

$$E = \int_0^\infty I \, dt = \frac{1}{\rho c} \int_0^\infty p^2 \, dt$$

where the integral, in a practical case, is taken over the duration of the wave. The units of energy flux density are ergs per square centimeter.

The term *spectrum level* refers to the level of a sound wave in a frequency band 1 Hz wide. It has meaning only for sounds having a "continuous" spectrum, that is, for those that have some sound, however small, in any frequency band. *Band level* refers to the level in a frequency band greater than 1 Hz wide. For a flat (white) spectrum, or as an approximation for any continuous spectrum if the band is not too wide, the band level BL is related to the spectrum level SPL at the middle of the band by

$$\text{BL} = \text{SPL} + 10 \log W$$

where W is the bandwidth in hertz. This expression implies that the intensities in all the adjacent 1-Hz bands can be added together to give the intensity in the W-hertz band.

1.6 The Micropascal Reference Unit and the Decibel

The *unit of intensity* in underwater sound is the intensity of a plane wave having an rms pressure equal to 1 micropascal (abbreviated 1 μPa) or 10^{-5} dyne per square centimeter. This reference intensity has replaced the two previous units in customary use, namely, 1 dyn/cm² and 0.000204 dyn/cm², and is now the American National Standard (17). In physical units this intensity amounts to 0.67×10^{-22} W/cm². Similarly the unit of energy flux density is the energy density of a plane wave of rms pressure equal to 1 μPa integrated over a period of one second; this unit amounts to 0.64×10^{-22} W · s/cm². In order to convert intensities from the old to this new reference, we simply multiply the intensity by the quantity $(1/10^{-5})^2 = 10^{10}$, or add 100 dB to the number of decibels referred to 1 dyn/cm². Similarly, in converting intensities from the reference intensity of a plane wave of rms pressure 0.000204 dyn/cm², we multiply by the quantity $(2.04 \times 10^{-3}/10^{-5})^2 = 416$, or equivalently, we add 26 dB to the number of decibels. The conversion of reference levels is given in more detail in Table 1.2. The micropascal reference standard serves the purpose of replacing the older two standards by a single new one. Incidentally, it is so low that negative decibels rarely, if ever, occur when referred to one micropascal.

table 1.2 Reference Level Conversion Table

dB re 1 dyn/cm²	dB re 0.0002 dyn/cm²	dB re 1 μPa	Plane wave rms pressure, dyn/cm²
20	94	120	10
0	74	100	1
−20	54	80	10^{-1}
−40	34	60	10^{-2}
−60	14	40	10^{-3}
−80	−6	20	10^{-4}
−100	−26	0	10^{-5}

The *decibel* has, historically, long been in use for reckoning acoustic quantities. Decibel units provide a convenient way of handling large changes in variables and, in addition, permit quantities to be multiplied together simply by adding their decibel equivalents, as we shall do in the next chapter when we write, and later on solve, the sonar equations. If I_1 and I_2 are two intensities, or two energy flux densities, the number N denoting their ratio I_1/I_2 is $N = 10 \log_{10} I_1/I_2$ dB. In decibel units, I_1 and I_2 are said to differ by N dB.

The *level* of a sound wave is the number of decibels by which its intensity, or energy flux density, differs from the intensity of the reference sound wave. That is, if I_2 in the above expression is the reference intensity, then a sound wave of intensity I_1 has a *level* equal to N dB. To make it quite clear what reference is used, the level would be written "N dB re 1 μPa." Thus, a wave having an intensity 100 times that of a plane wave of rms pressure 1 μPa would be said to have a *level* of $10 \log_{10} 100/1 = 2$ dB re 1 μPa.

It should be emphasized that the decibel is a comparison of intensities or energy densities, rather than directly of acoustic pressures, even though "20 dB re 1 μPa" appears to refer to a pressure. What is omitted from a statement of this kind are the words re "*the intensity of a plane wave of pressure equal to*" 1 μPa. Similarly, in electrical engineering, the decibel, strictly speaking, refers to a power ratio instead of to a ratio of voltage or current.

In the past, many different units have been used for the quantities occurring in oceanography and underwater sound. Horizontal distances have been stated in feet, yards, meters, and nautical miles, vertical distances in feet or fathoms, sound speeds in feet per second, etc. At present, there is, in this subject as well as in others, a clear tendency to "go metric," and to refer distances to the meter and its decade multiples, such as centimeters and kilometers, rather than the yard. However, in this book we will continue to use 1 yard as the reference distance for range calculations because of the persistence of this unit in naval usage and the existing range scales of sonar equipments, together with the fact that the bulk of the quantitative data in underwater sound has been reported in English units. Those wishing to make a range calculation in meters rather than yards will find a correction to the sonar parameters in Sec. 2.4 in the following chapter.

REFERENCES

1. Bell, T. G.: Sonar and Submarine Detection, *U.S. Navy Underwater Sound Lab. Rep.* 545, 1962.
2. Hunt, F. V.: "Electroacoustics," chap. 1, John Wiley & Sons, Inc., New York, 1954.
3. Listening Systems, *Nat. Def. Res. Comm. Div. 6 Sum. Tech. Rep.* 14, chap. 2, 1946.
4. Wood, A. B.: From the Board of Invention and Research to the Royal Naval Scientific Service, *J. Roy. Nav. Sci. Serv.*, **20**(4):185 (1965).
5. Hayes, H. C.: Detection of Submarines, *Proc. Am. Phil. Soc.*, **59**:1 (1920).
6. Lichte, H.: On the Influence of Horizontal Temperature Layers in Sea Water on the Range of Underwater Sound Signals, *Physik. Z.*, **17**:385 (1919). Translated by A. F. Wittenborn, Tracor, Inc., Rockville, Md.
7. Klein, E.: Underwater Sound and Naval Acoustical Research and Applications before 1939, *J. Acoust. Soc. Am.*, **43**:931 (1968).
8. Holt, L. E.: The German Use of Sonic Listening, *J. Acoust. Soc. Am.*, **19**:678 (1947).
9. Batchelder, L. B.: When Sonar Was Called Submarine Signalling, *J. Acoust. Soc. Am.*, **31**:832 (1959).
10. Wood, A. B.: From the Board of Invention and Research to the Royal Naval Scientific Service, *J. Roy. Nav. Sci. Serv.*, **20**:185 (1965).
11. Lasky, M.: Review of World War I Acoustic Technology, *U.S. Navy J. Underwater Acoust.*, **24**:363 (1973).
12. Lasky, M.: A Historical Review of Underwater Technology, 1916–1939 With Emphasis on ASW, *U.S. Navy J. Underwater Acoust.*, **24**:597 (1974).
13. Lasky, M.: Review of Undersea Acoustics to 1950, *J. Acoust. Soc. Am.*, **61**:283 (1977).
14. Hackman, W. D.: Underwater Acoustics and the Royal Navy, 1893–1930, *Ann. Sci.*, **36**:255 (1979).
15. "Present and Future Civil Uses of Underwater Sound," National Academy of Sciences, Washington, D.C., 1970.
16. Volovov, V. I., and others: Acoustical Determination of the Course, Speed, and Lateral Drift of a Ship, *Sov. Phys. Acoust.* **25**:162 (1979).
17. Preferred Reference Quantities for Acoustical Levels, ANSI S1.8-1969 (R1974), American National Standards Institute, New York.

two

the sonar equations

The many phenomena and effects peculiar to underwater sound produce a variety of quantitative effects on the design and operation of sonar equipment. These diverse effects can be conveniently and logically grouped together quantitatively in a small number of units called the *sonar parameters,* which, in turn, are related by the *sonar equations.* These equations are the working relationships that tie together the effects of the medium, the target, and the equipment; they are among the design and prediction tools available to the engineer for underwater sound applications.

The sonar equations were first formulated during World War II (1) as the logical basis for calculations of the maximum range of sonar equipments. In recent years, they have seen increasing use in the optimum design of sonars for new applications. Essentially the same relationships are employed in radar (2), though with linear instead of logarithmic units and with slightly different definitions of the parameters.

The essentially simple sonar equations serve two important practical functions. One is the *prediction of performance* of sonar equipments of known or existing design. In this application the design characteristics of the sonar set are known or assumed, and what is desired is an estimate of performance in some meaningful terms such as detection probability or search rate. This is achieved in the sonar equations by a prediction of range through the parameter transmission loss. The equations are solved for transmission loss, which is then converted to range through some assumption concerning the propagation characteristics of the medium.

The other general application of the equations is in *sonar design,* where a preestablished range is required for the operation of the equipment being designed. In this case the equation is solved for the particular troublesome parameter whose practical realization is likely to cause difficulty. An example would be the directivity required, along with other probable values of sonar parameters, to yield a desired range of detection in a detection sonar or the range of actuation by a passing ship of an acoustic mine mechanism. After the directivity needed to obtain the desired range has been found, the design continues through the "trade-offs" between directivity index and other parameters. The design is finally completed through several computations using the equations and the design engineer's intuition and experience.

2.1 Basic Considerations

The equations are founded on a basic equality between the desired and undesired portions of the received signal at the instant when some function of the sonar set is just performed. This function may be detection of an underwater target, or it may be the homing of an acoustic torpedo at the instant when it just begins to acquire its target. These functions all involve the reception of acoustic energy occurring in a natural acoustic background. Of the total acoustic field at the receiver, a portion may be said to be *desired* and is called the *signal*. The remainder of the acoustic field is *undesired* and may be called the *background*. In sonar the background is either *noise,* the essentially steady-state portion not due to one's own echo ranging, or *reverberation,* the slowly decaying portion of the background representing the return of one's own acoustic output by scatterers in the sea. The design engineer's objective is to find means for *increasing* the overall response of the sonar system to the signal and for *decreasing* the response of the system to the background—in other words, to *increase* the signal-to-background ratio.

Let us imagine a sonar system serving a practical purpose such as *detection, classification* (determining the nature of a target), *torpedo homing, communication,* or *fish finding*. For each of these purposes there will be a certain signal-to-background ratio that will depend on the functions being performed and on the performance level that is desired in terms of percentages of successes and "false alarms," such as an apparent detection of a target when no target is present. If the signal is imagined to be slowly increasing in a constant background, the desired purpose will be accomplished when the *signal level equals the level of the background which just masks it*. That is to say, when the sonar's purpose is *just* accomplished,

Signal level = background masking level

The term "masking" implies that not all the background interferes with the signal, but only a portion of it—usually that portion lying in the frequency band of the signal. The word "masking" is borrowed from the theory of audition, where it refers to that part of a broadband noise background that masks out a pure tone or a narrow-band signal presented to a human listener.

We should note that the equality just stated will exist at only *one* instant of time when a target approaches, or recedes from, a sonar receiver. At short ranges, its signal level will exceed the background masking level; at long ranges, the reverse will occur. But the instant of equality is the moment of greatest interest to the sonar engineer or designer, for it is at this instant that the sonar system *just* performs its assigned function. It is this instant to which the engineer or designer will often focus attention in a sonar calculation.

2.2 *The Active and Passive Equations*

The next step is to expand the basic equality in terms of the *sonar parameters* determined by the *equipment,* the *medium,* and the *target.* We will denote these parameters by two-letter symbols in order to avoid Greek and subscripted symbols as much as possible in the writing of the equations. These parameters are levels in units of decibels relative to the standard reference intensity of a 1-μPa plane wave. They are as follows:

Parameters Determined by the *Equipment*
    Projector Source Level: SL
    Self-Noise Level: NL
    Receiving Directivity Index: DI
    Detection Threshold: DT
Parameters Determined by the *Medium*
    Transmission Loss: TL
    Reverberation Level: RL
    Ambient-Noise Level: NL
Parameters Determined by the *Target*
    Target Strength: TS
    Target Source Level: SL

Two pairs of the parameters are given the same symbol because they are essentially identical. It should be mentioned in passing that this set of parameters is not unique. Others, which could be employed equally well, might be more fundamental or might differ by a constant. For example, sound velocity could be adopted as a parameter, and TS could be replaced by the parameter "backscattering cross section" expressed in decibels, as is done in radar. The chosen parameters are therefore arbitrary; those employed here are the ones conventionally used in underwater sound.* It should also be noted that they may all be expanded in terms of fundamental quantities like frequency, ship speed, and bearing—a subject that will be of dominant importance in the descriptions of the parameters that will follow. The units of the parameters are decibels, and they are added together in forming the sonar equations.

* There is, however, no conventional symbolism for many of the parameters.

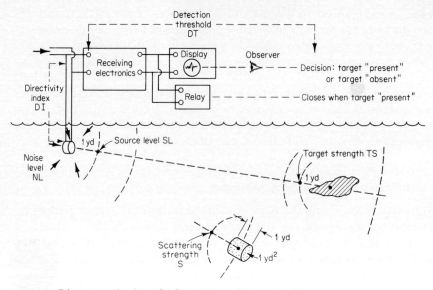

fig. 2.1 *Diagrammatic view of echo ranging, illustrating the sonar parameters.*

The meaning of these quantities can best be illustrated through some simple considerations for an active (echo-ranging) sonar (Fig. 2.1). A sound source acting also as a receiver (a *transducer*) produces by some means a *source level* of SL decibels at a unit distance (1 yd) on its axis. When the radiated sound reaches the target (if the axis of the sound source points toward the target), its level will be reduced by the *transmission loss,* and becomes SL − TL. On reflection or scattering by the target of target strength TS, the reflected or backscattered level will be SL − TL + TS at a distance of 1 yd from the acoustic center of the target in the direction back toward the source. In traveling back toward the source, this level is again attenuated by the *transmission loss* and becomes SL − 2TL + TS. This is the echo level at the transducer. Turning now to the background and assuming it to be isotropic noise rather than reverberation, we find that the *background level* is simply NL. This level is reduced by the *directivity index* of the transducer acting as a receiver or *hydrophone* so that at the terminals of the transducer the relative noise power is NL − DI. Since the axis of the transducer is pointing in the direction from which the echo is coming, the relative echo power is unaffected by the transducer directivity. At the transducer terminals, therefore, the echo-to-noise ratio is

SL − 2TL + TS − (NL − DI)

Let us now assume that the function that this sonar is called upon to perform is *detection,* that is, that its principal purpose is to give an indication of some sort on its *display* whenever an echoing target is present. When the input signal-to-noise ratio is above a certain detection threshold fulfilling certain probability criteria, a decision will be made by a human observer that a target is *present**; when the input signal-to-noise ratio is less than the detection threshold, the decision will be made that the target is *absent*. When the target is *just* being detected, the signal-to-noise ratio *equals* the detection threshold, and we have

$$SL - 2TL + TS - (NL - DI) = DT$$

We have here the active-sonar equation as an equality in terms of the *detection threshold*, called in audition and in much of the older underwater sound literature *recognition differential*. In terms of the basic equality described above, we could equally well consider that only that part of the noise power lying above the detection threshold masks the echo, and we would then have

$$SL - 2TL + TS = NL - DI + DT$$

a more convenient arrangement of the parameters, since the echo level occurs on the left-hand side and the noise-masking background level occurs on the right.

This is the active-sonar equation for the *monostatic case* in which the source and receiving hydrophone are coincident and in which the acoustic return of the target is back toward the source. In some sonars, a separated source and receiver are employed and the arrangement is said to be *bistatic;* in this case, the two transmission losses to and from the target are not, in general, the same. Also in some modern sonars, it is not possible to distinguish between DI and DT, and it becomes appropriate to refer to DI − DT as the increase in signal-to-background ratio produced by the entire receiving system of transducer, electronics, display, and observer (if one is used).

A modification is required when the background is reverberation instead of noise. In this case, the parameter DI, defined in terms of an isotropic background, is inappropriate, inasmuch as reverberation is by no means isotropic. For a reverberation background we will replace the terms NL − DI by an *equivalent plane-wave reverberation level* RL observed at the hydrophone terminals. The active-sonar equation then becomes

$$SL - 2TL + TS = RL + DT$$

where the parameter DT for reverberation has in general a different value than DT for noise.

In the passive case, the target itself produces the signal by which it is detected, and the parameter source level now refers to the level of the radiated noise of the target at the unit distance of 1 yd. Also, the parameter target strength becomes irrelevant, and one-way instead of two-way transmission is involved. With these changes, the *passive-sonar equation* becomes

---

* If the human observer is replaced by a relay at the output of the detector, then the detection threshold is the input signal-to-background ratio at the transducer terminals which *just* closes the relay to indicate "target present."

$$SL - TL = NL - DI + DT$$

Table 2.1 is a list of parameters, reference locations, and short definitions in the form of ratios. More complete definitions of the parameters will be given near the beginnings of the chapters dealing with the parameters.

2.3 Names for Various Combinations of Parameters

In practical work it is convenient to have separate names for different combinations of the terms in the equations. Methods exist for measuring some of these on shipboard sonars as a check on system operation. Table 2.2 is a listing of these names and the combination of terms that each represents. Of these, *the figure of merit* is the most useful, because it combines together the various equipment and target parameters so as to yield a quantity significant for the performance of the sonar. Since it equals the transmission loss at the instant when the sonar equation is satisfied, the figure of merit gives an immediate indication of the range at which a sonar can detect its target, or more generally, perform its function. However, when the background is reverberation instead of noise, the figure of merit is not constant, but varies with range and so fails to be a useful indicator of sonar performance.

table 2.1 The Sonar Parameters, Their Definitions, and Reference Locations

Parameter	symbol	Reference	Definition
Source level	SL	1 yd from source on its acoustic axis	$10 \log \dfrac{\text{intensity of source}}{\text{reference intensity*}}$
Transmission loss	TL	1 yd from source and at target or receiver	$10 \log \dfrac{\text{signal intensity at 1 yd}}{\text{signal intensity at target or receiver}}$
Target strength	TS	1 yd from acoustic center of target	$10 \log \dfrac{\text{echo intensity at 1 yd from target}}{\text{incident intensity}}$
Noise level	NL	At hydrophone location	$10 \log \dfrac{\text{noise intensity}}{\text{reference intensity*}}$
Receiving directivity index	DI	At hydrophone terminals	$10 \log \dfrac{\text{noise power generated by an equivalent nondirectional hydrophone}}{\text{noise power generated by actual hydrophone}}$
Reverberation level	RL	At hydrophone terminals	$10 \log \dfrac{\text{reverberation power at hydrophone terminals}}{\text{power generated by signal of reference intensity*}}$
Detection threshold	DT	At hydrophone terminals	$10 \log \dfrac{\text{signal power to just perform a certain function}}{\text{noise power at hydrophone terminals}}$

* The reference intensity is that of a plane wave of rms pressure 1 μPa.

table 2.2 Terminology of Various Combinations of the Sonar Parameters

Name	Parameters	Remarks
Echo level	SL − 2TL + TS	The intensity of the echo as measured in the water at the hydrophone
Noise-masking level	NL − DI + DT	Another name for these two combinations is *minimum detectable echo level*
Reverberation-masking level	RL + DT	
Echo excess	SL − 2TL + TS − (NL − DI + DT)	Detection just occurs, under the probability conditions implied in the term DT, when the echo excess is zero
Performance figure	SL − (NL − DI)	Difference between the source level and the noise level measured at the hydrophone terminals
Figure of merit	SL − (NL − DI + DT)	Equals the maximum allowable one-way transmission loss in passive sonars, or the maximum allowable two-way loss for TS = 0 dB in active sonars

2.4 The Parameters in Metric Units

Some of the parameters of Table 2.1 have 1 yard as their reference distance. These are SL, TL, TS, and (as a determining quantity for RL) the scattering strength S. If, instead, 1 meter is taken as the reference distance, and it is desired to use metric units in a calculation involving the sonar equations, these quantities should be *decreased* by the amount 20 log (1 meter/1 yard) = 0.78 dB. In addition, the attenuation coefficient, commonly expressed in English units in decibels per kiloyard, must be multiplied by 1.094 to convert it to decibels per kilometer. No other sonar quantities are affected by a choice of units; nor are the quantities echo level, noise-masking level, and echo excess (defined in Table 2.2). For finding sonar ranges in metric units, it is often more convenient to find the range first in kiloyards, and then to divide by the factor 1.094 to obtain the range in kilometers.

2.5 Echo, Noise, and Reverberation Level as Functions of Range

The sonar equations just written are no more than a statement of an equality between the desired portion of the acoustic field called the *signal*—either an echo or a noise from a target—and an undesired portion, called the *background* of noise or reverberation. This equality, in general, will hold at only one range; at other ranges, one or the other will be the greater, and the equality will no longer exist.

This is illustrated in Fig. 2.2, where curves of echo level, noise-masking level, and reverberation-masking level are shown as a function of range. Both the echo and reverberation fall off with range, whereas the noise remains

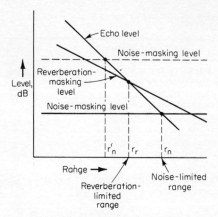

fig. 2.2 *Echo, noise, and reverberation as functions of range.*

constant. The echo-level curve will generally fall off more rapidly with range than the reverberation-masking level curve and will intersect it at the *reverberation-limited range* r_r given by the sonar equation for reverberation. The curve of echo level will also intersect the noise-masking level at the range of the sonar equation for noise r_n. If the reverberation is high, the former will be less than the latter, and the range will be said to be *reverberation-limited*. If for any reason the noise-masking level rises to the level shown by the dashed line in the figure, the echoes will then die away into a background of noise rather than reverberation. The new noise-limited range r'_n will then be less than the reverberation-limited range r_n, and the range will become *noise-limited*. Both ranges are given by the appropriate form of the sonar equation.

A knowledge of whether a sonar will be noise- or reverberation-limited is necessary for both the sonar predictor and the sonar designer. In general, the curves for echo and reverberation will not be straight lines because of complications in propagation and in the distribution of reverberation-producing scatterers. For a new sonar system, such curves should always be drawn from the best information available for the conditions most likely to be encountered in order to demonstrate visually to the design engineer the behavior of the signals and background with range.

For passive sonars, a convenient graphical way of solving the sonar equations is called SORAP, denoting "sonar overlay range prediction." It consists of two plots that are laid one upon the other (Fig. 2.3). The overlay (solid lines) is a plot of SL versus frequency for a particular passive target or class of targets; the underlay (dashed lines) is a plot of the sum of the parameters TL + NL − DI + DT for a particular passive sonar and for a number of different ranges. The range and frequency at which the target can be detected can be readily read off by inspection. In Fig. 2.3, where the two plots are superposed, the target will be detected first at a range of 10 miles by means of the line component at frequency f_1. However, if the criterion is adopted that *three* spectral lines must appear on the display, whatever it may be, before a detec-

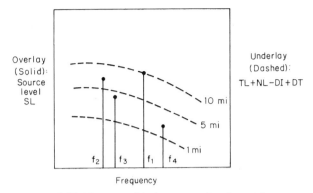

fig. 2.3 *SORAP: a graphical way to solve the passive sonar equation.*

tion is called, the range would be reduced to 4 miles, and the display would show the lines at frequencies $f_1, f_2,$ and f_3. The method is particularly useful for separating the target parameter SL from the equipment parameters and medium parameters at the location where it is used, while at the same time accommodating a wide range of frequencies. Thus, targets can be compared for the same equipment and locations, or, alternatively, locations can be compared for the same target, and so forth.

2.6 Transient Form of the Sonar Equations

The equations thus far have been written in terms of *intensity*, or the average acoustic power per unit area of the sound emitted by the source or received from the target. The word "average" implies a time interval over which the average is to be taken. This time interval causes uncertain results whenever short transient sources exist or, generally, whenever severe distortion is introduced by propagation in the medium or by scattering from the target.

A more general approach is to write the equations in terms of *energy flux density*, defined as the acoustic energy per unit area of wavefront (see Sec. 1.5). If a plane acoustic wave has a time-varying pressure $p(t)$, then the energy flux density of the wave is

$$E = \frac{1}{\rho c} \int_0^\infty p^2(t)\, dt$$

If the units of pressure are dynes per square centimeter and the acoustic impedance of the medium is in cgs units (for water, $\rho c \approx 1.5 \times 10^5$), then E will be expressed in ergs per square centimeter. The intensity is the mean-square pressure of the wave divided by ρc and averaged over an integral of time T, or

$$I = \frac{1}{T}\int_0^T \frac{p^2(t)}{\rho c}\,dt$$

so that over the time interval T,

$$I = \frac{E}{T}$$

The quantity T is accordingly the time interval over which the energy flux density of an acoustic wave is to be averaged to form the intensity. For long-pulse active sonars, this time interval is the duration of the emitted pulse and is very nearly equal to the duration of the echo. For short transient sonars, however, the interval T is often ambiguous, and the duration of the echo is vastly different from the duration of the transient emitted from the source. Under these conditions, however, it can be shown (3) that the intensity form of the sonar equations can be used, provided that the source level is defined as

$$SL = 10 \log E - 10 \log \tau_e$$

where E is the energy flux density of the source at 1 yd and is measured in units of the energy flux density of a 1-μPa plane wave taken over an interval of 1 second, and τ_e is the duration of the echo in seconds for an active sonar. For explosives, E is established by measurements for a given charge weight, depth, and type of explosive (Sec. 4.4). For pulsed sonars emitting a flat-topped pulse of constant source level SL' over a time interval τ_0, then, since the energy density of a pulse is the product of the average intensity times its duration,

$$10 \log E = SL' + 10 \log \tau_0$$

By combining the last two equations, the effective source level SL for use in the sonar equations is therefore

$$SL = SL' + 10 \log \frac{\tau_0}{\tau_e}$$

Here τ_0 is the duration of the emitted pulse of source level SL', and τ_e is the echo duration. For long-pulse sonars, $\tau_0 = \tau_e$ and SL = SL'. For short-pulse sonars, $\tau_e > \tau_0$, and the effective source level SL is less than SL' by the amount $10 \log (\tau_0/\tau_e)$. The effect of time stretching on source level may be visualized as shown in Fig. 2.4. A short pulse of duration τ_0 and source level SL' is replaced in a sonar calculation by an effective or equivalent pulse of longer duration τ_e and lower source level SL. The two source levels are related so as to keep the energy flux-density source levels the same, namely:

$$SL + 10 \log \tau_e = SL' + 10 \log \tau_0$$

or

$$SL = SL' + 10 \log \frac{\tau_0}{\tau_e}$$

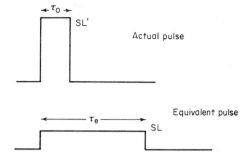

fig. 2.4 *Equivalent source level in short-pulse sonars.*

In effect, the pulse emitted by the source is stretched out in time and thereby reduced in level by the multipath effects of propagation and by the processes of target reflection. The appropriate values of other sonar parameters in the equations, such as TS and TL, are those applying for long-pulse or CW conditions, in which the effects of multipaths in the medium and on the target are added up and accounted for.

For active short-pulse sonars, the echo duration τ_e is, accordingly, a parameter in its own right. Figure 2.5 illustrates a pulse as a short exponential tran-

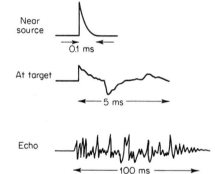

fig. 2.5 *Diagrams of the pressure of an explosive pulse near the source, on arrival at an extended target, and as an echo back in the vicinity of the source.*

sient at the source, as a distorted pulse at the target, and as an echo received back in the vicinity of the source. An exponential pulse, similar in form to the shock wave from an explosion of about 1 lb of TNT and having an initial duration of 0.1 ms, becomes distorted into an echo 1,000 times as long. Two actual examples of explosive echoes are shown in Fig. 2.6.

The echo duration can be conceived as consisting of three components: τ_0, the duration of the emitted pulse measured near the source; τ_m, the additional duration imposed by the two-way propagation in the sea; and τ_t, the additional duration imposed by the extension in range of the target. In this view, the echo duration is the sum of the three components, or

$$\tau_e = \tau_0 + \tau_t + \tau_m$$

Typical examples of the magnitude of these three components of the echo duration under different conditions are given in Table 2.3. Thus, with these

28 / principles of underwater sound

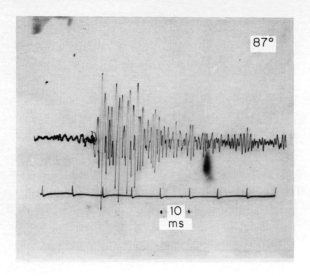

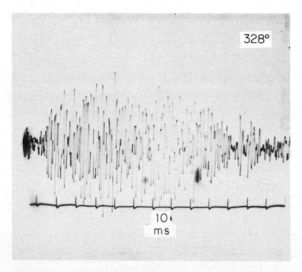

fig. 2.6 *Oscilloscope photographs of explosive echoes from a submarine at aspect angles 87° (near-beam aspect) and 328° (near-bow aspect). The time ticks are 10 ms apart.*

table 2.3 Components of Echo Duration

Component	Typical values, ms
Duration of the emitted pulse at short ranges	Explosives: 0.1 Sonar: 100
Duration produced by multiple paths	Deep water: 1 Shallow water: 100
Duration produced by a submarine target	Beam aspect: 10 Bow-stern aspect: 100

values, the time duration of an explosive echo from a bow-stern aspect submarine in shallow water would be 0.1 + 100 + 100 = 200.1 ms.

2.7 Statement of the Equations

A condensed statement of the equations is as follows.

Active sonars (monostatic):

Noise background

$$SL - 2TL + TS = NL - DI + DT$$

Reverberation background

$$SL - 2TL + TS = RL + DT_R$$

Passive sonars:

$$SL - TL = NL - DI + DT_N$$

where DT has been subscripted to make it clear that this parameter is quantitatively different for noise and for reverberation.

2.8 Limitations of the Sonar Equations

The sonar equations written in terms of intensities are not always complete for some types of sonars. We have already seen how short-pulse sonars require the addition of another term, the echo duration, to account for the time-stretching produced by multipath propagation. Another such addition is a *correlation loss* in correlation sonars to account for the decorrelation of the signal relative to a stored replica; such decorrelation occurs on bottom reflection and scattering in bottom-bounce sonars. Other terms may be conceivably required for other sophisticated sonars whose operation does not depend on intensity alone.

A limitation of another kind is produced by the nature of the medium in which sonars operate. The sea is a moving medium containing inhomogeneities of various kinds, together with irregular boundaries, one of which is in motion. Multipath propagation is the rule. As a result, many of the sonar parameters fluctuate irregularly with time, while others change because of unknown changes in the equipment and the platform on which it is mounted. Because of these fluctuations, a "solution" of the sonar equations is no more than a best-guess time average of what is to be expected in a basically stochastic problem.

Precise calculations, to tenths of decibels, are futile; a predicted sonar range is an average quantity about which the observed values of range are likely to congregate. We can hope that as our knowledge of underwater sound and its fluctuations improves, the accuracy of the predictions of the sonar equations can be expected to increase.

REFERENCES

1. Principles of Underwater Sound, *Nat. Def. Res. Comm. Div. 6 Sum. Tech. Rep.* 7, pp. 175–199 (1946).
2. Kerr, D. E. (ed.): "Propagation of Short Radio Waves," M.I.T. Radiation Laboratory Series, pp. 18–21, McGraw-Hill Book Company, New York, 1951.
3. Urick, R. J.: Generalized Form of the Sonar Equations, *J. Acoust. Soc. Am.*, **34:**547 (1962).

three

properties of transducer arrays: directivity index

A human being requires a great deal of assistance in the perception and localization of underwater sound. The unaided human ear is for many reasons completely useless for the purpose. Underwater sound equipments include a number of devices of varying degrees of complexity to make evident to an observer the existence of an underwater sound wave. These devices normally include a *hydrophone array* that transforms or *transduces* acoustic energy to electric energy, followed by some form of *signal processing* to feed some kind of aural or visual display suitable for the human observer.

Devices that convert sound and electric energy into each other—or, more generally, any two forms of energy—are called *transducers*, a hybrid Latin word meaning to "lead across." A transducer that converts sound into electricity is called a receiver or *hydrophone;* a transducer that converts electric energy into sound is called a *projector*. Some sonars use the same transducer for generating and receiving sound; others use a separate projector and hydrophone.

The ability to transduce, or interconvert, these two forms of energy rests on the peculiar properties of certain materials called *piezoelectricity* (and its variant, electrostriction) and *magnetostriction*. Some crystalline substances, like quartz, ammonium dihydrogen phosphate (ADP), and Rochelle salt, acquire a charge between certain crystal surfaces when placed under pressure; conversely, they acquire a stress when a voltage is placed across them. These crystalline substances* are said to be piezoelectric. Electrostrictive materials exhibit the same effect, but are polycrystalline ceramics that have to be properly polarized by being subjected to a high electrostatic field; examples are barium titanate and lead zirconate titanate. A magnetostrictive material is one that changes dimensions when placed in a magnetic field, and, conversely, changes the magnetic field within and around it when it is stressed. Magnetostrictive materials are also polarized in order to avoid frequency doubling and to achieve a higher efficiency.

In air, other kinds of transducers are commonly used. Among these are moving-coil, moving-armature, and electrostatic types. In water, piezoelectric and magnetostrictive materials are particularly suitable because of their better impedance match to water. Ceramic materials have become increasingly popular in underwater sound because they can be readily molded into desirable shapes. Present-day sonars use ceramic transducer elements almost exclusively because of their inexpensiveness and availability.

Other transducer types have been the subject of recent research studies, such as thin-films (1) and fiber optic hydrophones (2), both having possible applications, especially to towed-line arrays.

The ability to use these properties of matter for the reception or generation of underwater sound is based on the art of transducer design. It is a specialized science and technology of its own, which is too vast to be described here. The interested reader may find the basic principles of electroacoustic transduction in a book by Hunt (3). The design and construction of actual transducers are well described in two volumes of the NDRC Summary Technical Reports (4).

Single piezoelectric or magnetostrictive elements are normally utilized in hydrophones for research or measurement work. Nearly all other uses of hydrophones require hydrophone *arrays* in which a number of spaced elements are employed. The advantages of an array over a single hydrophone element are severalfold. First, the array is more sensitive, since a number of elements will generate more voltage (if connected in series) or more current (if connected in parallel) than a single element exposed to the same sound field. Second, the array possesses directional properties that enable it to discriminate between sounds arriving from different directions. Third, the array has an improved signal-to-noise ratio over a single hydrophone element since it discriminates against isotropic or quasi-isotropic noise in favor of a signal arriving in the direction that the array is pointing. These advantages dictate the use of hydrophone arrays in most practical applications of underwater sound. The first and second of the above three benefits exist for projectors as well as for hydrophones.

Examples of a plane and a cylindrical array are shown in Fig. 3.1. Figure 3.2 shows a spherical array for installation on a modern submarine. Figure 3.3

---

* V. A. Bazhenov has written a book on the piezoelectric properties of wood, a noncrystalline material (54).

fig. 3.1 *A plane and a cylindrical transducer array. Array elements are 45° Z-cut crystals cemented to a steel backing plate. Resonant frequency near 24 kHz. (Ref. 4, p. 237.)*

shows a cylindrical array placed inside a dome on a modern destroyer or frigate. In all such arrays the individual elements are connected to phasing or time-delay networks to form beams pointing in different directions.

3.1 Array Gain

Of the various benefits that result from using an array of hydrophones, the most important for the detection of underwater targets is the improvement in the signal-to-noise ratio. This improvement is measured by the *array gain* of

34 / *principles of underwater sound*

the array, or, alternatively for a special case of signal and noise, by its *directivity index*.

The gain of an array, AG, in decibels is defined as

$$AG = 10 \log \frac{(S/N)_{\text{array}}}{(S/N)_{\text{one element}}}$$

where the numerator is the signal-to-noise ratio at the array terminals, and the denominator is the signal-to-noise ratio at a single element of the array, assumed for simplicity to be the same over all the elements of the array.

For a linear, additive array—that is, one comprising transducer elements whose outputs are summed either in series or in parallel—two approaches exist for computing or estimating the array gain.

One approach involves the directional patterns of the signal and noise fields in which the array is placed, together with the beam pattern of the array. Let the signal and noise fields be characterized by the directional functions $S(\theta,\varphi)$ and $N(\theta,\varphi)$, representing the signal and noise power per unit solid angle, respectively, incident on the array from the polar directions θ and φ, and let $b(\theta,\varphi)$ be the beam pattern of the array. Then the array gain as just defined becomes

fig. 3.2 Skeleton of a large array installed on a number of U.S. submarines. The size of the array may be noted by comparison with the height of the workman standing below it. (Ref. 55.)

fig. 3.3 A modern cylindrical echo-ranging array, a part of the DE 1160 sonar manufactured by the Raytheon Company. The array is 32 in. high and comprises 192 elements. (Courtesy, Raytheon Co., Portsmouth, R.I.)

$$AG = 10 \log \frac{\int_{4\pi} S(\theta,\varphi) b(\theta,\varphi)\, d\Omega \Big/ \int_{4\pi} N(\theta,\varphi) b(\theta,\varphi)\, d\Omega}{\int_{4\pi} S(\theta,\varphi)\, d\Omega \Big/ \int_{4\pi} N(\theta,\varphi)\, d\Omega}$$

In this cumbersome expression, each integral is merely the directional pattern of signal or noise, weighted or multiplied by the beam pattern and integrated over all solid angle. For a single, nondirectional array element, $b(\theta,\varphi) = 1$. For deep-sea ambient noise at a bottomed location, some observed values of the directional pattern $N(\theta,\varphi)$ may be found in Sec. 7.9.

An alternative and more useful approach to array gain involves the *coherence* of signal and noise across the dimensions of the array. By coherence is meant the degree of similarity of waveform of signal and noise between any two elements of the array. Coherence is measured by the *crosscorrelation coefficient* of the outputs of different elements of the array. If $v_1(t)$ and $v_2(t)$ are voltages generated by two array elements as functions of time, then the crosscorrelation coefficient between them is defined as

$$\rho_{12} = \frac{\overline{v_1(t)v_2(t)}}{[\overline{(v_1)^2}\,\overline{(v_2)^2}]^{1/2}}$$

where the bars indicate time averages and the denominator serves as a normalization factor. The crosscorrelation coefficient of two functions of time is therefore the normalized time-averaged product of the two functions.

The array gain depends on the crosscorrelation coefficients of signal and noise between the elements of the array. Consider a linear additive array of n elements of equal sensitivity. Let the individual output voltages, as functions of time, including any phase shifts or delays incorporated for steering, be denoted by $s_1(t), s_2(t), \ldots, s_n(t)$. If the array elements are in series, the voltage appearing across the array terminals will be $s_1(t) + s_2(t) + \cdots + s_n(t)$. The average signal power into the load across the array terminals will be

$$\overline{S^2} = m\overline{[s_1(t) + s_2(t) + \cdots + s_n(t)]^2}$$

where m is a proportionality factor. Similarly, the noise power will be

$$\overline{N^2} = m\overline{[n_1(t) + n_2(t) + \cdots + n_n(t)]^2}$$

where the n's are the outputs of the array elements produced by noise. The average array signal-to-noise ratio will therefore be

$$\frac{\overline{S^2}}{\overline{N^2}} = \frac{\overline{[s_1(t) + s_2(t) + \cdots + s_n(t)]^2}}{\overline{[n_1(t) + n_2(t) + \cdots + n_n(t)]^2}}$$

Expanding and dropping the t's, one obtains

$$\frac{\overline{S^2}}{\overline{N^2}} = \frac{(\overline{s_1 s_1} + \overline{s_1 s_2} + \cdots + \overline{s_1 s_n}) + (\overline{s_2 s_1} + \overline{s_2 s_2} + \cdots + \overline{s_2 s_n}) + \cdots}{(\overline{n_1 n_1} + \overline{n_1 n_2} + \cdots + \overline{n_1 n_n}) + (\overline{n_2 n_1} + \overline{n_2 n_2} + \cdots + \overline{n_2 n_n}) + \cdots}$$

For simplicity, let the signal powers be equal for each element:

$$\overline{s_1^2} = \overline{s_2^2} \cdots = \overline{s_n^2} = \overline{s^2}$$

and similarly let the noise powers be equal for each element:

$$\overline{n_1^2} = \overline{n_2^2} \cdots = \overline{n_n^2} = \overline{n^2}$$

Then by the definition of the crosscorrelation coefficient, the array signal-to-noise power becomes

$$\frac{\overline{S^2}}{\overline{N^2}} = \frac{\overline{s^2}}{\overline{n^2}} \frac{[(\rho_s)_{11} + (\rho_s)_{12} + \cdots + (\rho_s)_{1n}] + [(\rho_s)_{21} + (\rho_s)_{22} + \cdots + (\rho_s)_{2n}] + \cdots}{[(\rho_n)_{11} + (\rho_n)_{12} + \cdots + (\rho_n)_{1n}] + [(\rho_n)_{21} + (\rho_n)_{22} + \cdots + (\rho_n)_{2n}] + \cdots}$$

$$= \frac{\overline{s^2}}{\overline{n^2}} \frac{\sum_j \sum_i (\rho_s)_{ij}}{\sum_j \sum_i (\rho_n)_{ij}}$$

Here $(\rho_s)_{ij}$ and $(\rho_n)_{ij}$ are the crosscorrelation coefficients between the ith element and the jth element of the signal and of the noise, respectively. The *array gain* is by definition the ratio, in decibel units, of the signal to noise of the array to the signal to noise of a single element, so that

$$AG = 10 \log \frac{\overline{S^2/N^2}}{\overline{s^2/n^2}} = 10 \log \frac{\sum_i \sum_j (\rho_s)_{ij}}{\sum_i \sum_j (\rho_n)_{ij}}$$

The gain of the array therefore depends on the sum of the crosscorrelation coefficients between all pairs of elements of the array, for both noise and signal. When the mean-square outputs of the individual array elements are not all the same, as when amplitude shading is used, the array gain becomes

$$AG = 10 \log \frac{\sum_i \sum_j a_i a_j (\rho_s)_{ij}}{\sum_i \sum_j a_i a_j (\rho_n)_{ij}}$$

where a_i is the rms voltage produced by the ith element due to the signal or the noise.

The quantities ρ_s and ρ_n are basic properties of the signal and noise sound field in which the array is placed. Accordingly, the same array will have a different array gain in different signal and noise fields.

Both ρ_s and ρ_n will depend on the electrical time delays that may be introduced into the array for steering; indeed, the purpose of these electrical delays is to maximize ρ_s when the array is steered in the direction from which the signal is coming.

Table 3.1 gives expressions for ρ_s and ρ_n for a unidirectional signal and for isotropic noise, respectively, for a signal frequency and for a frequency band, with and without electrical delays. Computed curves of the correlation of isotropic noise as a function of electrical delay and bandwidth may be found in a report by Jacobson (5). For deep-water ambient noise originating from random sources distributed over the sea surface, expressions for the crosscorrelation coefficient between separated receivers have been derived by Cron and Sherman (6).

It is instructive to compute the array gain for some simple situations. When both the signal and noise are either completely coherent or completely incoherent across the array, the array gain becomes zero decibels; the array is unable to distinguish signal from noise. On the other hand, for a *perfectly coherent* signal in *incoherent* noise, such as when

$$(\rho_n)_{ij} = 0 \quad i \neq j$$
$$(\rho_n)_{ij} = 1 \quad i = j$$

it is easy to show, by expansion of the double summations of the array-gain expression, that *an array of n elements has a gain of 10 log n* for incoherent noise. However, with a perfectly coherent signal in a background of only partly coherent noise, such that

table 3.1 Crosscorrelation Coefficients of Signals and Noise

	Unidirectional signal ρ_s	Isotropic noise ρ_n
Single frequency, zero time delay	$\cos \omega \tau_w$	$\dfrac{\sin(\omega d/c)}{\omega d/c}$
Single frequency, time delay	$\cos \omega(\tau_w + \tau_e)$	$\dfrac{\sin(\omega d/c)}{\omega d/c} \cos \omega \tau_e$
Flat bandwidth, zero time delay	$\dfrac{\sin\left[\frac{1}{2}(\omega_2 - \omega_1)\tau_w\right]}{\frac{1}{2}(\omega_2 - \omega_1)\tau_w} \cos \dfrac{(\omega_2 + \omega_1)}{2}\tau_w$	$\dfrac{1}{(\omega_2 - \omega_1)d/c}\left[\operatorname{Si}\!\left(\omega_2 \dfrac{d}{c}\right) - \operatorname{Si}\!\left(\omega_1 \dfrac{d}{c}\right)\right]$
Flat bandwidth, time delay	$\dfrac{\sin\left[\frac{1}{2}(\omega_2 - \omega_1)(\tau_w + \tau_e)\right]}{\frac{1}{2}(\omega_2 - \omega_1)(\tau_w + \tau_e)} \cos \dfrac{(\omega_2 + \omega_1)}{2}(\tau_w + \tau_e)$	$\dfrac{1}{2(\omega_2 - \omega_1)d/c}\left[\operatorname{Si}\!\left(\omega_2\!\left(\dfrac{d}{c} + \tau_e\right)\right) - \operatorname{Si}\!\left(\omega_1\!\left(\dfrac{d}{c} + \tau_e\right)\right)\right.$ $\left. + \operatorname{Si}\!\left(\omega_2\!\left(\dfrac{d}{c} - \tau_e\right)\right) - \operatorname{Si}\!\left(\omega_1\!\left(\dfrac{d}{c} - \tau_e\right)\right)\right]$

Symbols for table:
$\omega = 2\pi$ (frequency)
τ_w = travel time of signal between array elements
    = $(d/c) \cos \theta$
d = separation of array elements
θ = angle to line joining the two elements
c = velocity of sound
τ_e = electrical (steering) delay
$\omega_2 = 2\pi$ (upper frequency of band)
$\omega_1 = 2\pi$ (lower frequency of band)
$\operatorname{Si}(x) = \int_0^x \dfrac{\sin u}{u}\,du$

$$(\rho_n)_{ij} = \rho \quad i \neq j$$
$$(\rho_n)_{ij} = 1 \quad i = j$$

then, by expansion, the array gain becomes

$$10 \log \frac{n}{1 + (n-1)\rho}$$

and the array gain is less than $10 \log n$. It is obvious therefore that, in general, the gain of an array degrades as the signal coherence decreases and as the noise coherence increases. When, as in the actual ocean, the medium in which the array operates is not statistically time-stationary, but causes correlated amplitude and phase fluctuations in the received signal, the array performance is further degraded (7–10). The gain of an array, in short, depends on the statistics of the desired and undesired portions of the sound field in which the array operates.

As a simple example of a computation, we compute the gain of a three-element, equally spaced, unshaded line array. The coherence of signal and noise, plotted as a continuous function of separation d in the direction of the line, is shown in Fig. 3.4. The values of ρ_s and ρ_n corresponding to the spacings of the elements are those of the following matrixes:

ρ_s

j \ i	1	2	3
1	1	0.7	0.3
2	0.7	1	0.7
3	0.3	0.7	1

ρ_n

j \ i	1	2	3
1	1	0	0
2	0	1	0
3	0	0	1

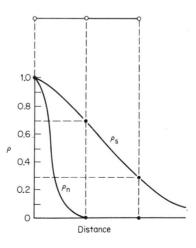

fig. 3.4 *Example of signal and noise coherence along a three-element array.*

The array gain is 10 times the logarithm of the ratio of the sums of the matrix elements, or

$$AG = 10 \log \frac{(1 + 0.7 + 0.3) + (0.7 + 1 + 0.7) + (0.3 + 0.7 + 1)}{(1 + 0 + 0) + (0 + 1 + 0) + (0 + 0 + 1)}$$

$$= 10 \log \frac{6.4}{3}$$

$$= 3.3 \text{ dB}$$

The effect of the degradation of signal coherence in this example is to degrade the array gain from $10 \log 3 = 4.8$ dB, when all the elements of ρ_s are unity, to 3.3 dB.

An interesting special case, investigated by Cox (11), is that of a line array of equally spaced elements in incoherent noise having a constant signal correlation coefficient ρ_s between adjacent elements. In this case, the signal correlation coefficient of the signal between the first and second elements is ρ_s; between the first and third ρ_s^2; between the first and fourth ρ_s^3, and so on. It is shown that the array gain reaches a maximum value of $10 \log (1 + \rho_s)/(1 - \rho_s)$ for arrays long enough for the coefficient to fall to 0.01 or less between the end elements of the array. Arrays longer than this will not achieve a greater array gain, since adding elements to the end of the array adds effectively only noise into the array output. An extension to the case of a linear, rather than an exponential, decay of coherence shows (12) that the maximum achievable array gain for a line of equally spaced elements in uncorrelated noise is $10 \log D_s/d$, where D_s is the distance for the signal correlation coefficient to fall to zero, and d is the separation distance between elements. Thus, if the signal correlation coefficient falls linearly to zero in a distance of 100 meters, a line array of elements one meter apart would have an array gain of $10 \log 100/1 = 20$ dB, provided the noise is uncorrelated between elements. Filling in the array with more elements will increase the gain as long as the noise remains uncorrelated; making the array longer will not.

On the other hand, occasionally very large array gains can be achieved by tailoring the array to the prevailing signal and noise fields. The simplest example is a vertical array used to discriminate against undesired sound arriving vertically (such as the bottom return, at times corresponding to normal incidence) in favor of a signal arriving horizontally. If a narrow frequency band can be tolerated, a half-wavelength spacing between elements will markedly reduce the denominator in the array gain expression, while leaving the numerator unaffected, and so yield a large array gain. The array, in a more customary view, has canceled out the noise. In general, array design for maximum gain in a particular signal and noise field is best achieved by evaluating on a digital computer the gain of a number of selected arrays and choosing the best. However, it should be noted that in some applications beamwidth and side-lobe level will be more important than array gain and will govern the acoustical design.

Signal fluctuations degrade the gain of arrays. For example, it was shown (13) that phase fluctuations of ±80 degrees rms cause a degradation of 6 dB in the gain of an 8-element array, whereas amplitude fluctuations of ±6 dB produce a degradation of only 0.8 dB. Thus, it would appear that phase changes are more important than amplitude changes in decreasing the array gain.

3.2 Measuring the Coherence of Sound Fields

The coherence of a signal or noise field can be determined by a pair of small probe hydrophones whose separation and orientation can be systematically varied to explore the space in which an array is to be used. This space might consist of the inside of a sonar dome, the length of ocean behind a ship pulling a towed-line array, or the volume of ocean near the sea bed where a three-dimensional array will be installed. The outputs of the pair of hydrophones, which could be part of an array, are fed into a *time-delay correlator*, whose basic components are shown in Fig. 3.5, for generating a *time-delay correlogram*. From such correlograms, after digitizing and storing in a computer, the ρ's for signal and noise can be read and summed at the time delays required for beam-forming at the separation distances and orientation appropriate for a trial array, and array gain obtained. If clipped processing is to be employed, then clipping or hard limiting should be done at the outputs of the two hydrophones before correlating, and the final result is the *clipped correlation coefficient*. Since amplitude changes are removed by clipping, this is a correlation coefficient of the phase between the two hydrophones. If, for some reason, phase can be ignored and amplitude is important, the *envelope*

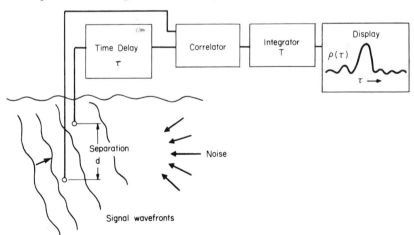

fig. 3.5 *Component parts of a correlogram generator. Two hydrophones are located a distance d apart. The device inserts a rapidly varying time delay into the output of one and produces a synchronized trace—the time delay correlogram—on a display, usually an oscilloscope screen.*

correlation coefficient will be of interest; here the time-delay correlogram is that for the envelope of the two hydrophone outputs after rectifying and low-pass filtering.

3.3 Receiving Directivity Index

When the signal is a unidirectional *plane wave* and is therefore *perfectly coherent*, and when the noise is *isotropic*, that is, when the noise power per unit solid angle is the same in all directions so that $N(\theta,\varphi) = 1$, the array gain reduces to the quantity called *directivity index*, which has had a long historical use in underwater sound. With a coherent signal in isotropic noise, and for an array steered in the direction of the signal, the expression for array gain in terms of the directional functions of the signal and noise becomes

$$AG = DI = 10 \log \frac{\int_{4\pi} d\Omega}{\int_{4\pi} b(\theta,\varphi)\, d\Omega} = 10 \log \frac{4\pi}{\int_0^{2\pi}\int_{-\pi/2}^{\pi/2} b(\theta,\varphi) \cos\theta\, d\theta\, d\varphi}$$

If, further, the beam pattern has rotational symmetry and is nondirectional in the plane in which φ is reckoned, then the above expression simplifies to

$$DI = 10 \log \frac{4\pi}{2\pi \int_{-\pi/2}^{\pi/2} b(\theta) \cos\theta\, d\theta}$$

DI is always a positive number of decibels.*

For simple arrays, such as the line and circular plane, DI can be evaluated mathematically. Expressions for the DI of some simple transducer arrays are given in Table 3.2 in terms of the array dimensions. For arrays that cannot be approximated by these simple forms, the DI can be found by integration of the beam pattern, for which some graphical aids have been developed (14).

A nomogram for finding the directivity index of line and circular-piston arrays in terms of their dimensions and the frequency is given in Fig. 3.6. The DI is the point of intersection of the line joining the dimension and frequency scales with the center scale of the diagram. For example, the DI of a circular-piston array of 5 in. diameter at a frequency of 30 kHz is 18 dB.

3.4 Limitations of Directivity Index

Because it can be evaluated for some of the common array configurations of underwater sound, as by Fig. 3.6 or Table 3.2, directivity index is a useful parameter for providing at least a first-cut estimate of the gain of an array. Yet its restriction to the special case of a perfectly coherent signal in isotropic noise must not be overlooked. In the real ocean these ideal conditions seldom,

---

* Before about 1948, directivity index was regarded as a negative instead of a positive quantity.

properties of transducer arrays: directivity index / 43

table 3.2 Directivity of Simple Transducers

Type	Pattern function	DI = 10 log
Continuous line of length L $L \gg \lambda$	$\left[\dfrac{\sin(\pi L/\lambda)\sin\theta}{(\pi L/\lambda)\sin\theta}\right]^2$	$\dfrac{2L}{\lambda}$
Piston of diameter D in an infinite baffle $D \gg \lambda$	$\left[\dfrac{2J_1[(\pi D/\lambda)\sin\theta]}{(\pi D/\lambda)\sin\theta}\right]^2$	$\left(\dfrac{\pi D}{\lambda}\right)^2$
Line of n elements of equal spacing d	$\left[\dfrac{\sin(n\pi d\sin\theta/\lambda)}{n\sin[(\pi d/\lambda)\sin\theta]}\right]^2$	$\dfrac{n}{1 + \dfrac{2}{n}\sum\limits_{\rho=1}^{n-1}(n-\rho)\dfrac{\sin(2\rho\pi d/\lambda)}{2\rho\pi d/\lambda}}$
Two-element array; as above but $n = 2$	$\left[\dfrac{\sin(2\pi d\sin\theta/\lambda)}{2\sin[(\pi d/\lambda)\sin\theta]}\right]^2$	$\dfrac{2}{1 + \left[\dfrac{\sin(2\pi d/\lambda)}{2\pi d/\lambda}\right]}$

if ever, occur. For example, the noise background of the sea is known to be anisotropic, and to have directionality in both vertical and (at low frequencies) horizontal planes (Sec. 7.9). Moreover, signals transmitted in the sea are perfectly coherent only at short ranges; at long ranges, where the improvement in signal to noise provided by an array is most desired, transmitted signals commonly are received from a number of different vertical directions over a variety of refracted and reflected multipaths. Such multipath propagation causes the coherence of a low-frequency signal to decrease rapidly with range and hydrophone separation (15), so that, when an array is steered toward the

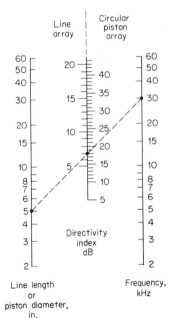

fig. 3.6 *Directivity-index nomogram for line and circular-piston arrays in an infinite baffle.*

signal arriving along one of the multipaths, the signals arriving from the others act as noise and produce a degraded array gain. Because of this, directivity index, although it is still a useful sonar parameter for approximate calculations, should be used with caution in the complicated signal and noise environment of the real ocean. Whenever the coherence characteristics of signal and noise are known, or can be guessed, array gain should replace DI in realistic sonar calculations.

3.5 Transducer Responses

Since a hydrophone is a device which (normally) linear transforms sound into electricity, there exists a proportionality factor, called the *response* of the hydrophone, that relates the generated voltage to the acoustic pressure of the sound field.

The *receiving response* of a hydrophone is the voltage across its terminals produced by a plane wave of unit acoustic pressure (before the introduction of the hydrophone into the sound field). It is customary to express the receiving response as the *open-circuit response* obtained when the hydrophone works into an infinite impedance. The receiving response is usually stated as the number of decibels relative to 1 volt produced by an acoustic pressure of 1 micropascal (10^{-5} dyn/cm^2), and written as decibels re 1 V/μPa. Thus, a response stated as -80 dB re 1 volt means that the hydrophone generates an open-circuit rms voltage of 10^{-4} volt when placed in a plane-wave sound field having an rms pressure of 1 micropascal.

The *transmitting-current response* of a projector is the pressure produced at a point 1 meter from the projector in the direction of the axis of its beam pattern by a unit current into the projector. The transmitting response is usually stated as the number of decibels relative to 1 micropascal as measured at the reference distance, produced by 1 ampere into the electric terminals of the projector, and expressed as a certain number of decibels relative to 1 μPa/A. Transmitting responses are customarily referred to a point distant 1 *meter* from the source; for a reference distance of 1 yard, a correction of 20 log 39.4/36 or +0.78 dB is required to convert the transmitting response to the source level required for the sonar equations expressed in yards. As an example, a projector having a response of 100 dB re 1 μPa/A (referred to 1 meter) will produce an rms pressure of 10^5 dyn/cm^2 at 1 meter when driven with a current of 1 rms ampere. The corresponding source level would be 100.78 dB re 1 μPa/A at 1 yard.

3.6 Calibration Methods

The determination of the response as a function of frequency and direction constitutes the calibration of a transducer. A large number of methods for determining transducer responses are available. Table 3.3 is a summary of calibration methods as compiled from the literature by T. F. Johnston,

table 3.3 Comparison of Frequency Response Calibration Methods

No.	Method	Advantages	Disadvantages	References (List at foot of table)
1	Reciprocity	Standards not required; absolute method	Lengthy; complex; requires reciprocal transducer	
	(a) Spherical			1
	(b) Cylindrical			2
	(c) Plane			3
	(d) Diffuse			4
	(e) Self-reciprocity			5
	(f) General theory			6–10
2	Substitution or comparison	Simple; rapid	Not absolute method; requires calibrated standard transducer	11
3	Static	Simple	Indirect computation required	12, 13
4	Hydrostatic	Large body of water not required	Low frequency only	14, 25
5	Two-projector null method	Absolute method	Low frequency only; requires sealed tank	1, 15
6	Near calibration of large transducers	Eliminates need for large body of water	Special instrumentation; complex data analysis	16–18
7	Pulse calibration	Reduces effect of reflections and standing waves	Elaborate equipment required	19–21
8	Noise calibration	Continuous wideband coverage in frequency	White (or "pink") noise generator required	22
9	Impedance calibration	Tank or body of water not required	Complex data reduction	23
10	Explosions	Wideband	Results are complex; noisy; unsafe	24
11	Pressure tube	For high-pressure calibrations	Special tube needed	26
12	Pressure-gradient calibrator	For pressure-gradient hydrophones	Special tank needed	27

1. Calibration of Underwater Electroacoustic Transducers, S1.20-1972, American National Standards Institute, New York, 1972.
2. Bobber, R. J., and G. A. Sabin: Cylindrical Wave Reciprocity Parameter, *J. Acoust. Soc. Am.*, **33:**446 (1961).
3. Simmons, B. D., and R. J. Urick: The Plane Wave Reciprocity Parameter and Its Application to the Calibration of Electroacoustic Transducers at Close Distances, *J. Acoust. Soc. Am.*, **21:**633 (1949).
4. Diestel, H. G.: Reciprocity Calibration of Microphones in a Diffuse Sound Field, *J. Acoust. Soc. Am.*, **33:**514 (1961).
5. Carstensen, E. G.: Self-Reciprocity Calibration of Electroacoustic Transducers, *J. Acoust. Soc. Am.*, **19:**702 (1947).
6. Ballantine, S.: Reciprocity in Electromagnetic, Mechanical, Acoustical and Interconnected Systems, *Proc. Inst. Radio Eng.*, **17:**929 (1929).
7. MacLean, W. R.: Absolute Measurement of Sound without a Primary Standard, *J. Acoust. Soc. Am.*, **12:**140 (1940).
8. Foldy, L. L., and H. Primakoff: General Theory of Passive Linear Electroacoustic Transducers and the Electroacoustic Reciprocity Theorem: I, *J. Acoust. Soc. Am.*, **17:**109 (1945).
9. Primakoff, H., and L. L. Foldy: General Theory of Passive Linear Electroacoustic Transducers and the Electroacoustic Reciprocity Theorem: II, *J. Acoust. Soc. Am.*, **19:**50 (1947).

10. McMillan, E. M.: Violation of the Reciprocity Theorem in Linear Passive Electromechanical Systems, *J. Acoust. Soc. Am.*, **18**:345 (1946).
11. Albers, V. M.: "Underwater Acoustics Handbook," 1st ed., chap. 18, p. 243, The Pennsylvania State University Press, University Park, Pa., 1960.
12. Snavely, B. L., and G. S. Bennett: Static Calibrations of Acoustic System MK1 Hydrophones, U.S. Nav. Ord. Lab., White Oak, Md., NOLM 8698, Aug. 19, 1946.
13. Raymond, F. W.: Nonacoustical Method for Measuring Sensitivity of a Piezoelectric Hydrophone, *J. Acoust. Soc. Am.*, **35**:70 (January 1963).
14. Bobber, R. J.: The Calibration of a Low-Frequency Calibrating System, *U.S. Navy Underwater Sound Res. Lab. Rep.* 23, Orlando, Fla.
15. Trott, W. J., and E. N. Lide: Two-Projector Null Method for Calibration of Hydrophones at Low Audio and Infrasonic Frequencies, *J. Acoust. Soc. Am.*, **27**:951 (September 1955).
16. Horton, C. W., and G. S. Innis: The Computation of Far-Field Radiation Patterns from Measurements Made near the Source, *J. Acoust. Soc. Am.*, **33**:877 (1961).
17. Baker, D. D.: Determination of Far-Field Characteristics of Large Underwater Sound Transducers from Near-Field Measurements, *J. Acoust. Soc. Am.*, **34**:1737 (1962).
18. Baker, D. D., and K. McCormack: Computation of Far-Field Characteristics of a Transducer from Near-Field Measurements Made in a Reflective Tank, *J. Acoust. Soc. Am.*, **35**:736 (1963).
19. Ginty, G. F.: Transducer Test and Calibration Equipment, 1954, U.S. Navy Underwater Sound Laboratory, Fort Trumbull, New London, Conn. (In its *Res. Develop. Rep.* 260, quarterly report, July 1–Sept. 30, 1954, pp. 49–52.)
20. Wallace, J. D., and E. W. McMorrow: Sonar Transducer Pulse Calibration System, *J. Acoust. Soc. Am.*, **33**:75 (1961).
21. Terry, R. L., and R. B. Watson: Pulse Technique for the Reciprocity Calibration of Microphones, *J. Acoust. Soc. Am.*, **23**:684 (1951).
22. McMorrow, E. W., J. D. Wallace, and J. J. Coop: Technique for Rapid Calibration of Hydrophones Using Random Noise Fields, *Tech. Memo* ADC-EL41-EWM:JDW:JJC, U.S. Nav. Air Develop. Center, Johnsville, Pa., 1953.
23. Sabin, G. A.: Transducer Calibration by Impedance Measurements, *J. Acoust. Soc. Am.*, **28**:705 (1956).
24. Carter, J. L., and M. F. M. Osborne: Hydrophone Calibration by Explosion Waves, *U.S. Nav. Res. Lab. Rep.* S-2179, Washington, D.C., 1944.
25. Schloss, F., and M. Strasberg: Hydrophone Calibrations in a Vibrating Column of Liquid, *J. Acoust. Soc. Am.*, **34**:958 (1962).
26. Beatty, L. G., R. J. Bobber, and D. L. Phillips: Sonar Transducer Calibration in a High-Pressure Tube, *J. Acoust. Soc. Am.*, **39**:48 (1966).
27. Bauer, B. B., L. A. Abbagnaro, and J. Schumann: Wide Range Calibration System for Pressure Gradient Hydrophones, *J. Acoust. Soc. Am.*, **51**:1717 (1972).

formerly of the Naval Ordnance Laboratory. These methods have different desirable and undesirable features depending upon transducer size and frequency range. Further information on transducer calibration techniques can be found in a book by Bobber (16).

Comparison method The most simple and straightforward of these is the comparison method, wherein the output of the unknown transducer is compared with that of a transducer that has been previously calibrated. This method is easily and rapidly carried out and is the most commonly used of all methods, but requires a comparison transducer of known and reliable calibration.

A large variety of standard transducers have been developed by, and are available from, the Underwater Sound Reference Division of the Naval Research Laboratory (17).

Reciprocity method At the other extreme is the reciprocity method of transducer calibration, which requires no reference standard, but requires a series of measurements on several transducers. Because of its basic importance, it will be described in some detail.

The practical importance of electroacoustic reciprocity was first pointed out by MacLean (18) and was applied to underwater sound calibration by Ebaugh and Meuser (19). The method was extended by Carstensen (20) to a self-reciprocity method involving reflection from a reflecting surface. The reciprocity method is based on the *electroacoustic reciprocity principle,* similar to the well-known electrical reciprocity principle for passive bilateral networks. This principle is sometimes expressed by saying that in any network composed of linear bilateral elements excited by a zero-impedance generator, the reading of a zero-impedance ammeter is unchanged when the ammeter and generator are interchanged. The corresponding theorem for electroacoustic networks can be shown to relate current and open-circuit voltage on the electrical side to the diaphragm velocity and impressed force on the rigid, or blocked, diaphragm on the acoustical side (Fig. 3.7). The relationship is

$$\frac{|E|}{|v|} = \pm \frac{|F|}{|I|} \quad \text{or} \quad \frac{|E|}{|F|} = \pm \frac{|v|}{|I|}$$

where $|E|$ = absolute magnitude of electric voltage
$|I|$ = absolute magnitude of electric current
$|v|$ = magnitude of velocity on diaphragm
$|F|$ = magnitude of force on diaphragm

The ± signs arise because of the different kinds of electromechanical coupling in different types of transducers (the + sign applies for piezoelectric, the − sign for magnetostrictive and electromagnetic transducers). The above expressions can be applied to the problem of transducer calibration by

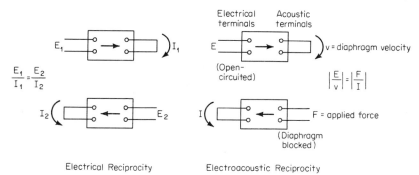

fig. 3.7 *Recriprocity in electric networks and electroacoustic transducers.*

relating the diaphragm velocity v to the pressure produced at a distance when the transducer is acting as a projector, and by relating the force F to the incident sound wave when the transducer is acting as a hydrophone. These relationships depend on the kind of sound field being considered, that is, whether the transducer generates spherical, cylindrical, or plane waves.

Spherical waves If the transducer is considered to be a pulsating sphere and to generate spherical waves, the pressure at a large distance can be shown from acoustic theory (21) to be

$$|P|_r = \frac{f\rho}{2r}|Q|$$

where $|P|_r$ = absolute magnitude of pressure at distance r
$|Q|$ = absolute magnitude of volume of source
ρ = fluid density
f = frequency

For a spherical source, the volume velocity $|Q|$ is the product of area A and surface velocity $|v|$ so that

$$|P|_r = \frac{f\rho}{2r}A|v|$$

When the transducer becomes a hydrophone, the force F of the incident sound field upon it is equal to its area multiplied by the incident sound pressure,* or

$$F = P_i A$$

Dropping the absolute magnitude signs from here on, and substituting for v and F in the reciprocity relationship, we obtain

$$\frac{E}{P_i A} = \frac{P_r}{I}\frac{2r}{\rho f A}$$

But the ratios E/P_i and P_r/I are the receiving and transmitting responses (for $r = 1$) as previously defined. Call these ratios M, for "microphone," and S, for "speaker." Then

$$M = S\frac{2r}{\rho f}$$

or

$$\frac{M}{S} = \frac{2r}{\rho f} \equiv J_s$$

---

* This is true only at low frequencies where diffraction effects are absent. However, a rigorous derivation of reciprocity, based on the electroacoustic equations involved in Fig. 3.7, is not limited by this restriction.

The quantity J_s is called the *spherical-wave-reciprocity parameter* and is the ratio of the response of a reciprocal transducer acting as a projector to its response as a hydrophone when the transducer is regarded as a source of spherical waves.

Cylindrical waves Now let the transducer be a pulsating cylinder generating cylindrical waves. If the surface velocity of the cylinder is v, the acoustic pressure at distance r can be shown from basic theory (22) to be

$$P_r = \pi \rho a v \left(\frac{cf}{r}\right)^{1/2}$$

where a = radius of cylinder and c = velocity of sound. If, as before, we change the transducer to a hydrophone, the force exerted upon it by an incident sound field of pressure P_i will be

$$F = AP_i = 2\pi a L P_i$$

where A = face area of cylindrical hydrophone and L = length of cylindrical hydrophone. Inserting in the reciprocity relationship, we obtain

$$\frac{E}{P_i 2\pi a L} = \frac{P_r}{I} \left(\frac{r}{cf}\right)^{1/2} \frac{1}{\pi \rho a}$$

Writing, as before, $E/P_i \equiv M$, $P_r/I \equiv S$, we obtain

$$\frac{M}{S} = \frac{2L}{\rho c} \left(\frac{cr}{f}\right)^{1/2} \equiv J_c$$

The ratio J_c is called the *cylindrical-wave-reciprocity parameter* and was first derived by Bobber and Sabin (23). It is the parameter appropriate for transducers generating cylindrical waves.

Plane waves When the transducer, acting as projector, is imagined to produce plane waves, the pressure produced at distance r when its diaphragm has the velocity v is simply

$$P_r = \rho c v$$

When the transducer is acting as a receiver of plane waves of pressure P_i, the force upon its surface of area A is

$$F = 2AP_i$$

where the factor 2 accounts for pressure doubling at the transducer face. Inserting in the reciprocity relationship, we obtain

$$\frac{E}{2AP_i} = \frac{P_r}{\rho c I}$$

so that

$$\frac{M}{S} = \frac{E/P_i}{P_r/I} = \frac{2A}{\rho c} = J_P$$

This is the *plane-wave-reciprocity parameter* originally derived by Simmons and Urick (24) and is the appropriate parameter for transducers generating plane waves.

Summary The three reciprocity parameters can be written synoptically as follows:

$$\text{Plane: } J_p = \frac{2}{\rho c}(\lambda r)^0 A$$

$$\text{Cylindrical: } J_c = \frac{2}{\rho c}(\lambda r)^{1/2} L$$

$$\text{Spherical: } J_s = \frac{2}{\rho c}(\lambda r)$$

where the exponent of the term λr indicates the spreading law (Sec. 5.2) for the three kinds of acoustic radiation.

For a given transducer, the choice of parameters for a reciprocity calibration depends upon the geometry of the transducer and the distance at which the calibration is done. A long cylindrical transducer to be calibrated at a short distance $r \leq L^2/\lambda$ would require the cylindrical parameter, whereas a flat-faced pistonlike transducer calibrated at a distance $r \leq A/\lambda$ would require the plane-wave parameter. When large projectors must be calibrated at close distances, however, special techniques are needed (25, 26) in order to appropriately sample the complex sound field at distances comparable to the dimensions of the radiating faces of the projectors.

The reciprocity parameter J is the ratio of the receiving response of a reciprocal transducer to its transmitting response. When one of these is known, the other can be readily computed by means of the reciprocity parameter, provided the transducer is reciprocal and the type of waves for which the calibration applies is known. For spherical waves, the relation between the two responses is

$$20 \log S = 20 \log M + 20 \log f + 354$$

where $20 \log S$ is the transmitting response in decibels re $1 \mu\text{Pa/A}$ at 1 meter, $20 \log M$ is the receiving response in decibels re $1 \text{ V}/\mu\text{Pa}$, and f is the frequency in kilohertz. This expression is useful for finding one of the responses of a transducer when the other is known without resorting to a separate measurement, provided the transducer is indeed reciprocal. As an example, suppose we have a reciprocal transducer with a *receiving* response at 1 meter of $M = -77.3$ dB re $1 \text{ V/(dyn)(cm}^2) = -177.3$ dB re $1 \text{ V}/\mu\text{Pa}$. What is its *transmitting* response at a frequency of 41 kHz? The answer is $S = -177.3 + 32.3 + 354.0 = 209.0$ dB re $1 \mu\text{Pa} = 109.0$ dB re $1 \text{ dyn/(cm}^2)$ (A) at 1 meter, or 109.8 dB at 1 yard.

3.7 Reciprocity Calibration

Electroacoustic reciprocity makes possible the calibration of transducers without reference to a transducer of known response. A reciprocity calibration requires three transducers. One must be reciprocal (in the sense of obeying the basic reciprocity condition); another must be a projector; and the third must be a hydrophone. The response of either the projector or the hydrophone is desired. The transducer being calibrated need not be reciprocal.

Let the three units be nondirectional and placed at equal distances r from one another, as shown in Fig. 3.8. The method in its simplest form involves two steps. In step 1, the projector P is driven with current i, and the voltages appearing at the output terminals of the transducer T (acting as hydrophone) and the hydrophone H are measured. In step 2, the transducer T acts as projector and is driven with the same current i as before. The voltage v'_H generated by the hydrophone H is measured. No other measurements are required.

Let the desired quantity be the receiving response M_H of the hydrophone H. Then, from step 1,

$$\frac{M_H}{M_T} = \frac{v_H}{v_T} \qquad (1)$$

where M_T is the unknown receiving response of T. If the projector P, observed at distance r, is assumed to radiate spherical waves, then, from the definitions of transmitting and receiving responses, we have

$$S_P = \frac{P_r r}{i}$$

and

$$M_H = \frac{v_H}{P_r}$$

where P_r is the acoustic pressure produced by P at distance r. Multiplying these two equations together, we obtain

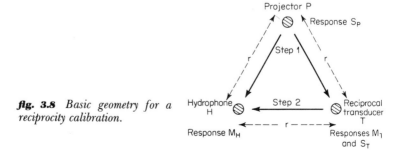

fig. 3.8 Basic geometry for a reciprocity calibration.

$$S_P M_H = \frac{v_H r}{i} \tag{2}$$

From the measurement of step 2,

$$\frac{S_P}{S_T} = \frac{v_H}{v'_H} \tag{3}$$

Solving (1) for M_H, solving (3) for S_P, and inserting into (2), we obtain

$$S_T M_T = \frac{v_T v'_H}{v_H} \frac{r}{i}$$

But by reciprocity, the two responses of the transducer T are related by

$$M_T = J_S S_T$$

Proceeding to eliminate S_T, we find

$$M_T = \left(J_S \frac{v_T v'_H}{v_H} \frac{r}{i}\right)^{1/2}$$

and on substituting in (1), we obtain the desired receiving response of the hydrophone H:

$$M_H = \left(J_S \frac{v_H v'_H}{v_T} \frac{r}{i}\right)^{1/2}$$

The calibration is therefore obtained by three measured output voltages, the distance, the driving current, and the appropriate reciprocity parameter (in this case J_S). The other responses M_T, S_T, and S_P can be computed, if desired, without additional measurements. Standard procedures for making a reciprocity calibration are available (27).

Self-Reciprocity Only a single reciprocal transducer is required for a *self-reciprocity* calibration. Here the transducer to be calibrated is pulsed, and the amplitude of the pulse after reflection from a plane surface is measured. The image of the transducer in the surface acts as the reciprocal transducer (T in Fig. 3.8) for the calibration. In the preceding expression, v_H and v_T become equal, and the expression for the receiving sensitivity reduces to

$$M_H = \left(\frac{J_S v'_H r}{i}\right)^{1/2}$$

Patterson (28) has used this method to obtain a calibration of a transducer at sea. Referring to Fig. 3.9, the unit was lowered to a depth d, and the voltage v'_H of a series of short pulses, generated with current i and reflected from the sea surface, was measured. The distance r is twice the depth, or $2d$. An accuracy of 2 dB was said to have been obtained. This method is particularly useful for obtaining or checking the calibration of a transducer while at sea,

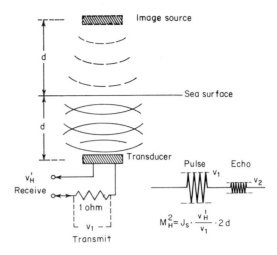

fig. 3.9 *Self-reciprocity at sea. The transducer to be calibrated is suspended at a depth d and sends out short pulses reflected from the surface. The receiving response M_H is given in terms of v_1, the voltage of the transmitted pulse across a 1-ohm resistor, and v_H', the open-circuit voltage of the reflected pulse.*

but it requires the presence of a calm surface, the absence of a prominent supporting platform over the unit, and equipment for generating and receiving short pulses.

3.8 Calibration of Large Arrays

The emergence of large arrays for sonar applications requires some special techniques for calibration at short distances where the array cannot be considered to be a point source. Two solutions to the problem of calibration in the "near field" have been proposed. One method (25, 29, 30) utilizes a probe hydrophone to measure the pressure amplitude and phase at many points in the vicinity of the array. The measurements are entered into a computer program based on theory relating the near-field and "far-field" pressures produced by an arbitrary distribution of sound sources. The result is the pressure at any arbitrary point in the far field. The other method (26, 31) employs a plane array of small sound sources having just the right amplitude shading needed to produce a plane-wave sound field throughout a volume in front of it. The transducer to be calibrated is placed within this volume, and conventional measurements are made of its receiving response and beam pattern. By reciprocity, the special array can be used as a receiver for finding the transmitting response of a large source. Both methods, in effect, integrate the sound field in the vicinity of the large array being measured in order to find the pressure at a point a large distance away.

Another way of producing a plane-wave sound field is to use a parabolic reflector such as those manufactured for use in radar, with a small sound source at its focus. This method (32) was successfully applied to the calibration of a number of transducers and should be generally useful for the high-frequency (>100 kHz) calibration and measurement of beam patterns of transducers that are large compared to a wavelength.

3.9 Beam Patterns

The response of a transducer array varies with direction relative to the array. This property of array *directionality* is highly desirable, for it enables the direction of arrival of a signal to be determined and enables closely adjacent signals to be resolved. At the same time, directionality reduces noise, relative to the signal, arriving in other directions. In projector arrays, directionality serves to concentrate the emitted sound in a desired direction.

Directionality in hydrophone arrays results from the fact that sinusoidal signals arriving in some one direction tend to be in phase at all the array elements, whereas the noise background is out of phase. Similarly, broadband signals arriving in this direction tend to be correlated between pairs of array elements, whereas the noise is not. This direction of maximum inphase condition or maximum correlation is that of the *acoustic axis* of the array and is the direction of maximum sensitivity.

The response of an array varies with direction in a manner specified by the *beam pattern* of the array. If the response is written $R(\theta,\varphi)$ to indicate that it is a function of the angles θ and φ in polar coordinates, then we may write

$$R(\theta,\varphi) = R(0,0)v(\theta,\varphi)$$

where $R(0,0)$ = response in direction $\theta = 0$, $\varphi = 0$
$v(\theta,\varphi)$ = a response function normalized so that $v(0,0) = 1$

The beam pattern, or pattern function, is the square of v or

$$b(\theta,\varphi) = v^2(\theta,\varphi)$$

For hydrophones, $b(\theta,\varphi)$ is the mean-square voltage produced by an array of unit response when sound of unit pressure is incident on it in the direction θ, φ. For projectors, $b(\theta,\varphi)$ is the mean-square pressure produced at unit distance when unit current is fed into the projector. The direction $(0,0)$ is arbitrary, but is ordinarily taken to be the direction of maximum response. In regular arrays, the beam pattern is symmetrical, and the direction $(0,0)$ to which the axial response $R(0,0)$ refers is the acoustic axis of the array. Beam patterns, like responses, are commonly expressed in decibels.

Arrays can normally be *steered*, either mechanically by physically rotating the array, or electrically by inserting in series or in parallel with each array element appropriate phasing networks (for narrow-band arrays) or time-delay networks (for broadband arrays) that effectively rotate the direction of maximum response into some desired direction. For example, the direction of maximum sensitivity of a plane array of elements located at the positions of the open circles in Fig. 3.10 can be rotated into a direction lying at angle θ_0 to a reference direction by delaying the output of each element by the time for sound to travel the distances l. In this way, the array is effectively converted into a line array along the line AB. The output voltage of an array steered in the (θ,φ) direction can be written as

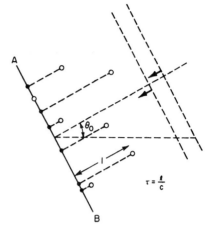

fig. 3.10 Beam steering of an irregular plane array. Electrical time delays equal to l/c are inserted in series with each array element, where l is the perpendicular distance from the reference line AB and c is the velocity of sound. By so doing, the elements are brought to lie, in effect, along the line AB.

$$v(\theta,\theta_0, \varphi,\varphi_0) = R(\theta_0,\varphi_0)v(\theta_0 - \theta, \varphi_0 - \varphi)$$

where $R(\theta_0,\varphi_0)$ is the response when the array is steered along (θ_0,φ_0) direction and $v(\theta_0 - \theta, \varphi_0 - \varphi)$ is the square root of the beam pattern referred to this direction and again normalized so that $v(0,0) = 1$. For steered arrays, the beam pattern is not usually symmetrical, so that $b(\theta_0 - \theta, \varphi_0 - \varphi) \neq b(\theta - \theta_0, \varphi - \varphi_0)$.

In their simplest form, many arrays are arranged with elements along a line or distributed along a plane. The acoustic axis of such *line* or *plane* arrays, when unsteered, lies at right angles to the line or plane. The beam pattern of a line array may be visualized as a doughnut-shaped figure having supernumerary attached doughnuts formed by the side lobes of the pattern. The three-dimensional pattern of a plane array is a searchlight type of figure with rotational symmetry about the perpendicular to the plane plus side lobes. Three-dimensional views of such patterns are drawn in Fig. 3.11.

Line of equally spaced elements The beam pattern of a line of equally spaced, equally phased (i.e., unsteered) elements can be derived as follows. Let a plane sinusoidal sound wave of unit pressure be incident at an angle θ to a line of n such elements. As seen in Fig. 3.12, the output of the mth element relative to that of the zeroth element will be delayed by the amount of time necessary for sound to travel the distance $l_m = md \sin \theta$; the corresponding phase delay for sound of wavelength λ will be, at frequency $\omega = 2\pi f$,

$$u_m = \frac{2\pi}{\lambda} l_m = mu$$

where the phase delay between adjacent elements is in radians,

$$u = \frac{2\pi d}{\lambda} \sin \theta$$

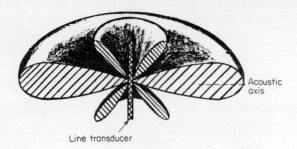

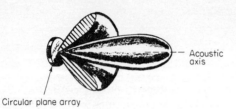

fig. 3.11 *Three-dimensional view of the beam pattern of a line and a circular-plane array.*

The output voltage of the mth element of voltage response R_m will be

$$v_m = R_m \cos(\omega t + mu)$$

and the array voltage will be the sum of such terms:

$$V = R_0 \cos \omega t + R_1 \cos(\omega t + u) + \cdots \\ + R_m \cos(\omega t + mu) + \cdots + R_{n-1} \cos[\omega t + (n-1)u]$$

In the complex notation, the array voltage will be

$$V = (R_0 + R_1 e^{iu} + R_2 e^{2iu} + \cdots + R_{n-1} e^{(n-1)iu}) e^{i\omega t}$$

If the array elements are all of unit response, we can drop the R's and write

$$V = (1 + e^{iu} + e^{2iu} + \cdots + e^{(n-1)iu}) e^{i\omega t}$$

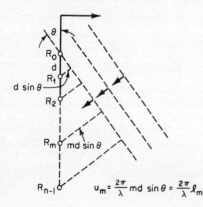

fig. 3.12 *Line array of equally spaced elements.*

Multiplying by e^{iu} and subtracting, we find

$$V = \frac{e^{inu} - 1}{e^{iu} - 1} e^{i\omega t}$$

which, after manipulating, and neglecting the time dependence, becomes

$$V = \frac{\sin(nu/2)}{\sin(u/2)}$$

Finally, expressing u in terms of θ, we obtain the desired beam pattern, which is the square of this function normalized to unity at $\theta = 0$

$$b(\theta) = \left(\frac{V}{n}\right)^2 = \left[\frac{\sin(n\pi d \sin\theta/\lambda)}{n \sin(\pi d \sin\theta/\lambda)}\right]^2$$

Arrays can be *steered* by introducing appropriate phase or time delays in the outputs of the various elements, as mentioned above, in order to rotate the main lobe of the pattern to a desired direction. When this is done, the beam width and side-lobe structure of the beam pattern are changed. Figure 3.13

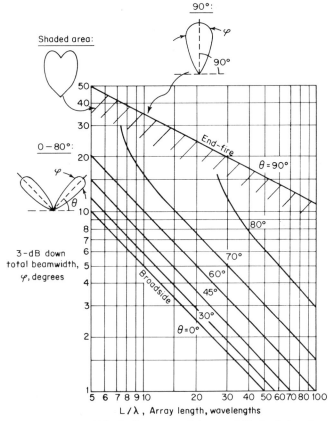

fig. 3.13 Beam width of a line array of elements $\lambda/2$ apart at various steering angles to broadside. (Ref. 33.)

gives the angular width between the −3-dB down points of the main lobe of a line array, with steering angle as a parameter (33). In the "end-fire" orientation ($\theta = 90°$), the beam pattern is searchlight-shaped, instead of doughnut-shaped; in a transition region near 90°, shown shaded in Fig. 3.13, the 3-dB-down beam width is ambiguous. Although the beam width widens as the beam is steered from the broadside to end-fire, the directivity index of the steered line remains, strangely enough, approximately the same.

In a *synthetic aperture* line array, the hydrophones of a multielement conventional line array are replaced by a *single* hydrophone towed, or otherwise moved, through the water. The principle of the synthetic aperture array as applied to a side-looking sonar is shown in Fig. 3.14. A small transducer, acting as both a pinging source and a receiver, is towed through the water so as to be at positions T_1 to T_5 at successive instants of time. By means of suitable electronics, the backscattered return, along with the echo from any fixed object P, is recorded, stored, and later range-gated and summed at the appropriate time intervals corresponding to the lateral range r of the object P. The sum of the gated samples will be seen to be equivalent to the output of a five-element conventional array "focused" on the object P. Very long virtual arrays can be realized in this way with only a single transducer, but they require that the echo be coherent over the distance and time interval required for beam formation. Synthetic arrays have been extensively employed in side-scan radars for ground mapping from aircraft (34).

Continuous-line and plane circular arrays When the array elements are so close together that they may be regarded as adjacent, the array becomes a

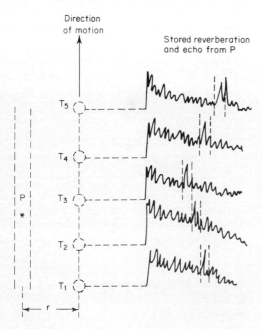

fig. 3.14 *Principle of the synthetic aperture array. A transducer is moved to successive positions, T_1 to T_5. The return from a short ping is stored and sampled at the time delays required to focus the virtual beam on P. The samples between the dotted lines are summed to form the virtual beam.*

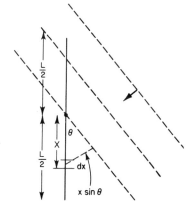

fig. 3.15 Geometry of the continuous-line array.

continuous-line transducer, and the beam pattern can be found by integration instead of by summation. For this case let the line transducer be of length L, and have a response per unit length of R/L. The contribution to the total voltage output produced by a small element of line length dx located a distance x from the center (Fig. 3.15) will be (neglecting the time dependence)

$$dV = \frac{R}{L} e^{(i2\pi/\lambda)x \sin \theta} \, dx$$

and the total voltage will be

$$V = \frac{R}{L} \int_{-L/2}^{L/2} e^{(i2\pi/\lambda)x \sin \theta} \, dx$$

$$= R \, \frac{e^{(i\pi L/\lambda) \sin \theta} - e^{-(i\pi L/\lambda) \sin \theta}}{(i2\pi L/\lambda) \sin \theta}$$

$$= R \, \frac{\sin\left[(\pi L/\lambda) \sin \theta\right]}{(\pi L/\lambda) \sin \theta}$$

The beam pattern will be the square of V normalized so that $b(0) = 1$, or

$$b(\theta) = \left(\frac{V}{R}\right)^2 = \left[\frac{\sin\left[(\pi L/\lambda) \sin \theta\right]}{(\pi L/\lambda) \sin \theta}\right]^2$$

In a similar manner, the beam pattern of a circular plane array of diameter D of closely spaced elements can be shown to be

$$b(\theta) = \left[\frac{2J_1\left[(\pi D/\lambda) \sin \theta\right]}{(\pi D/\lambda) \sin \theta}\right]^2$$

where $J_1[(\pi D/\lambda) \sin \theta]$ is the first-order Bessel function of argument $(\pi D/\lambda) \sin \theta$.

60 / principles of underwater sound

Generalized beam patterns for continuous-line and circular plane arrays are drawn in Fig. 3.16 in terms of the quantities $(L/\lambda) \sin \theta$ and $(D/\lambda) \sin \theta$.

Figure 3.17 is a nomogram for finding the angular width, between the axis and the -3-dB down and -10-dB down points, of the beam pattern of continuous-line and circular plane arrays. The dashed lines indicate how the nomogram is to be used. Thus, a circular plane array of 20 in. diameter at a wavelength of 4 in. (corresponding to a frequency of 15 kHz at a sound velocity of 5,000 ft/s) has a beam pattern 6° wide between the axis of the pattern and the -3-dB down point, and 10° wide between the axis and the -10-dB down point. For a line array 30 in. long at a wavelength of 10 in. (frequency 6 kHz), the corresponding angles are 8 and 15°.

In general, the angle in degrees to the -3-dB down point of the pattern of a continuous-line array of length L at wavelength λ is equal to 25.3 λ/L; for a circular plane array of diameter D, the angle equals 29.5 λ/D.

3.10 Product Theorem and the Mills Cross

So far, each element of an array of hydrophones has been nondirectional. When the array comprises identical hydrophones that are themselves directional, the beam pattern is given by the *product theorem,* which states that the beam pattern of an array of identical, equally spaced directional hydrophones is the product of (1) the pattern of each hydrophone alone, and (2) the pattern of an identical array of nondirectional hydrophones.

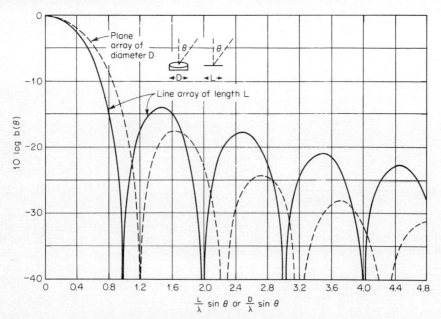

fig. 3.16 Beam patterns of a line array of length L and of a circular plane of diameter D in terms of the dimensionless parameters L/λ sin θ and D/λ sin θ.

properties of transducer arrays: directivity index / 61

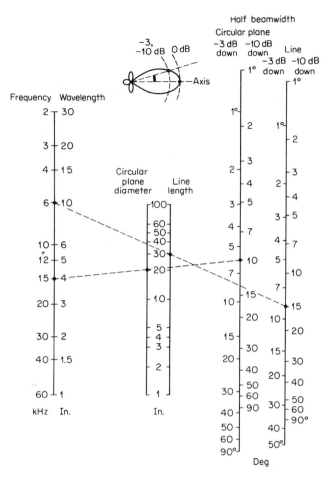

fig. 3.17 Nomogram for finding the width of the beam pattern of circular-plane and line transducers.

The theorem is almost self-evident. Consider a line array of equally spaced elements with a separation d, all of unit response. If they are nondirectional, the array voltage produced by a unit sound pressure will be, as before,

$$V_{Nond}(\theta) = 1 + e^{iu} + e^{2iu} + \cdots$$

where u is the angular function $u = (2\pi d/\lambda) \sin \theta$. If each element itself has a response that varies with direction according to the function $v(\theta,\varphi)$, the array output will be

$$V_D(\theta,\varphi) = v(\theta,\delta) + v(\theta,\varphi)e^{iu} + v(\theta,\varphi)e^{2iu} + \cdots = v(\theta,\varphi)V_{Nond}(\theta)$$

Hence

$$b_D(\theta,\varphi) = V_D{}^2(\theta,\varphi) = b(\theta,\varphi)b_{Nond}(\theta)$$

The beam pattern of the array of directional elements $b_D(\theta,\varphi)$ will therefore be the product of the beam pattern $b_{Nond}(\theta)$ of an identical array of nondirectional elements multiplied by the beam pattern $b(\theta,\varphi)$ of each element alone. The theorem provides a handy shortcut in array design involving directional hydrophone elements.

The *Mills Cross* is an array of two line transducers whose outputs are multiplied, or correlated, together. If the two lines are perpendicular to each other, the beam pattern, by the product theorem, is the same as that of a plane rectangular array of elements of the same overall dimensions. That is, if each of the two line transducers has n elements, the beam pattern of a Mills Cross containing $2n$ elements is the same as that of rectangular array of n^2 elements. The Mills Cross has the advantages of light weight and an economy of transducer elements, though at the expense of a lower array gain and a lower sensitivity than a rectangular array. Its usefulness occurs under conditions of high signal-to-noise ratio, where array gain is immaterial, and where factors such as weight, cost, and good directionality are important considerations.

3.11 Shading and Superdirectivity

Shading is a method by which some degree of control can be exercised over the pattern of an array having some particular geometry. Amplitude shading involves adjusting the responses of array elements to provide the most desirable pattern for some particular purpose. Nearly always, an array is shaded with maximum response at the center and least response at the ends or sides of the array so that the sensitivity is *tapered* from a high value on the inside toward a lower value on the outside.

The effects of adjusting the response of array elements may be appreciated by comparing the beam patterns of a six-element line array shaded in different ways, as shown in Fig. 3.18. A convenient starting point is an unshaded array, whose shading formula, expressing the responses of the array elements, may be written 1, 1, 1, 1, 1, 1. The beam pattern of an unshaded line array has been derived above. When the sensitivity is all at the ends, with formula 1, 0, 0, 0, 0, 1, the pattern consists of a series of narrow lobes all of equal amplitude. When, on the other hand, the array is shaded strongly toward the center, with formula 0, 0, 1, 1, 0, 0, the side lobes disappear, leaving a single, broad main lobe. Thus the effect of tapering from center outward (0, 0, 1, 1, 0, 0) is widening of the main beam and reduction of the side lobes; the effect of reverse tapering (1, 0, 0, 0, 0, 1) is narrowing of the main beam at the expense of increased side lobes. *Binomial* shading, with sensitivities proportional to the coefficients of a binomial expansion (0.1, 0.5, 1, 1, 0.5, 0.1), yields the narrowest main lobe, with a total absence of side lobes: *Dolph-Chebyshev* shading (0.30, 0.69, 1, 1, 0.69, 0.30) gives the narrowest main lobe, with side lobes of a preassigned level. The procedures involved in a Dolph-Chebyshev design may be found in a paper by Davids, Thurston, and Meuser (35) and in a book by Albers (36). In general, the selection of

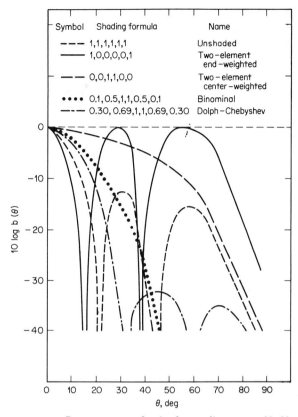

fig. 3.18 Beam patterns of a six-element line array of half-wave spacing between elements for different shading formulas.

shading factors to achieve a desired beam pattern can be done by the method of Lagrangian multipliers, as pointed out originally in a paper by Pritchard (37).

Amplitude shading is a convenient way to tailor a beam pattern to some desired shape. However, with a coherent signal in uniform incoherent noise, shading results in a lower array gain than with constant element sensitivity. It follows that heavy shading should not be used under conditions of low signal-to-noise ratio, where the array gain is an important consideration in the design of the array.

Sometimes, as in the self-noise environment in sonar domes on shipboard, or with an array of hydrophones placed along the length of the ship, the level of the noise is different at different elements of the array. Under this condition it is beneficial to reduce the sensitivity of the noisy elements. It can be shown that if the amplitude shading factor of the ith element is a_i, and if the noise intensity at the ith element is I_i, the maximum array gain results when the a_i's are chosen so as to keep the products $a_i^2 I_i$ constant over all the elements of the array. This shading adjustment can be made adaptively by varying the gain of preamplifiers at the outputs of the array elements so as to take care of any changes that may occur in the noise levels I_i at the array elements.

An extreme form of shading is called *superdirectivity*, in which narrow beams may be obtained with arrays of limited size. In a superdirective array, the elements are spaced less than one-fourth wavelength apart, with the signs, or polarities, of adjacent elements reversed. The properties of superdirective arrays were pointed out by Pritchard (38). For example, the directivity index of an array of five elements one-eighth wavelength apart—and thus only one-half wavelength long—and having the shading factors $+1, -4, +6, -4, +1$ was computed to be 5.6 dB. The price to be paid for superdirectivity is a low array sensitivity due to the phase reversals of the elements, together with relatively high side lobes—a price usually greater than the gain in directivity index that it buys.

Phase shading may also be used in arrays. In this kind of shading, the spacing of the array elements is varied to obtain the desired pattern. This method has received considerable attention in the design of radio antennas (39–41) but has not been extensively employed in underwater sound.

3.12 Adaptive Beam Forming

When a source of noise exists in a certain direction, a null can be placed in the beam pattern to cancel out the noise. This cancellation can be done *adaptively* by using appropriate digital techniques that allow automatically for changes in the angle of the interfering noise. One such adaptive technique is called *dicanne* (42). In dicanne processing, suitable time delays are inserted to form a so-called *estimator beam* in the direction of the interference. This beam is then subtracted from the outputs of the array elements. After subtraction, complementary delays are inserted that cancel out, in effect, the delays used for forming the estimator beam. Finally, the array element outputs are fed into a conventional beam former. The result of this process is a modified beam pattern having a dip in the direction of the source of the interference.

Another form of adaptive processing is to allow for differing noise levels between array elements. In shipboard arrays, the self-noise may increase along the length of the array, one end being quiet, the other noisy. On submarines, for example, boundary layer noise increases with distance aft from the bow, while in towed arrays, the forward array elements tend to be more noisy than the elements that are farther away from the noise of the towing vessel. In such arrays, wherein the signal level is constant over the array but the noise level is not, it has been shown (43) that for uncorrelated noise the maximum gain results if amplitude shading is used such that products $a_i^2 I_i$ are kept constant. Here a_i is the amplitude shading coefficient of the ith element and I_i is the noise intensity at that element. This amounts to reducing the sensitivity of the noisy elements relative to the quiet elements, a process that can be readily done automatically by simple electronics.

3.13 Multiplicative Arrays

Thus far we have been concerned with arrays that are both *linear* and *additive*—that is, they comprise elements that have an output linearly proportional to the acoustic pressure and are simply added together, or summed, to give the array output. Instead of being added, the element outputs may be *multiplied* together so as to form a multiplicative or correlative array. Such arrays have properties different from those of linear arrays.

Figure 3.19a shows a four-element linear array in which the elements are simply added together. In Fig. 3.19b, the four elements are multiplied together, and the output of the multiplier is averaged or smoothed. This kind of multiplicative array, called an intraclass correlator, was originally studied by Faran and Hills (44) and more recently by Fakley (45). When the outputs of various multipliers are themselves multiplied together, a time-average-product (TAP) array [so called by Berman and Clay (46)] is obtained. One arrangement of a four-element TAP array is shown in Fig. 3.19c.

A multiplicative array of only two elements can be made to have the same beam pattern as a linear array of an arbitrary number of equally spaced elements. To see this, consider a linear array of an odd $(2n - 1)$ number of elements spaced a distance d apart. If the array is assumed to be amplitude-shaded symmetrically about the center element, the array output can be written as

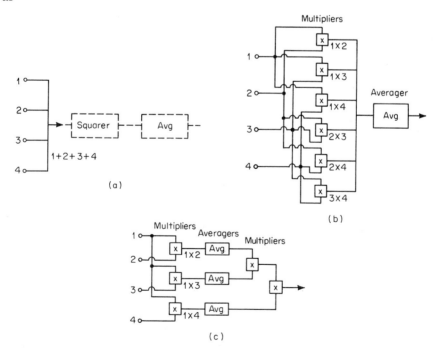

fig. 3.19 Schematic diagrams of (a) additive array, (b) intraclass correlator, and (c) TAP array (Class I of Berman and Clay).

$$v(u) = \sum_{k=1}^{n} a_k \cos 2ku$$

where a_k is the common shading factor of the two elements a distance kd away from the center element, and $u = (\pi d/\lambda) \sin \theta$. Similarly, for an even number of elements,

$$v(u) = \sum_{k=1}^{n} a_k \cos (2k - 1)u$$

where a_k is the common shading factor for elements distant $(k - \frac{1}{2})d$ from the center. In these expressions, each of the $\cos 2ku$ or $\cos (2k - 1)u$ terms can be expanded in a power series of $\cos u$, thus

$\cos u = \cos u$
$\cos 2u = 2 \cos^2 u - 1$
$\cos 3u = 4 \cos^3 u - 3 \cos u$
$\cos 4u = 8 \cos^4 u - 8 \cos^2 u + 1$
. . . .

These polynomial expansions of $\cos mu$ in powers of $\cos u$ are, in the general case, the Chebyshev polynomials, written

$$T_m(\cos u) = \cos mu$$

These were used originally by Dolph (47) in antenna design, as mentioned above. The array output can therefore be written as a sum of terms like $\cos^m u$, appropriately weighted by the shading factors of the particular array. This means that the beam pattern of an array of any number of elements can be synthesized by an array of only two elements working into appropriate power-law amplifiers and combined in accordance with the shading factors and the terms of the $\cos mu$ expansions. An example, taken from Brown and Rowlands (48), of how the beam pattern of a linear array of $2n + 1$ elements can be achieved, in principle, with only two elements plus nonlinear amplifiers is shown in Fig. 3.20.

This technique is appealing in applications where a reduction in the number of hydrophone elements is necessary. Here one may trade physical size, or number of elements, or both, for complexity in processing. The method is valid only at high signal-to-noise ratios, since the effect of noise (so far neglected) is to destroy the beam pattern unless a long averaging time is used.

The saving in number of array elements is coupled with a degradation of the signal-to-noise ratio when the signal-to-noise ratio is less than unity. Using communication theory, Brown and Rowlands (48) showed that for improving the signal-to-noise ratio of small signals buried in noise, one cannot do better with n elements than to form an additive array of the n elements. If the n elements are split into two additive arrays of $n/2$ elements each and the outputs are multiplied together, it was shown by Faran and Hills (44) and Jacob-

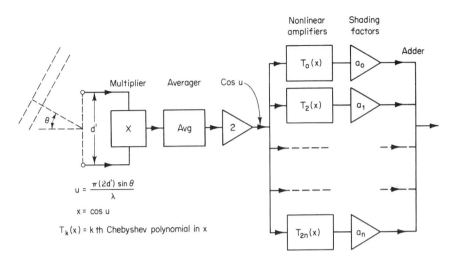

fig. 3.20 Synthesis of the beam pattern of a linear array of an odd number of equally spaced elements by two elements plus amplifiers generating Chebyshev polynomials of cos u. (After Ref. 48.)

son (49) that the array gain in uncorrelated noise is $10 \log(n/2)$, or a loss of 3 dB over simple addition of the elements. The advantages of multiplicative processing lie in narrower beams and in better angular resolution of closely spaced sources. In one study (50), for example, it was shown that a four-element array two wavelengths in overall length and having a band-width-time product of 9 has the same resolution as a linear array approximately five wavelengths long. Another benefit lies in the flexibility in design that multiplicative processing provides. Some examples of possible nonlinear arrays are described in an article by Tucker (51).

A number of different beams may be obtained with a single array of hydrophones by means of *dimus* processing of the outputs of the array elements. In the dimus (*di*gital *mu*ltibeam *s*teering) technique (52, 53), the time delays needed to form different beams are obtained with digital shift registers using successive infinitely clipped samples of the output of the elements. The advantage of dimus is that the various directional beams are formed simultaneously; the array can "look" acoustically in different directions at the same time, though with a loss of about 1 dB in signal-to-noise ratio.

With linear arrays, the beam width, array gain, and resolution are all interconnected. With multiplicative arrays there is no longer any direct relation among these array properties. They must be determined mathematically for particular configurations of multipliers and averages. Indeed, with multiplicative arrays, the terms DI and DT in the sonar equations must be considered together because it is impossible to separate clearly the array from the signal processing which accompanies it.

In summary, it may be said that multiplicative arrays find applications in conditions of high signal-to-noise ratio where narrow beams or high resolution are desirable or where a reduction in size or number of elements over a corresponding linear array is mandatory. In these applications, the trade-offs in processing complexity and signal-to-noise ratio must be seriously considered.

REFERENCES

1. Hennion, C., and J. Lewiner: Condenser Electret Hydrophones, *J. Acoust. Soc. Am.*, **63:**279 (1978).
2. Bucaro, J. A., and others: Fiber-optic Hydrophone, *J. Acoust. Soc. Am.*, **62:**1302 (1977).
3. Hunt, F. V.: "Electroacoustics," John Wiley & Sons, Inc., New York, 1954.
4. Design and Construction of Crystal Transducers, Design and Construction of Magnetostriction Transducers, *Nat. Def. Res. Comm. Div. 6 Sum. Tech. Reps.* 12 and 13, 1946.
5. Jacobson, M. J.: Space-Time Correlation in Spherical and Circular Noise Fields, *J. Acoust. Soc. Am.*, **34:**971 (1962).
6. Cron, B. F., and C. H. Sherman: Spatial Correlation Functions for Various Noise Models, *J. Acoust. Soc. Am.*, **34:**1732 (1962).
7. Bourret, R. C.: Directivity of a Linear Array in a Random Transmission Medium, *J. Acoust. Soc. Am.*, **33:**1793 (1961).
8. Berman, H. G., and A. Berman: Effect of Correlated Phase Fluctuation on Array Performance, *J. Acoust. Soc. Am.*, **34:**555 (1962).
9. Brown, J. L.: Variation of Array Performance with Respect to Statistical Phase Fluctuations, *J. Acoust. Soc. Am.*, **34:**1927 (1962).
10. Lord, G. E., and S. R. Murphy: Reception Characteristics of a Linear Array in a Random Transmission Medium, *J. Acoust. Soc. Am.*, **36:**850 (1964).
11. Cox, H.: Line Array Performance When the Signal Coherence is Spatially Dependent, *J. Acoust. Soc. Am.*, **54:**1744 (1973).
12. Green, M. C.: Gain of a Linear Array for Spatially Dependent Signal Coherence, *J. Acoust. Soc. Am.*, **60:**129 (1976).
13. Kleinberg, L. I.: Array Gain for Signals and Noise Having Amplitude and Phase Fluctuations, *J. Acoust. Soc. Am.*, **67:**572 (1980).
14. Basic Methods for the Calibration of Sonar Equipment, *Nat. Def. Res. Comm. Div. 6 Sum. Tech. Rep.* 10, pp. 21–26, 1946.
15. Urick, R. J.: Measurements of the Vertical Coherence of the Sound from a Near-Surface Source in the Sea and the Effect on the Gain of an Additive Vertical Array, *J. Acoust. Soc. Am.*, **54:**115 (1973).
16. Bobber, R. J.: "Underwater Electroacoustic Measurements," U.S. Naval Research Laboratory, 1970. Available from U.S. Government Printing Office, Washington, D.C.
17. Groves, I. D.: Twenty Years of Underwater Electroacoustic Standards, *U.S. Nav. Res. Lab. Rep.* 7735, 1974.
18. MacLean, W. R.: Absolute Measurement of Sound without a Primary Standard, *J. Acoust. Soc. Am.*, **12:**140–146 (1940).
19. Ebaugh, P. E., and R. Meuser: Practical Application of the Reciprocity Theorem in the Calibration of Underwater Sound Transducers, *J. Acoust. Soc. Am.*, **19:**695–700 (1947).
20. Carstensen, E. L.: Self-Reciprocity Calibration of Electroacoustic Transducers, *J. Acoust. Soc. Am.*, **19:**961–965, 1947.
21. Morse, P. M.: "Vibration and Sound," 2d ed., eq. 27.4, McGraw-Hill Book Company, New York, 1948.
22. Morse, P. M.: "Vibration and Sound," 2d ed., eq. 26.2, McGraw-Hill Book Company, New York, 1948.
23. Bobber, R. J., and G. A. Sabin: Cylindrical Wave Reciprocity Parameter, *J. Acoust. Soc. Am.*, **33:**446 (1961).

24. Simmons, B. D., and R. J. Urick: The Plane Wave Reciprocity Parameter and Its Applications to the Calibration of Electroacoustic Transducers at Close Distances, *J. Acoust. Soc. Am.*, **21:**633 (1949).
25. Horton, C. W., and G. S. Innis: The Computation of Far-Field Radiation Patterns from Measurements Made near the Source, *J. Acoust. Soc. Am.*, **33:**877 (1961).
26. Trott, W. J.: Underwater Sound Transducer Calibration from Nearfield Data, *J. Acoust. Soc. Am.*, **36:**1557 (1964).
27. Calibration of Underwater Electroacoustic Transducers, S1.20–1972, American National Standards Institute, New York, 1972.
28. Patterson, R. B.: Using the Ocean Surface on a Reflector for a Self-Reciprocity Calibration of a Transducer, *J. Acoust. Soc. Am.*, **42:**653 (1967).
29. Baker, D. D.: Determination of the Far-Field Characteristics of Large Underwater Sound Transducers from Near-Field Measurements, *J. Acoust. Soc. Am.*, **34:**1737 (1962).
30. Baker, D. D., and K. McCormack: Computation of Far-Field Characteristics from Near-Field Measurements in a Reflective Tank, *J. Acoust. Soc. Am.*, **35:**736 (1963).
31. Trott, W. J., and I. D. Groves: Application of Near-Field Array Technique to Sonar Evaluations, *U.S. Nav. Res. Lab. Rep.* 6734, 1968.
32. McKemie, M. J., and C. M. McKinney: An Experimental Investigation of the Parabolic Reflector as a Nearfield Calibration Device For Underwater Sound Transducers, *J. Acoust. Soc. Am.*, **67:**523 (1980).
33. Elliott, R. S.: Beamwidth and Directivity of Large Scanning Arrays, *Microwave J.*, **6:**53 (1963).
34. Harger, R. O., "Synthetic Aperture Radar Systems: Theory and Design," Academic Press, Inc., New York, 1970.
35. Davids, N., E. G. Thurston, and R. E. Meuser: The Design of Optimum Directional Acoustic Arrays, *J. Acoust. Soc. Am.*, **24:**50 (1952).
36. Albers, V. M.: "Underwater Acoustics Handbook," chap. 11, The Pennsylvania State University Press, University Park, Pa., 1960.
37. Pritchard, R. L.: Optimum Directivity Patterns for Linear Point Arrays, *J. Acoust. Soc. Am.*, **25:**879 (1953).
38. Pritchard, R. L.: Maximum Directivity Index of a Linear Point Array, *J. Acoust. Soc. Am.*, **26:**1034 (1954).
39. Sandler, S. S.: Some Equivalence between Equally and Unequally Spaced Arrays, *Inst. Radio Eng. Trans. Antennas Propagation,* **8:**496 (1960).
40. Harrington, R. F.: Sidelobe Reduction by Non-Uniform Element Spacing, *Inst. Radio Eng. Trans. Antennas Propagation,* **9:**187 (1961).
41. Ishimaru, A., and Y. S. Chen: Thinning and Broadbanding Antenna Arrays by Unequal Spacings, *Inst. Radio Eng. Trans. Antennas Propagation,* **13:**34 (1965).
42. Anderson, V. C.: DICANNE, a Realizable Adaptive Process, *J. Acoust. Soc. Am.*, **45:**398 (1969).
43. W. A. Von Winkle: Increased Signal-to-Noise Ratio of a Summed Array Through Dynamic Shading, *U.S. Nav. Underwater Sound Lab. Rep.* 579, 1963.
44. Faran, J. J., and R. Hills: Application of Correlation Techniques to Acoustic Receiving Systems, *Harvard Univ. Acoust. Res. Lab. Tech. Memo* 27, September 1952. Also, *Tech. Mem.* 28, November 1952.
45. Fakley, D. C.: Time Averaged Product Array, *J. Acoust. Soc. Am.*, **31:**1307 (1959).
46. Berman, A., and C. S. Clay: Theory of Time-Average Product Arrays, *J. Acoust. Soc. Am.*, **29:**805 (1957).
47. Dolph, C. L.: A Current Distribution of Broadside Arrays Which Optimizes the Relationship between Beam Width and Side-Lobe Level, *Proc. Inst. Radio Eng.*, **34:**335 (June 1946).
48. Brown, J. L., and R. O. Rowlands: Design of Directional Arrays, *J. Acoust. Soc. Am.*, **31:**1638 (1959).
49. Jacobson, M. J.: Analysis of a Multiple Receiver Correlation System, *J. Acoust. Soc. Am.*, **29:**1342 (1957).

50. Linder, I. W.: Resolution Characteristics of Correlation Arrays, *J. Res. Nat. Bur. Stand.*, **6JD**(3):245 (1961).
51. Tucker, D. G.: Sonar Arrays, Systems and Displays, in V. M. Albers (ed.), "Underwater Acoustics," Plenum Press, New York, 1963.
52. Anderson, V. C.: Digital Array Phasing, *J. Acoust. Soc. Am.*, **32**:867 (1960).
53. Rudnick, P.: Small Signal Detection in the DIMUS Array, *J. Acoust. Soc. Am.*, **32**:871 (1960).
54. Bazhenov, V. A.: "Piezoelectric Properties of Wood" (translation), Consultants Bureau, New York, 1961.
55. Booda, L. L.: ASW: Challenges and Bold Solutions, *Sea Technol.*, November 1973, p. 25.

four

generation of underwater sound: projector source level

All active sonars utilize some form of *projector* to generate acoustic energy. Sonar projectors normally, but not always, generate sound through a process of converting electric energy. They usually consist of arrays of individual elements, by means of which a directional beam is formed to send the generated energy into directions where it is wanted. In many sonars, the projector is also used as a hydrophone in order to save cost, weight, and space.

In the sonar equations, the parameter *source level* specifies the amount of sound radiated by a projector. It is defined as the intensity of the radiated sound in decibels relative to the intensity of a plane wave of rms pressure 1 μPa, referred to a point 1 yd from the acoustic center of the projector in the direction of the target. Because a directional projector ordinarily "points" in the direction of the target, the reference point for source level lies along the axis of the projector's beam pattern.

The *transmitting directivity index* of a projector is the difference, measured at a point on the axis of the beam pattern, between the level of the sound generated by the projector and the level that would be produced by a nondirectional projector radiating the same total amount of acoustic power. Figure 4.1 shows the beam patterns of a directional projector and of the equivalent nondirectional projector radiating the same total acoustic power. If the intensity represented by the directional pattern along its axis is I_D, and if the intensity represented by the nondirectional pattern is I_{Nond},

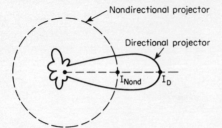

fig. 4.1 *Beam patterns of a directional projector and the equivalent nondirectional projector.*

then the transmitting directivity index is

$$\mathrm{DI}_T = 10 \log \frac{I_D}{I_{Nond}}$$

where the subscript T emphasizes that the transmitting, rather than the receiving, directivity index is referred to. In terms of the geometry of the projector array and the wavelength, the same expressions apply for both the transmitting and the receiving directivity indexes.

At short distances a projector does not radiate like a point source of sound. In the near field the intensity at any point is the sum of contributions from different parts of the radiating surface. The sound field is therefore irregular and does not fall off smoothly with distance, as it does in the far field. As shown in Fig. 4.2, the two regions are separated by a transition region centered about a transition range r_0 which may be taken to be approximately the

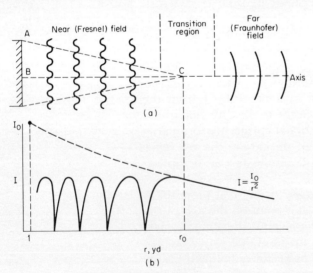

fig. 4.2 *The near and far sound fields of a projector: (a) The near or Fresnel field of a circular piston projector separated from the far or Fraunhofer field by a transition region. In the near field the wavefronts are crenulated and nondivergent; in the far field the wavefronts are smooth and spherical. (b) Intensity versus range along the axis.*

beginning of the far field. If, in Fig. 4.2, this range is taken to be that for which the differential distance AC − BC equals $\lambda/2\pi$, where λ is the wavelength, then it can be shown by trigonometry that $r_0 = A/\lambda$, where A is the radiating face area of a circular piston projector. A calibration of a projector is always made in the far field beyond r_0. The measured level is "reduced" to one yard by adding the spherical spreading correction $20 \log r$ to obtain the source level of the projector. Large arrays present special problems and special techniques have to be used (Sec. 3.8). The source level of the radiated noise of ships, submarines, and torpedoes is obtained in the same manner: by making a measurement at a long enough range so that the vessel is essentially a point source of sound and then applying the $20 \log r$ correction back to one yard.

When an electric voltage is impressed upon a material having the property of being piezoelectric or magnetostrictive, the material is caused to change its shape and thereby creates sound in the medium in which it is placed. This process of converting electric energy into sound, and, also, the design and construction of electroacoustic projectors, involves specialized theory and engineering practice. Much useful material on these subjects may be found in the World War II Summary Technical Reports of Division 6 of the NDRC on crystal and magnetostrictive transducers (1). Figure 4.3 shows the construction and beam pattern of a 24-kHz circular projector 19 in. in diameter, as constructed and measured during the World War II years. It is a mosaic of 728 ADP (ammonium dihydrogen phosphate) crystals cemented to a backing plate with an air cavity behind it. This unit was estimated to have an efficiency of 50 percent (ratio of acoustic-power output to electric-power input).

Not all projectors are electroacoustic. An explosive charge detonated in water converts chemical energy into sound and has interesting acoustical properties, which will be described in Sec. 4.4.

4.1 Relation between Source Level and Radiated Acoustic Power

The source level of a projector is related in a simple way to the acoustic power that it radiates and to its directivity index. Let a nondirectional projector be located in a homogeneous, absorption-free medium. At a large distance r, let the intensity of the sound emitted by the projector be I_r. If r is taken to be very large, the intensity is related to the rms pressure p_r in dynes per square centimeter by the plane-wave expression

$$I_r = \frac{p_r^2}{\rho c} \times 10^{-7} \quad \text{W/cm}^2$$

where ρ = density, g/cm³
c = velocity of sound, cm/s

Substituting $\rho = 1$ g/cm³ and $c = 1.5 \times 10^5$ cm/s = 4,920 ft/s and converting from centimeters to yards, we find

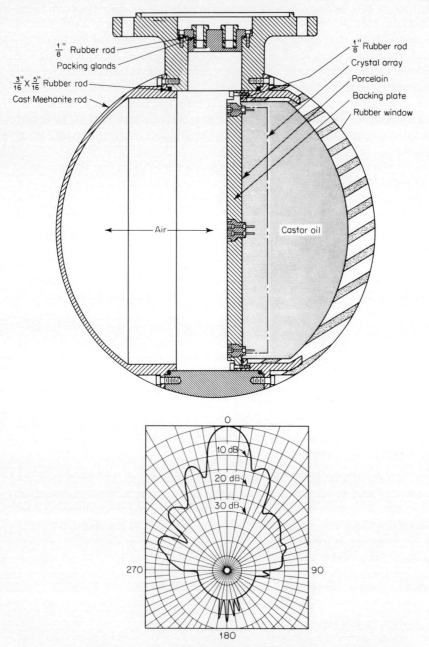

fig. 4.3 *Construction and beam pattern of a piezoelectric projector comprising a crystal array cemented to a backing plate. (After Ref. 1, report 12, pp. 243, 246.)*

$$I_r = 5.58 \times 10^{-9} p_r^2 \quad \text{W/yd}^2$$

For a nondirectional projector, this intensity corresponds to a radiated power output of

$$P_r = 4\pi r^2 I_r = 70.0 \times 10^{-9} p_r^2 r^2 \quad \text{W}$$

At a distance of 1 yd, the power is

$$P = 70.0 \times 10^{-9} p_1^2$$

where p_1 is the rms pressure at 1 yd in dynes per squate centimeter. Converting to decibels and remembering that $10 \log p_1^2$ is the source level SL, when $p_1 = 1 \ \mu\text{Pa}$, we obtain

$$10 \log P = -171.5 + \text{SL}$$

or

$$\text{SL} = 171.5 + 10 \log P$$

If the projector is directional with transmitting directivity index DI_T, we obtain, by the definition of DI_T,

$$\boxed{\text{SL} = 171.5 + 10 \log P + \text{DI}_T}$$

This function is graphed for convenience in Fig. 4.4. It should be noted that the value of SL computed above is not necessarily the value that would be measured by a calibrated hydrophone placed 1 yd from the projector. As in the case of the source level of the radiated noise of ships (Chap. 10), SL is a *reduced* quantity obtained through measurements made at large distances,

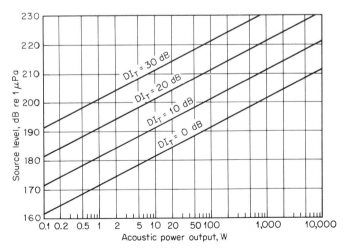

fig. 4.4 *Plots of the function* $\text{SL} = 171.5 + 10 \log P + \text{DI}_T$ *relating source level, power output, and transmitting directivity index.*

with 1 yd merely an arbitrary standard distance to which the sonar parameters are referred.

The quantity P is the total *acoustic* power radiated by the projector. For an electroacoustic projector, P is less than the *electric* power fed into the projector. If the electric power is P_e the ratio of these powers is the efficiency E of the projector, or

$$E = \frac{P}{P_e}$$

In terms of the electric power and efficiency, the source level expression becomes

$$\boxed{SL = 171.5 + 10 \log P_e + 10 \log E + DI_T}$$

The efficiencies of most sonar projectors range from 20 to 70 percent and depend upon the bandwidth employed; sharply tuned high-Q projectors have higher efficiencies than broadband projectors.

The radiated acoustic powers of shipboard sonar equipments range from a few hundred watts to some tens of kilowatts and have transmitting directivity indexes between 10 and 30 dB (2). It follows that the source levels of active shipboard sonars range from about 210 to about 240 dB.

4.2 Limitations on Sonar Power

To achieve the maximum range with active sonars, it is desirable to generate the maximum amount of acoustic power—at least up to the point where the just-detectable echo occurs in a background of reverberation rather than noise. In attempting to do so, two peculiar limitations, other than electric failures, are encountered. One, caused by *cavitation*, is primarily a property of the fluid medium into which the acoustic power is radiated; the other is caused by *interaction effects* between radiating elements of densely packed arrays of projector elements.

Cavitation limitation When the power applied to a sonar projector is increased, it is found that cavitation bubbles begin to form on the face and just in front of the projector. These bubbles are a manifestation of the rupture of the water caused by the negative pressures of the generated sound field. These negative pressures tear the liquid apart, so to speak, when they exceed a certain value called the *cavitation threshold*. The cavitation threshold may be expressed as a peak pressure in atmospheres or as a plane-wave intensity in watts per square centimeter. Since 1 atm is almost equal to 10^6 dyn/cm^2, the relation between the two is

$$I_c = \frac{[(0.707)10^6 p_c]^2}{\rho c} \times 10^{-7} = 0.3 p_c^2 \quad \text{W/cm}^2$$

where I_c = cavitation threshold, W/cm²
p_c = peak pressure of sound wave causing cavitation, atm
$\rho = 1.0$ g/cm³
$c = 1.5 \times 10^5$ cm/s

A cavitation threshold of 1 atm is therefore equivalent to a plane-wave intensity of 0.3 W/cm². When multiplied by the face area of the projector (in square centimeters), the cavitation threshold represents the maximum working value for the power output (in watts) of the projector. When this limit is exceeded by driving the projector harder, a number of deleterious effects begin to occur, such as erosion of the projector surface, a loss of acoustic power in absorption and scattering by the cavitation bubble cloud, a deterioration in the beam pattern of the projector, and a reduction in the acoustic impedance into which the projector must operate. In all these ways, the cavitation threshold represents the onset of a gradual deterioration of projector performance.

The physics of cavitation in liquids and the many phenomena accompanying it have received a great deal of attention in the literature. The reader interested in the processes of phenomena of cavitation is referred to a comprehensive review by Blake (3), to subsequent literature quoted by Strasberg (4), and to a summary article in a book edited by Mason (5).

The onset of cavitation is associated with the presence of cavitation nuclei in the liquid. As first postulated by Blake (6), these are believed to be extremely small microscopic bubbles of air that occur in cracks or cavities of small solid particles suspended in the liquid. Alternatively, cavitation nuclei may exist as free air bubbles surrounded by skins of organic impurities that hinder the escape of air into solution. In a sound field, these tiny bubbles form the nuclei into which dissolved air diffuses under the influence of the negative-going portions of the sound wave. When the negative pressure begins to exceed the cavitation threshold, the nuclei begin to grow in size by a process called (by Blake) *rectified diffusion,* by which more dissolved air diffuses into the bubble than out of it.

The occurrence of cavitation is accompanied by a variety of interesting physical and chemical phenomena, such as luminescence and the breakdown of chemical compounds. In laboratory experiments, cavitation is produced by focusing sound on a small region of the liquid under study, such as by exciting a spherical container into vibrational resonance. At the center of such a pulsating sphere, the occurrence of cavitation can be seen as the appearance of a bubble cloud or heard as the hiss of collapsing cavitation bubbles.

Practically speaking, the cavitation threshold may be said to occur at the point of departure from linearity of the input-output curve of the projector illustrated in Fig. 4.5. This point also marks the beginning of harmonic distortion in the output of a sinusoidal wave. The cavitation threshold of liquids is notably sensitive to a variety of factors, such as temperature, dissolved-air content, and previous pressure history, as well as to the particular criterion used for establishing the onset of cavitation.

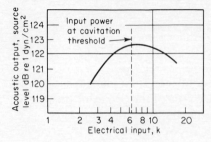

fig. 4.5 *Input versus output curve for a projector in the vicinity of the cavitation threshold. (After Ref. 9.)*

The cavitation threshold of a projector may be raised, and more acoustic power may be radiated, by (1) increasing the frequency, (2) decreasing the pulse length, and (3) increasing the depth (hydrostatic pressure) of the projector. Figure 4.6 shows the variation of cavitation threshold of fresh water with frequency as measured and compiled by Esche (7). The shaded area is the range of observed values, and the dashed curve is an estimated average curve. Below 10 kHz, the method of cavitation detection used by Esche led to thresholds lower than 1 atm, the value indicated by the dashed curve. A number of other measurements may be mentioned. At 25 kHz, Strasberg (4) found values ranging from 2.5 atm in tap water saturated with air at atmospheric pressure to 6.5 atm in water devoid of dissolved air. At 60 kHz, Blake (6) measured thresholds ranging from 3.5 to 4.6 atm for air-saturated and de-

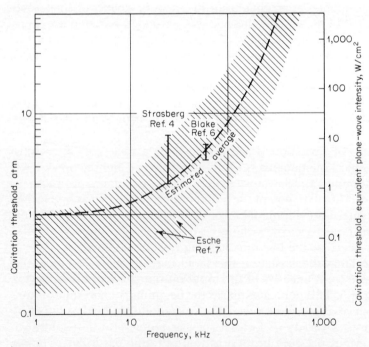

fig. 4.6 *Frequency dependence of the cavitation threshold. CW data on fresh water at atmospheric pressure.*

gassed water. Finally, at 550 kHz, Rosenberg (8) measured thresholds between 150 and 380 atm, a range of values lying close to the extension of the curve of Fig. 4.6.

Figure 4.7 shows the cavitation threshold of five full-scale sonar projectors measured by Liddiard (9) in open water as a function of pulse length employing the criterion shown in Fig. 4.5. The projectors used were flat-faced piston magnetostrictive and crystal projectors resonant in the frequency range 14.2 to 31.6 kHz. For water, the cavitation limit is seen to increase with a shortening of the pulse length below about 5 ms—in keeping with the idea that cavitation nuclei require a finite time interval to grow to a size where their effects are observable. As an interesting comparison, the laboratory measurements on castor oil by Briggs, Johnson, and Mason (10) may be mentioned. Castor oil, as well as other liquids of high viscosity and free of dissolved air, can tolerate much higher cavitation thresholds and is, accordingly, useful as filling liquid in transducer construction.

The effect of an increased depth of operation is to increase the cavitation threshold by the amount of the hydrostatic pressure. Since each 33 ft of depth represents an increased pressure of 1 atm, the cavitation threshold at depth h (in feet) becomes (in atmospheres)

$$P_c(h) = P_c(0) + \frac{h}{33}$$

where $P_c(h)$ = threshold at depth h
$P_c(0)$ = threshold at zero depth

The amount by which the cavitation threshold exceeds the ambient hydrostatic pressure may be thought of as the *tensile strength* of the fluid. Since the pressure at depth h equals $1 + h/33$ atm, we may write

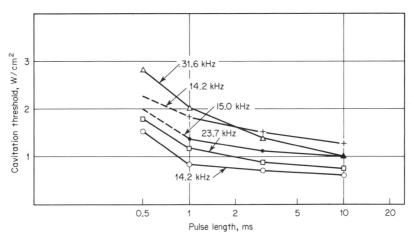

fig. 4.7 Variation of cavitation threshold with pulse duration for five projectors. (After Ref. 9.)

$$P_c(h) = 1 + \frac{h}{33} + T$$

where T is the tensile strength in atmospheres.

When actual sonar projector arrays are used, cavitation limits lower than those measured by other methods are found. These lower values are the result of "hot spots," or regions of high acoustic intensity, near the face of the projector. The effect of the nonuniform sound field in front of projectors on the cavitation limit has been investigated theoretically by Sherman (11), who defined a factor γ to express the near-field effect on the cavitation limit. The factor γ is the ratio of the cavitation limit for a particular transducer type to its value in a plane acoustic wave. For circular pistons, plane and hemispherical, γ was found to range between 0.3 and 0.6, with an average of about 0.5.

In summary, the cavitation threshold for long-pulse sonar projectors can be written

$$I_c = 0.3\gamma \left(P_c(0) + \frac{h}{33}\right)^2 \quad \text{W/cm}^2$$

$$= 0.15 \left(P_c(0) + \frac{h}{33}\right)^2 \quad \text{W/cm}^2$$

for a value of $\gamma = 0.5$ and with h in feet. The threshold P_c approximates 1 atm at low frequencies and increases with frequency, as indicated in Fig. 4.6. At a depth of 33 ft, the low-frequency cavitation limit is thus 0.6 W/cm² for pulse lengths in excess of about 5 ms. For shorter pulses, higher limits exist, as illustrated in Fig. 4.7.

It should be noted that the cavitation limit so predicted for sonar projectors is a conservative estimate since it corresponds (with $\gamma = 0.5$) to the *onset* of cavitation at a *single point* on the transducer surface. In practice, it is likely that considerably higher powers, higher perhaps by a factor of 2 or 3, can be usefully radiated without incurring excessively deleterious effects on the operation and performance of the projector.

Interaction effects A near-field effect, in many ways similar to that just described, occurs in large arrays of closely spaced resonant projector elements. When such an array is driven electrically, it is found that the velocity of motion of each element is not uniform, but varies from element to element of the array in a complex manner. This erratic behavior is caused by acoustic interactions between one element and another. For example, of two elements of the array, one may act as a sink for the acoustic output of another. In doing so, it absorbs, rather than radiates, sound and may possibly be driven to destruction. Interaction effects, unless compensated for in design, reduce the power output of a projector array, and deteriorate its beam pattern. They are important for arrays intended to radiate large amounts of acoustic power.

From a theoretical standpoint, interaction effects can be considered in terms of the mutual radiation impedance of a pair of sound sources, as has been done in a number of papers (12–19). Various control measures are available, as pointed out by Carson (20). First, and most simply, the mutual radiation impedance can be reduced by separating the elements of the array, though at a great cost in reduction of acoustic efficiency and power and in deterioration of the beam pattern. A second way of reducing the mutual impedance between elements is insertion of a reactance in series or in parallel with each element—a treatment effective at only a single frequency. A third method is to make the elements of the array individually large enough so that their self-radiation impedance overwhelms the mutual radiation impedance between elements—a "cure" which is costly from several engineering aspects. Finally, and most elegantly, individual amplifiers may be used to drive each array element in just the right amplitude and phase to yield the desired uniform inphase velocity of motion; a clever method of doing this, according to Carson, might be to place an accelerometer on the radiating face of each element to provide feedback to the driver amplifiers.

4.3 Nonlinear Effects in Sonar

Water, with or without contaminants such as air bubbles, is to some extent nonlinear. That is, the change in density caused by a change of pressure of a sound wave in water is not linearly proportional to the change in pressure. More generally, in any nonlinear system, such as a mixer stage in electronics, frequencies different from the input frequency occur at the output. For sinusoidal acoustic waves, a variety of additional frequencies in water are found to be generated. These include the harmonic and subharmonic frequencies when a single sinusoidal wave is generated and the sum and difference frequencies when two sinusoids of different frequencies are made to propagate together through the medium. These extraneous frequencies occur, practically speaking, only at high amplitudes of the primary wave; hence, the term "finite amplitude acoustics" to refer to such phenomena. The subject of nonlinear or finite-amplitude acoustics is theoretically complex with a rich literature; unfortunately, much of it has little content of interest to the sonar engineer.

Harmonic generation   One effect of nonlinearity is a dependence of propagation velocity on the pressure amplitude of the sound wave; high positive or negative pressures travel faster than low ones. The result is that an initially sinusoidal waveform progressively steepens as it propagates, so as to eventually acquire, in the absence of dissipation, a sawtoothlike waveform having a steep beginning and a more gradually sloping tail. An illustration of this progressive waveform distortion may be seen in Fig. 4.8.

Implicit in the acquisition of this sawtooth waveform is the generation of harmonics of the fundamental frequency. This creation of harmonics occurs at the expense of the fundamental. A portion of the power in the fundamen-

82 / principles of underwater sound

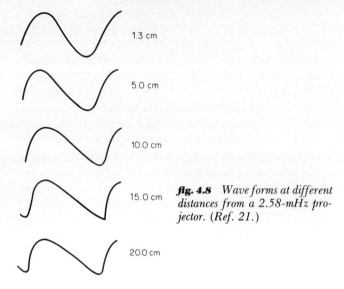

fig. 4.8 *Wave forms at different distances from a 2.58-mHz projector. (Ref. 21.)*

tal is converted into harmonics, where it is more rapidly lost because of the greater absorption at higher frequencies. This harmonic conversion process is greater at high amplitudes than at low ones so that, as the source level increases, the harmonic content increases also. A kind of saturation effect occurs, whereby an increase of source level does not give rise to a proportional increase in the level of the fundamental frequency. This saturation effect forms another limitation on the use of high acoustic powers in water. To the interested engineer, formulas are available for finding the fraction of power existing in the second and third harmonics at different ranges and frequencies under conditions such that absorption can be neglected (22).

Fortunately, for high-power sonars, the deleterious effects of harmonic conversion are lessened by spreading, which, by reducing the intensity of the primary wave, delays in range the onset of a stable distorted waveform and the saturation effect of the fundamental just mentioned. Yet at any given range from a sound source, there must exist, in principle at least, a maximum acoustic level which the source can produce at a particular frequency. This effect, however, remains conjectural since the source levels required are higher than those occurring in existing sonars.

Subharmonic generation Subharmonics are generated at frequencies lower than the fundamental in water whenever gas bubbles are present. In the sea, examples of such bubbles are those in marine organisms and in ships' wakes, as well as the microbubbles speculated to occur on surface indentations of tiny solid particles suspended in the sea.

The motion of a gas bubble in a sound field is slightly nonlinear. When the acoustic driving frequency is an integral multiple of the resonant frequency of the bubble, the bubble tries, in effect, to vibrate at its own resonant frequency;

in so doing, it gives rise to sound at a subharmonic of the driving frequency. This occurs most readily at a driving frequency equal to twice the bubble resonance frequency, when the half-frequency is generated (23).

Because of damping of the bubble's motion, a threshold exists for subharmonic generation; the driving pressure must be greater than a certain value in order that the subharmonic frequency may occur. The physical process is one of vibrational mode conversion, wherein a part of the energy of the driving sound wave is converted into energy in the first mode at the resonant frequency of the bubble.

Bubbles created by projector cavitation also give rise to subharmonics. In cavitation studies (24), the onset of the subharmonic frequency has been found to be a sensitive indicator of incipient cavitation at the face, or in front, of a sound source; the half-frequency is found to be more abrupt in its onset than the broadband noise created by collapse of the bubbles.

Difference-frequency generation (parametric sonar) When two sound waves of differing frequencies f_1 and f_2 are made to propagate in the same direction through water, they interact with each other to form the sum and difference frequencies $(f_1 + f_2)$ and $(f_1 - f_2)$. The sum frequency $(f_1 + f_2)$ is of no practical interest since it is more highly absorbed than either f_1 or f_2. However, the difference frequency $(f_1 - f_2)$, generated through nonlinear interaction, has a number of peculiar properties that make it useful for practical applications in depth sounding and communication.

This kind of projector has several valuable advantages:

1. *It has no side-lobe radiation outside the main beam at the difference frequency*, a property useful in communications where secure communication is important, as from one submarine to another.

2. *It has a narrower beam width* than can be achieved by direct generation of the difference frequency in a projector of the same physical size. This is useful where high directional resolution is required in a system of limited size, as in a small high-resolution echo sounder.

3. *It has an inherently broad bandwidth*, because a large proportional change in $f_1 - f_2$ can be achieved by making only a small proportional change in f_1, f_2, or both. Thus, if f_2 is kept fixed at 1.0 MHz, then $f_1 - f_2$ can be made to vary over the range 0 to 100 kHz by changing f_1 only from 1.0 to 1.1 MHz; this is a range easily accommodated by a projector of high Q and therefore of great efficiency, resonant in the region 1.0 to 1.1 MHz.

4. *The beam width is nearly constant* over a broad frequency band.

5. *Projector cavitation is not a problem*, since the primary frequencies f_1 and f_2 can be made so high that cavitation will not take place at achievable power outputs. The price to be paid for these desirable features is poor conversion efficiency, because only a minute fraction of the acoustic power generated at the primary frequencies appears at the difference frequency.

The beam width at the difference frequency is approximately the same as that at the primary frequencies f_1 and f_2 if these two frequencies are close together, as is normally the case.

A parametric or finite-amplitude projector is illustrated in Fig. 4.9a. The two primary frequencies are applied through a power amplifier to a conventional transducer. As these frequencies travel together through the water away from the projector, they interact to produce the difference frequency $f_2 - f_1$. All other harmonically generated frequencies, along with the two primary frequencies, are rapidly absorbed in the medium, leaving only $f_2 - f_1$ at a large distance. Generation of $f_2 - f_1$ continues out to several hundred yards or more, depending on the primary frequencies and source level. The result is equivalent to radiation by an end-fire line projector having an exponentially tapered shading.

A design nomogram for a parametric projector is given in Fig. 4.10. An example of its use is illustrated by the dashed lines. The beam width θ_d resulting from using a 100-kHz primary frequency and a 10-kHz difference frequency ($f_2 = 105$ kHz, $f_1 = 95$ kHz) would be 1.1 degree; the source level at the difference frequency for a power of 100 watts at one of the primaries would be 190 dB re 1 μPa.

The theory of the nonlinear interaction of two parallel-plane sound waves was originally worked out by Westervelt (25), and was experimentally verified by Bellin and Beyer (26). Its practical applications have received much attention by Berktay and others at the University of Birmingham, England (27).

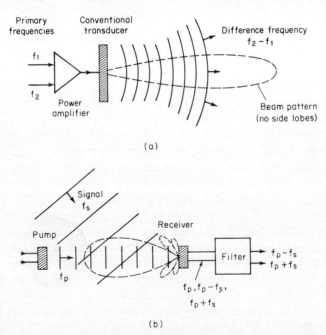

fig. 4.9 Schematic diagrams of a parametric or finite-amplitude projector (a) and receiver (b).

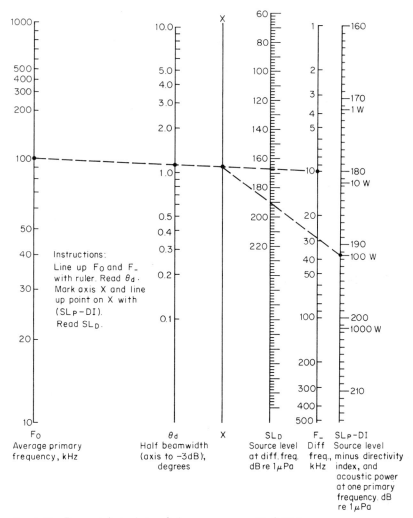

fig. 4.10 *Parametric projector design nomogram. (Ref. 28.)*

The production of sum and difference frequencies by high-amplitude sounds in water also makes possible *parametric receivers,* in which high-amplitude high-frequency plane-wave sound generated by a special *pump* projector is made to interact with the incoming low-frequency signal along a certain length (Fig. 4.9b). At the end of this interaction column is a conventional hydrophone that picks up the sum and difference frequencies between signal and pump, in addition to the pump frequency itself that is removed by filtering. Thus, if the pump frequency is 1 MHz and the signal frequency is 10 kHz, the output frequencies would be 1.0, 1.01, and 0.99 MHz. The beam pattern of a parametric receiver is like that of an end-fire line array, with a beam width at the signal frequency given by

$$\theta_{-3} = 139(f_s R)^{-1/2}$$

where θ_{-3} is the angle in degrees between the -3-dB down points, f_s is the signal frequency in kHz, and R is the interaction distance, or pump-receiver separation, in meters. Parametric receivers have the advantages of a narrow beam width, though with side lobes, at the signal frequency, as well as a broad bandwidth. They are also insensitive to sources of locally generated self-noise, since such sources will not coherently interact with the pump frequency (29, 30). However, parametric receivers have not been extensively employed in sonar applications.

4.4 Explosions as Sources of Underwater Sound

A kind of underwater sound source that does not suffer from the difficulties described in the preceding section, but has other limitations, is the underwater explosion. Underwater explosive sound sources are commonly small charges of explosive material ranging from a few grains to a few pounds in weight. A photograph of a standard United States Navy explosive sound source is shown in Fig. 4.11. This is an air-dropped pressure-detonated "sound signal" Mk 64 containing a small charge of TNT and detonating at 800 ft. It is used primarily for explosive echo ranging against submarine targets. A variety of similar charges detonating at other depths and containing other weights of explosive charge are in use in the United States Navy.

The basic phenomena of underwater detonation were established during World War II and are summarized in a book by Cole (31), as well as in a three-volume compendium of papers published by the Office of Naval Research (32).

The pressure signature When an explosion is initiated in a mass of explosive material, a pressure wave, initiated inside the material, propagates into the surrounding medium. This pressure wave is part of a complex series of phe-

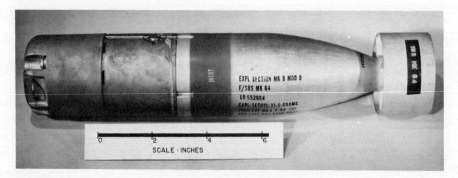

fig. 4.11 *A standard United States Navy explosive sound signal.*

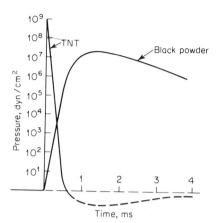

fig. 4.12 *Pressure pulses at 1 yd from a 1-lb charge of TNT and from a 1-lb charge of black powder in a heavy metal container.*

nomena occurring during the conversion of a solid explosive material into gaseous reaction products. The conversion process, and the accompanying pressure wave, proceeds in the material at a speed depending on the kind of material. In "high" explosives like TNT, the velocity of detonation lies between 15,000 and 30,000 ft/s, and a *shock wave* that propagates in all directions is produced in the medium. In other materials, such as black powder, the explosion process is one of *deflagration,* or burning, rather than *detonation* and occurs at a velocity of only 0.1 to 1 ft/s. The pressure pulses produced by these two processes are greatly different, as Fig. 4.12 illustrates. The detonating material creates a shock wave characterized by an infinitely steep front, a high peak pressure, and a rapid decay. The deflagrating material gives rise (33) to a relatively low, broad pressure blob. Only "high" TNT-like explosives are used as sources of underwater sound because they produce a much greater amount of acoustic energy at the higher frequencies useful in sonar.

The shock wave from a detonation is normally followed by a series of pressure pulses called *bubble pulses*. These are caused by successive oscillations of the globular mass of gaseous materials called the *gas globe* that remains after the detonation is completed. At the instant of minimum volume of the oscillating gas globe, a positive pressure pulse is generated; successive pulsations generate additional pulses, with each successive pulse being weaker than the preceding one. No bubble pulses occur at depths so shallow that the gas globe breaks the sea surface; neither do they occur when small charges are contained in a case that does not rupture when the charge is fired.

The pressure "signature" of a detonating explosion, as observed at short ranges, accordingly consists of the shock wave followed by a small number of bubble pulses, as illustrated in Fig. 4.13. At longer ranges, the signature is complicated by refraction and multipath-propagation effects in the sea. These complications range from the simple addition of a negative-going reflected pulse from the sea surface, when a surface-reflected path exists between the explosion and a receiver, to the distorted pulses seen by Brockhurst, Bruce,

88 / *principles of underwater sound*

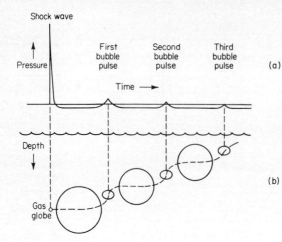

fig. 4.13 (a) Shock wave and bubble pulses from an explosion. (b) Pulsations and migration in depth of the gas globe. (After Ref. 47.)

and Arons (34) in shadow zones, to the long, drawn-out complex sofar signature of shots fired at great ranges in the deep sea. Yet the bubble period tends to persist in the signature even at long ranges. In one experiment (35), the depth of detonation of 1.8 lb SUS charges like that of Fig. 4.9, with a nominal detonation depth of 60 feet, could be determined within one foot by analyses of recordings made at ranges of hundreds of miles.

Of the total chemical energy of the explosive, about 40 percent is radiated as acoustic energy to distances beyond 1 yd for a 1-lb charge. Table 4.1 is a breakdown given by Arons and Yennie (36) of the dissipation of the chemical energy on the detonation of TNT. If the ratio of acoustic to chemical energy is taken as a measure of acoustic efficiency, explosives may be said to be comparable in efficiency to electroacoustic sonar projectors.

The Shock Wave The characteristics of the shock wave from explosions of various sizes over a wide range of distances in the sea were established by Arons and Yennie (37) by means of field measurements made in 1948 and earlier. Using charges ranging in size from ½ to 55 lb, they observed the amplitude and shape of the shock wave in 40 ft of water in Vineyard Sound, Massachusetts, at distances from a few feet to nearly 1 mile. It was found that the shock wave, expressed as a relationship between pressure and time, could be approximated by the function

table 4.1 Partition of the Energy of an Explosion of TNT (From Ref. 36)

Total acoustic radiation at $R = w^{1/3}/0.352$ ft (shock wave plus first and second bubble pulses)	410 cal/g
Shock-front dissipation at short ranges, up to $R = w^{1/3}/0.352$ ft	200
Energy remaining at third bubble maximum	95
Unaccounted for (turbulence, viscosity, heat conduction, chemical changes in the gaseous products)	345
Total	1,050 cal/g

$$p = p_0 e^{-t/t_0}$$

where p is the instantaneous pressure at time t after the onset of the shock front, p_0 is the peak pressure occurring at time $t = 0$, and t_0 is the time constant of the exponential pulse, or the time for the pressure to decay to $1/e = 0.368$ of its initial value p_0. Both p_0 and t_0 were found to be functions of the charge weight and range r. Arons, Yennie, and Cotter (38) found that for a charge of weight w pounds at a distance r feet, the peak pressure p_0 in pounds per square inch could be written as

$$p_0 = 2.16 \times 10^4 \left(\frac{w^{1/3}}{r}\right)^{1.13}$$

and the time constant in microseconds as

$$t_0 = 58 w^{1/3} \left(\frac{w^{1/3}}{r}\right)^{-0.22}$$

Figures 4.14 and 4.15 show p_0 and t_0 as a function of range for a 1-lb charge ($w = 1$), using the more customary underwater sound units of pressure in dynes per square centimeter and range in yards. For other charge weights, the peak pressure at a given range r may be found by reading off its value from Fig. 4.14 at a "reduced" range $rw^{-1/3}$; the time constant may be found by reading off a value from Fig. 4.15 at the "reduced" range $rw^{-1/3}$ and then multiplying by $w^{1/3}$.

The occurrence of the factor $w^{1/3}$ in the peculiar manner shown by the above expressions is a result of the *similarity principle* for shock waves (31, pp. 110–114), which states that the parameters of the shock wave are the same for all charge weights when the distances are reckoned in units of charge radii, or in units of $w^{1/3}$ for a spherical charge. The other peculiarity of the expressions for p_0 and t_0 is the occurrence of the exponents 1.13 and -0.22. In an absorption-free linear fluid, one would expect the exponents to be unity and zero, respectively, corresponding to spherical spreading and a time constant independent of distance. The excess attenuation of the peak pressure with range, as well as the broadening of the shock-wave decay, is due to a combination of the effects of nonlinearity of the underwater medium and greater attenuation in the sea at higher frequencies, with the latter effect overwhelming the former at the ranges of interest to sonar. The effects of absorption and high amplitude on the form of the shock wave were first studied by Arons, Yennie, and Cotter (38) and Arons (39). In addition to the effects of range on p_0 and t_0, long-range shock pulses have a finite time of rise to the peak pressure p_0 (amounting to about 20 μs at 1 kyd)—an effect attributable to high-frequency absorption, but not observable with certainty with the instrumentation used to obtain the empirical expressions for the form of the shock wave. A more modern study of the effects peculiar to the propagation of high-amplitude waves to long ranges has been made by Marsh, Mellen, and Konrad (40).

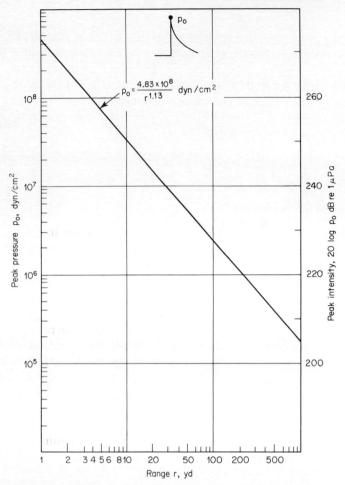

fig. 4.14 Peak pressure of a 1-lb charge of TNT as a function of range. To find the peak pressure of a charge of weight w lb, multiply r by $w^{-1/3}$. (Adapted from Ref. 38.)

Bubble pulses Following the shock wave occur a series of positive pressure pulses emitted by the pulsating gas globe at the successive instants of minimum volume. The amplitude of these pulses decreases progressively in size as the energy of the gas globe is dissipated. Measurements show that the peak pressure of the second bubble pulse is only about one-fifth that of the first; accordingly, only one or two bubble oscillations are practically significant. Similar bubble pulses are also radiated by the sudden collapse of a gaseous cavity under pressure, as when a glass bottle containing air breaks, or *implodes,* at a depth in the sea (41).

For explosives the time interval T, in seconds, between the shock wave and the first bubble pulse is found by theoretical analysis (42) to be

$$T = \frac{Kw^{1/3}}{(d+33)^{5/6}}$$

where K = a proportionality constant
w = charge weight, lb
d = depth of detonation below sea surface, ft

K depends slightly on the type of explosive and is equal to 4.36 for TNT (31, p. 282). Figure 4.16 is a plot of the above relationship for a number of different weights of charge. The reciprocal of T is the frequency of the first bubble maximum in the acoustic-energy spectrum of the charge.

With increasing depth of detonation, the interval T decreases and the pressure amplitude of the bubble pulses relative to the shock wave increases. These effects are illustrated by Fig. 4.17, which shows the pressure signature of a 1-lb charge fired at three depths and recorded near the sea surface overhead. Another effect of depth is a shortening of the apparent duration of the shock wave. This shortening is caused by the increased negative-going portions of the bubble oscillation at the deeper depths.

Acoustic-energy flux density and energy spectrum In order to avoid the complication caused by the propagation effects just mentioned, it is necessary to describe an explosion as a source of sound in terms of its acoustic-energy flux density and the distribution of the acoustic energy in frequency.

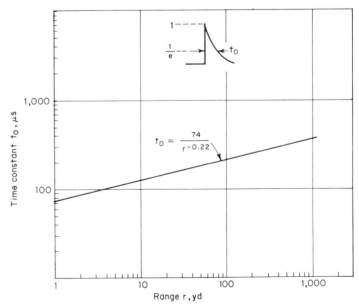

fig. 4.15 *Time constant of the shock wave for a 1-lb charge. To find t_0 for a charge of weight w lb, read off the value for a reduced range $rw^{-1/3}$ and multiply by $w^{-1/3}$. (Adapted from Ref. 38.)*

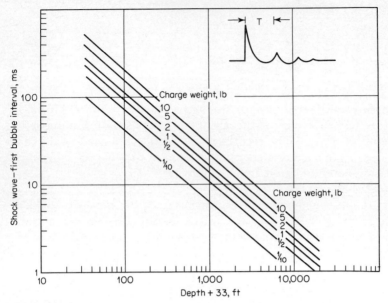

fig. 4.16 *Interval between the shock wave and the first bubble pulse as a function of the quantity (depth + 33) ft for various charge weights.*

In units of ergs per square centimeter, the energy-density of a plane wave is (Secs. 1.5 and 2.6)

$$E = \frac{1}{\rho c} \int_0^\infty p^2(t)\, dt$$

where $p(t)$ = pressure signature, dyn/cm²
ρ = density of medium
c = sound velocity of medium

For sonar purposes, a more practical unit than ergs per square centimeter is (Sec. 1.5) the energy-density of a plane wave of rms pressure of 1 dyn/cm² integrated over a period of 1 second. The two units differ by the factor $1/\rho c$, amounting to 52 dB.

The total acoustic energy in the shock wave can be found by substituting the expression $p(t) = p_0 e^{-t/t_0}$ in the integral expression for E. The result is

$$E = \frac{p_0^2 t_0}{2\rho c} \quad \text{ergs/cm}^2$$

which is the same as that of a square-topped sinusoidal pulse of peak pressure p_0 lasting for t_0 seconds. At 100 yd, we conclude that the shock wave from a 1-lb charge, using values for p_0 and t_0 from Figs. 4.14 and 4.15, has an energy flux density of +188 dB re (1 μPa)²s. In comparison, a pinging sonar of source level 230 dB and a pulse length of ½ second has an energy flux density at 100 yd of $230 - 40 - 3 = 187$ dB re (1 μPa)²s.

Since broadband reception is nearly always employed with explosive sources, it is necessary to consider the frequency distribution of the acoustic flux energy-density of an explosive source. This subject has been considered theoretically and experimentally by Weston (43). For an exponential pulse of peak pressure p_0 and time constant t_0, the energy flux spectral density $E_0(f)$ can easily be found by Fourier analysis to be

$$E_0(f) = \frac{2p_0^2}{\rho c (1/t_0^2 + 4\pi^2 f^2)} \quad \text{ergs/(cm}^2\text{)(Hz)}$$

where f is the frequency. For a single simulated bubble pulse of peak pressure p_1 having an exponential rise and an exponential decay of time constant t_1, Weston showed that the energy flux spectral density is given by

$$E_0'(f) = \frac{8}{\rho c} \left[\frac{p_1/t_1}{1/t_1^2 + 4\pi^2 f^2} \right]^2 \quad \text{ergs/(cm}^2\text{)(Hz)}$$

For values of p_0, t_0, p_1, and t_1 appropriate for a 1-lb charge, the functions $E_0(f)$ and $E_0'(f)$ are plotted, following Weston, as the dashed lines in Fig. 4.18. The solid curve is the combined spectrum of shock wave and first two bubble pulses. At the higher frequencies, the combined spectrum is given by

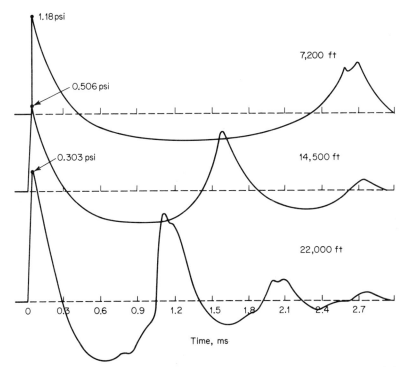

fig. 4.17 Pressure signature of a 1-lb charge at different depths, as recorded near the sea surface above. (After Ref. 48.)

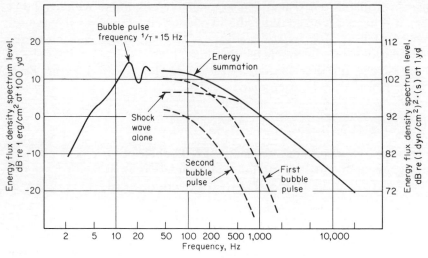

fig. 4.18 *Energy spectrum of a 1-lb charge at a depth of 20 fathoms, showing addition of shock-wave and bubble-pulse energies at frequencies greater than 1/T. [After Weston (43).]*

simple energy addition and is determined principally by the energy of the shock wave; at low frequencies, the combined spectrum falls off with decreasing frequency owing to phase cancellation of shock wave and bubble pulses; in between occur a number of oscillations representing interference between the two, with the principal peak occurring at a frequency $1/T$ equal to the reciprocal of the shock wave–first bubble interval. The number and amplitude of these oscillations depend upon the bandwidth being considered.

Figure 4.19 shows explosive spectra for a number of charge weights, all

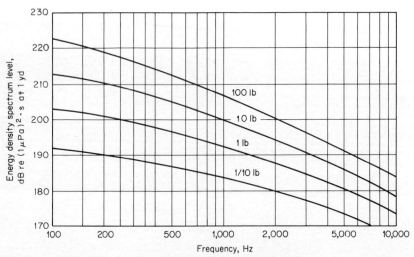

fig. 4.19 *Explosive source spectra for charges of different weights. [Based on data of Weston (43) at 100 yd and reduced to 1 yd by adding 40 dB.]*

based on the measurements of Weston at a distance of 100 yd and a depth of 120 ft. Generally confirming measurements were subsequently made by Stockhausen (44). The energy spectrum of a charge of any weight w pounds at frequency f may be obtained from that of a 1-lb charge by entering the 1-lb spectrum at a frequency equal to $w^{1/3}f$ and multiplying the value as read by $w^{4/3} = 13.3 \log w$.

At depths greater than 120 ft the spectrum is modified by the changing characteristics of the bubble pulses. The characteristics of deep charges have been investigated in papers by Blaik and Christian (45) and Christian and Blaik (46). Figure 4.20 shows octave-band spectra compiled from data in these papers, along with the shallow-depth spectrum of a 1-lb charge from the preceding figures. Different and more complicated scaling laws must be employed to find the spectrum when both charge weight and depth are varied.

Figure 4.21 may be used to find the energy–flux density level produced by an explosive charge under other conditions of charge weight and detonation depth. Instructions for its use and an example are given on the figure.

Advantages and disadvantages of explosive sound sources Explosives have some clear advantages, as well as some real disadvantages, in comparison with conventional-pulsed electroacoustic sonar sources. On the desirable side is the fact that they are mobile—in the sense of being free of connecting cables—and therefore can be easily launched and detonated at any depth. They yield a short high-power broadband pulse that is useful when range resolution is

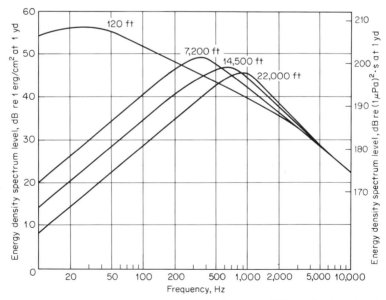

fig. 4.20 *Energy-density spectra of a 1-lb charge at different depths reduced from measurements made near the surface in octave frequency bands. [120-ft spectrum after Weston (43); deep spectra after Christian and Blaik (46).]*

96 / principles of underwater sound

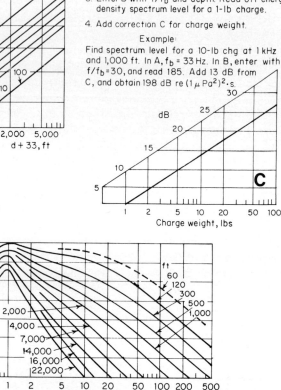

fig. 4.21 Curves for finding the energy-density spectrum level of an explosive charge (TNT or tetryl) of arbitrary weight and depth. (Ref. 49.)

important. They are nondirectional and need not be beamed to send an adequate amount of acoustic energy in the direction of the target.

On the negative side are the disadvantages that explosives cannot be made repeatable so as to yield repeated "looks" at the target and that the short duration makes processing of the received signal difficult. Also, the combination of high power and nondirectionality yields a high background of reverberation which, in echo ranging, obscures the desired echo. Finally, the explosive signature cannot easily be coded, or its spectrum altered, so as to take advantage of propagation and target characteristics or to make signal processing easier.

For these reasons, explosive sources in sonar are restricted to research studies and to air-dropped applications where their mobility and depth flexibility are of paramount importance.

REFERENCES

1. Design and Construction of Crystal Transducers, Design and Construction of Magnetostriction Transducers, *Nat. Def. Res. Comm. Div. 6 Sum. Tech. Reps.* 12 and 13, 1946.
2. Sahlein, H. F.: Sonar Equipment Summary, *U.S. Navy Electron. Lab. Rep.* 379, 1953.
3. Blake, F. G.: The Tensile Strength of Liquids: A Review of the Literature, *Harvard Univ. Acoust. Res. Lab. Tech. Memo* 9, 1949.
4. Strasberg, M.: Onset of Ultrasonic Cavitation in Tap Water, *J. Acoust. Soc. Am.*, **31**:163 (1959).
5. Mason, W. P. (ed.): Physics of Acoustic Cavitation in Liquids, in H. G. Flynn, "Physical Acoustics," vol. 1, pt. B, chap. 9, Academic Press, Inc., New York, 1964.
6. Blake, F. G.: The Onset of Cavitation in Liquids, *Harvard Univ. Acoust. Res. Lab. Tech. Memo* 12, 1949.
7. Esche, R.: Schwingungskavitation in Flussigkeiten, *Akust. Beih. (Acustica)*, **4**:AB208 (1952).
8. Rosenberg, L. D.: La Génération et l'étude des vibrations ultrasonares, *Acustica*, **12**:40 (1962).
9. Liddiard, G. E.: Short Pulse Cavitation Thresholds for Projectors in Open Water, *U.S. Navy Electron. Lab. Rep.* 376, 1953.
10. Briggs, H. B., J. B. Johnson, and W. P. Mason: Properties of Liquids at High Sound Pressures, *J. Acoust. Soc. Am.*, **19**:664 (1947).
11. Sherman, C. H.: Effect of the Near-Field on the Cavitation Limit of Transducers, *J. Acoust. Soc. Am.*, **35**:1409 (1963).
12. Karnovskii, M. L.: Calculations of the Radiation Resistance of Several Types of Distributed Radiator Systems, *Sov. Phys. Acoust.*, **2**:200 (1956).
13. Rusby, J. S. M.: Investigation of a Mutual Radiation Impedance Anomaly between Sound Projectors Mounted in an Array, *Acustica*, **14**:127 (1964).
14. Eichler, E.: Calculation of Coupling Effects in Pritchard's Array, *J. Acoust. Soc. Am.*, **36**:1393 (1964).
15. Arase, E. M.: Mutual Radiation Impedance of Square and Rectangular Pistons in a Rigid Infinite Baffle, *J. Acoust. Soc. Am.*, **36**:1521 (1964).
16. Stumpf, F. B., and F. J. Lukman: Radiation Resistance of Magnetostrictive-Stack Transducer in Presence of Second Transducer at Air-Water Surface, *J. Acoust. Soc. Am.*, **32**:1420 (1960).
17. Waterhouse, R. V.: Radiation Impedance of a Source near Reflectors, *J. Acoust. Soc. Am.*, **35**:1144 (1963).
18. Sherman, C. H.: Theoretical Model for Mutual Radiation Resistance of Small Transducers at an Air-Water Surface, *J. Acoust. Soc. Am.*, **37**:532 (1965).
19. Waterhouse, R. V.: Mutual Impedance of Acoustic Sources, Paper J21, 5th International Congress on Acoustics, September 1965.
20. Carson, D. L.: Diagnosis and Cure of Erratic Velocity Distributions in Sonar Projector Arrays, *J. Acoust. Soc. Am.*, **34**:1191 (1962).
21. Beyer, R. T.: "Non-linear Acoustics," U. S. Navy Sea Systems Command, 1974.
22. McDaniel, O. H.: Harmonic Distortion of Spherical Sound Waves in Water, *J. Acoust. Soc. Am.*, **38**:644 (1965).
23. Eller, A., and H. G. Flynn: Generation of Subharmonics of Order One-Half by Bubbles in Sound Field, *J. Acoust. Soc. Am.*, **46**:722 (1969).
24. Desantis, P., D. Sette, and F. Wanderlingh: Cavitation Detection: The Use of Subharmonics, *J. Acoust. Soc. Am.*, **42**:514(L) (1967).
25. Westervelt, P. J.: Scattering of Sound by Sound, *J. Acoust. Soc. Am.*, **29**:924 (1957).

26. Bellin, J. L. S., and R. T. Beyer: Experimental Investigation of an End-Fire Array, *J. Acoust. Soc. Am.*, **34:**1051 (1962).
27. Berktay, H. O.: Some Finite Amplitude Effects in Underwater Acoustics, in V. M. Albers (ed.), "Underwater Acoustics," vol. 2, chap. 12, Plenum Press, New York, 1967. See also several papers in *J. Sound Vib.*, **2:**435–480 (1965).
28. Lockwood, J. C.: Nomographs for Parametric Array Calculations, *Appl. Res. Lab., Univ. Tex. Tech. Memo* 73-3 (1973).
29. Berktay, H. O.: Parametric Amplification by the Use of Acoustic Non-Linearities and Some Possible Applications, *J. Sound Vib.*, **2:**462 (1965).
30. Barnard, G. R., and others: Parametric Acoustic Receiving Array, *J. Acoust. Soc. Am.*, **52:**1437 (1972).
31. Cole, R. H.: "Underwater Explosions," Princeton University Press, Princeton, N.J., 1948.
32. "Underwater Explosions Research," a compendium of British and American Reports, U.S. Navy Office of Naval Research, Washington, D.C., 1950.
33. Slifko, J. P.: Informal communication, U.S. Naval Ordnance Laboratory, 1962.
34. Brockhurst, R. R., J. G. Bruce, and A. B. Arons: Refraction of Underwater Explosion Shock Waves by a Strong Velocity Gradient, *J. Acoust. Soc. Am.*, **33:**452 (1961).
35. Mitchell, S. K., and others: Determination of Source Depth from the Spectra of Small Explosions Observed at Long Ranges, *J. Acoust. Soc. Am.*, **60:**825 (1976).
36. Arons, A. B., and D. R. Yennie: Energy Partition in Underwater Explosion Phenomena, *Rev. Mod. Phys.*, **20:**519 (1948).
37. Arons, A. B., and D. R. Yennie: Long Range Shock Propagation in Underwater Explosion Phenomena I, *U.S. Navy Dept. Bur. Ord. NAVORD Rep.* 424, 1949.
38. Arons, A. B., D. R. Yennie, and T. P. Cotter: Long Range Shock Propagation in Underwater Explosion Phenomena II, *U.S. Navy Dept. Bur. Ord. NAVORD Rep.* 478, 1949.
39. Arons, A. B.: Underwater Explosion Shock Wave Parameters at Large Distances from the Charge, *J. Acoust. Soc. Am.*, **26:**343 (1954).
40. Marsh, H. W., R. H. Mellen, and W. L. Konrad: Anomalous Absorption of Pressure Waves from Explosions in Water, *J. Acoust. Soc. Am.*, **38:**326 (1965).
41. Urick, R. J.: Implosions as Sources of Underwater Sound, *J. Acoust. Soc. Am.*, **35:**2026 (1963).
42. Arons, A. B.: Secondary Pressure Pulses Due to Gas Globe Oscillation in Underwater Explosions: II. Selection of Adiabatic Parameters in the Theory of Oscillation, *J. Acoust. Soc. Am.*, **20:**277 (1948).
43. Weston, D. E.: Underwater Explosions as Acoustic Sources, *Proc. Phys. Soc. London*, **76**(pt. 2):233 (1960).
44. Stockhausen, J. H.: Energy-per-Unit-Area Spectrum of the Shock Wave from 1-Lb TNT Charges Exploded Underwater, *J. Acoust. Soc. Am.*, **36:**1220 (1964).
45. Blaik, M., and E. A. Christian: Near Surface Measurements of Deep Explosions: I: Pressure Pulses from Small Charges, *J. Acoust. Soc. Am.*, **38:**50 (1965).
46. Christian, E. A., and M. Blaik: Near Surface Measurements of Deep Explosions: II: Energy Spectra of Small Charges, *J. Acoust. Soc. Am.*, **38:**57 (1965).
47. Snay, H. G.: Hydrodynamics of Underwater Explosions, *Nav. Hydrodyn. Publ.* 515, chap. XIII, National Academy of Sciences, 1957.
48. Genau, M. A.: Unpublished memorandum, U.S. Naval Ordnance Laboratory, 1965.
49. Urick, R. J.: Handy Curves for Finding the Source Level of an Explosive Charge Fired at a Depth in the Sea, *J. Acoust. Soc. Am.*, **49:**935 (1971).

five

propagation of sound in the sea: transmission loss, I

5.1 Introduction

Definition of transmission loss The sea, together with its boundaries, forms a remarkably complex medium for the propagation of sound. It possesses an internal structure and a peculiar upper and lower surface which create many diverse effects upon the sound emitted from an underwater projector. In traveling through the sea, an underwater sound signal becomes delayed, distorted, and weakened. Transmission loss expresses the magnitude of one of the many phenomena associated with sound propagation in the sea.

The sonar parameter *transmission loss* quantitatively describes the weakening of sound between a point 1 yd from the source and a point at a distance in the sea. More specifically, if I_0 is the intensity at the reference point located 1 yd from the "acoustic center" of the source ($10 \log I_0$ is the source level of the source) and I_1 is the intensity at a distant point, then the transmission loss between the source and the distant point is

$$\text{TL} = 10 \log \frac{I_0}{I_1} \quad \text{dB}$$

Since I_0 and I_1 are intensities, a time average is implied in the definition. For a CW or quasi-CW source, this time average is taken over a long enough interval of time to include all the effects of fluctuation in transmission; for a short transient pulse, the averaging time must be long enough to include the multipath transmission effects that cause distortion of the received signal. For short pulses, a transmission loss equivalent to that for CW is given by the ratio of the energy flux density at 1 yd from the source E_0 to the energy flux density at the distant point E_1, or

$$\text{TL} = 10 \log \frac{E_0}{E_1} \quad \text{dB}$$

If metric units of range are used and if the reference distance is 1 m, TL is 0.78 dB less than if the reference distance is 1 yd. However, this correction to TL is usually insignificant in practical work in comparison with the large uncertainties associated with TL as a sonar parameter.

Sources of loss Transmission loss may be considered to be the sum of a loss due to *spreading* and a loss due to *attenuation*. *Spreading loss* is a geometrical effect representing the regular weakening of a sound signal as it spreads outward from the source. As will be shown in Sec. 5.2, the spreading loss varies with range according to the logarithm of the range, and can be expressed as a certain number of *decibels per distance doubled*. *Attenuation loss* includes the effects of absorption, scattering, and leakage out of sound channels. As shown in Sec. 5.3, attenuation loss varies linearly with range, and is expressed as a certain number of *decibels per unit distance*. In addition to these two principal kinds of loss, other losses not readily identified with range may occur; examples are the *convergence gain* in sound channels or its converse, *refraction loss* in shadow zones.

Transmission loss, as a single number, summarizes the effects of a variety of propagation phenomena in the sea. These phenomena are many and varied; extensive studies of them can be found in the literature. Propagation has been, and continues to be, the most active aspect of underwater sound from a research standpoint and has attracted, because of its complexities, the attention of experimenters and theoreticians for many years. The number of papers has been increasing exponentially. Much of the recent attention is the result of theoretical modeling and computer programming, rather than at-sea observations, doubtless because of the difficulties and cost of doing work in the actual ocean.

In the following pages we shall approach the complex subject of sound propagation in the sea by beginning with the simplest form of loss—spreading in an ideal medium—and then proceeding to the complications caused by absorption, refraction, reflections from the boundaries, the various kinds of sound channels existing in the sea, and the multipath effects causing fluctuations in sound transmission.

5.2 Spreading Laws

Spherical spreading Referring to Fig. 5.1*a*, let a small source of sound be located in a homogeneous, unbounded, lossless medium. For this most simple propagation condition, the power generated by the source is radiated equally in all directions so as to be equally distributed over the surface of a sphere surrounding the source. Since there is no loss in the medium, the power P crossing all such spheres must be the same. Since power equals intensity times area, it follows that

$$P = 4\pi r_1^2 I_1 = 4\pi r_2^2 I_2 = \cdots$$

If r_1 be taken as 1 yd, the transmission loss to range r_2 is then

$$\text{TL} = 10 \log \frac{I_1}{I_2} = 10 \log r_2^2 = 20 \log r_2$$

This kind of spreading is called *inverse-square*, or *spherical, spreading*. The intensity *decreases* as the square of the range, and the transmission loss *increases* as the square of the range. Another way of describing spherical

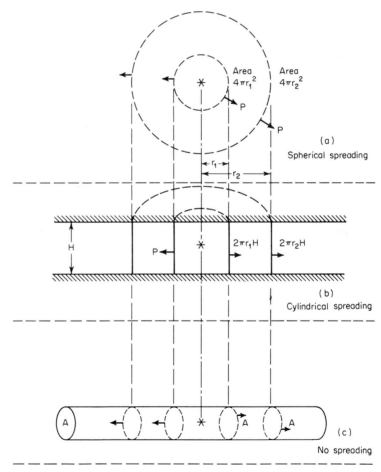

fig. 5.1 Spreading in (a) an unbounded medium, (b) a medium between two parallel planes, and (c) a tube.

spreading is to say that the product of the rms sound pressure and the range is a constant.

Cylindrical spreading When the medium has plane-parallel upper and lower bounds (Fig. 5.1*b*), the spreading is no longer spherical because sound cannot cross the bounding planes. Beyond a certain range, the power radiated by the source is distributed over the surface of a cylinder having a radius equal to the range and a height H equal to the distance between the parallel planes. The power crossing cylindrical surfaces at range r_1 and r_2 is

$$P = 2\pi r_1 H I_1 = 2\pi r_2 H I_2 = \cdots$$

If r_1 be taken as 1 yd, the transmission loss to r_2 then becomes

$$\text{TL} = 10 \log \frac{I_1}{I_2} = 10 \log r_2$$

and the spreading is said to be *inverse first power,* or cylindrical. In cylindrical spreading the product of the rms pressure and the square root of the range is a constant. This type of spreading exists at moderate and long ranges whenever sound is trapped by a sound channel in the sea.

No spreading As an academic case (Fig. 5.1*c*), we may consider propagation down a lossless tube or pipe of constant cross section. Beyond a certain range, the area over which the power is distributed is a constant, and the pressure, intensity, and transmission loss of a wave propagating down the tube are independent of range.

Time stretching A special kind of spreading occurs when the signal from a pulsed source is spread out in *time* as the pulse propagates. The pulse becomes elongated by multipath propagation effects and becomes smeared out in time as it travels in range. Such an effect is particularly important in long-range propagation in the deep-ocean sound channel. If the medium were unbounded, and time stretching proportional to range occurred, the intensity would fall off as the inverse cube of the range. In a sound channel, time stretching proportional to range causes the intensity to decrease as the *inverse square* of the range instead of as the *inverse first power*.

Table 5.1 is a summary of the various kinds of spreading and the spreading laws to which they give rise.

5.3 Absorption of Sound in the Sea

Absorption loss as a function of range Absorption is a form of loss that obeys a different law of variation with range than the loss due to spreading. It involves a process of conversion of acoustic energy into heat and thereby represents a true loss of acoustic energy to the medium in which propagation is taking place.

table 5.1 Spreading Laws

Type	Intensity varies as	Transmission loss is, dB	Propagation in
No spreading*	r^0	0	Tube
Cylindrical	r^{-1}	$10 \log r$	Between parallel planes
Spherical	r^{-2}	$20 \log r$	Free field
Hyperspherical*	r^{-3}	$30 \log r$	Free field with time stretching

* Hypothetical for sonar.

When a plane wave travels through an absorbing medium, a certain fraction of its intensity is lost in each unit small distance traveled. If the intensity at some distance is I, the loss of intensity dI in traveling a small additional distance dx is given by

$$\frac{dI}{I} = -n\, dx$$

where n is a proportionality constant, and the minus sign indicates that dI is a *negative* change of intensity. On integration between ranges r_2 and r_1, we find that the intensity I_2 at range r_2 is related to the intensity I_1 at range r_1 by

$$I_2 = I_1 e^{-n(r_2 - r_1)}$$

Taking 10 times the logarithm to the base 10, we obtain

$$10 \log I_2 - 10 \log I_1 = -10n(r_2 - r_1) \log_{10} e$$

Writing $\alpha = 10n \log_{10} e$, the change of level between range r_2 and r_1 becomes

$$10 \log I_2 - 10 \log I_1 = -\alpha(r_2 - r_1)$$

or

$$\alpha = \frac{10 \log I_1 - 10 \log I_2}{r_2 - r_1}$$

The quantity α is the *logarithmic absorption coefficient* (to the base 10), and is expressed in decibels per kiloyard (dB/kyd). It means that for each kiloyard traveled, the intensity is diminished by absorption by the amount α dB.

Measurements of absorption The first measurements of the absorption coefficient of seawater were made during the years 1931 to 1934 by Stephenson (1) using sinusoidal pulses transmitted through the sea between a surface ship and a submerged submarine. By the beginning of World War II, measurements had been made in different ocean areas in the frequency range then of interest to sonar (2). During the wartime years, extensive additional data were obtained, principally by the University of California Division of War Research (3). Later, laboratory measurements were made by Wilson and Leonard (4) using the decay of sound in a glass sphere of seawater excited into vibration in one of its natural modes; still later, a rectangular cavity was used at lower frequencies, both in the laboratory (5) and at sea (6).

Causes of absorption It was soon clear that the absorption of sound in the sea was unexpectedly high compared with that in pure water, and could not be attributed to scattering, refraction, or other anomalies attributable to propagation in the natural environment. For example, the absorption in seawater at frequencies between 5 and 50 kHz was found to be some 30 times that in distilled water. This excess absorption was attributed by Liebermann (7) to a kind of chemical reaction that occurs under the influence of a sound wave and involves one of the minor dissolved salts in the sea.

In seawater, the absorption of sound is caused by three effects. One is the effect of ordinary or *shear viscosity*, a classical effect investigated theoretically long ago by Rayleigh (8, p. 316). Rayleigh derived an expression for the absorption coefficient equivalent to

$$\alpha = \frac{16\pi^2 \mu_s}{3\rho c^3} f^2$$

where α = intensity absorption coefficient, cm^{-1}
μ_s = shear viscosity, poises (about 0.01 for water)
ρ = density, g/cm^3 (about 1 for water)
c = sound velocity (about 1.5×10^5 cm/s)
f = frequency, Hz

For pure water, the value of α computed by this formula is $6.7 \times 10^{-11} f^2$ dB/kyd and amounts to only about one-third of the absorption actually measured in pure (distilled) water. Figure 5.2 shows curves of measured absorption

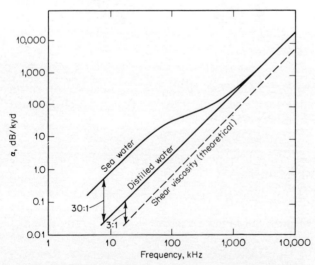

fig. 5.2 *Absorption coefficients in seawater and in distilled water and theoretical absorption due to shear viscosity alone.*

coefficients in seawater and distilled water and the theoretical absorption due to shear viscosity. The additional absorption in pure water over that due to shear viscosity is attributed to a second kind of viscosity, called *volume viscosity*, which produces absorption as a result of a time lag required for the water molecules to "flow" under pressure into lattice "holes" in the crystal structure. The effect adds an additional viscosity term to the above expression. The absorption coefficient due to both kinds of viscosity becomes (9)

$$\alpha = \frac{16\pi^2}{3\rho c^3}\left(\mu_s + \frac{3}{4}\mu_v\right)f^2$$

where μ_v is the volume-viscosity coefficient. The ratio μ_v/μ_s has the value 2.81 for water. The dominant cause of absorption in seawater below 100 kHz is due to *ionic relaxation* of the magnesium sulfate ($MgSO_4$) molecules in seawater. This is a dissociation-reassociation process, involving a finite time interval called the *relaxation time*, in which the $MgSO_4$ ions in solution dissociate under the pressure of the sound wave. Although $MgSO_4$ amounts to only 4.7 percent by weight of the total dissolved salts in seawater, this particular salt (rather than the principal constituent of seawater, NaCl) was found by Leonard, Combs, and Skidmore (10) to be the dominant absorptive constituent of seawater.

Variation with frequency Liebermann showed theoretically (11) that the ionic relaxation mechanism, together with viscosity, should yield a frequency dependence of the absorption coefficient of the form

$$\alpha = a\frac{f_T f^2}{f_T^2 + f^2} + bf^2$$

where a and b = constants
$\qquad f_T$ = relaxation frequency equal to reciprocal of relaxation time.

A modified form of this frequency law was subsequently fitted by Schulkin and Marsh (12) to some thirty thousand measurements made at sea between 2 and 25 kHz out to ranges of 24 kyd. The result was

$$\alpha = A\frac{Sf_T f^2}{f_T^2 + f^2} + B\frac{f^2}{f_T} \qquad \text{dB/kyd}$$

in which S is the salinity in parts per thousand (ppt); A and B are constants found to be equal to 1.86×10^{-2} and 2.68×10^{-2}, respectively; f is the frequency in kilohertz; and f_T is the temperature-dependent relaxation frequency given by

$$f_T = 21.9 \times 10^{6-1{,}520/(T+273)} \qquad \text{kHz}$$

where T is the temperature in degrees Celsius.

At frequencies below about 5 kHz the attenuation coefficients measured at sea are much higher than would be expected from the expressions just given.

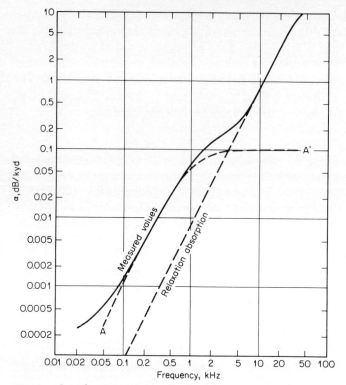

fig. 5.3 *Low-frequency attenuation coefficients. [Solid curve is the center of measured data compiled by Thorp (13).]*

At these frequencies an additional source of loss other than that caused by ionic relaxation of $MgSO_4$ must therefore be predominant.

A compilation of measured data made by Thorp (13) clearly shows that at frequencies below 1 kHz the attenuation coefficient is about 10 times higher than would be expected from a downward extrapolation of the high-frequency values. The data fall along the solid curve of Fig. 5.3, and are higher by an order of magnitude than the $MgSO_4$ relaxation mechanism would predict.

The cause, or causes, of this excess attenuation were a mystery for a number of years. Several possible processes were advanced. High-amplitude shock wave effects (14), some yet-undiscovered relaxation mechanism (15, 16), diffraction and scattering out of the deep sound channel (17, 18), eddy viscosity (19), solid particles in the sea (20)—all have been proposed as possible explanations. The presently accepted explanation is a boron-borate relaxation process discovered through laboratory techniques by Yeager, Fisher, et al. (21) to have the sufficiently long relaxation time needed to account for the observed attenuation; by this process, boric acid, a substance occurring in only minute concentrations in seawater, is the culprit responsible for the excess attenuation.

In an attempt to unravel the mystery, the Naval Underwater Systems Center has conducted measurements in various bodies of water all over the world, including Lake Superior, Lake Tanganyika, and the Red Sea (22). The results of this extensive series of measurements are concordant with the hypothesis of the boron-borate relaxation, together with, in fresh water and at low frequencies in seawater, scattering by inhomogeneities in the index of refraction of the medium (23). The latter contribution is a constant, independent of frequency, and amounts to about 0.003 dB/kyd.

The boric acid [B(OH)$_3$] ionization process is by no means a simple one, since the ionic interaction mechanism appears to depend in a complicated way on the other chemicals present in the seawater solution. Other ions, which can occur in greater or lesser amounts, affect the process, and hence the absorption coefficient. The resonator method, originally used by Leonard and co-workers (10), has been refined and extended downward in frequency as far as 10 kHz and has been used by Mellen, Browning and Simmons (24) to establish still another ionic process (involving MgCO$_3$) having a still smaller absorption. A second study by the same authors has further elaborated the complex boron relaxation mechanism. The practical result is that the boron process depends on the pH, or acidity factor, of the various seas and oceans. For example, the North Pacific Ocean has a different pH because of its chemical composition, and as a result has only half the absorption coefficient of the North Atlantic, at the frequencies where the boric acid relaxation is the principal absorption-producing process (25). Of course, at these frequencies, the coefficients involved are extremely small (between about 10^{-2} and 10^{-1} dB/kyd); yet such oceanic differences become important for long-range propagation.

At still lower frequencies, a compilation of measurements made by Kibblewhite and Hampton (26) shows a wide disparity of values, nearly all larger than the boron absorption process would anticipate. Figure 5.4 shows in the crosshatched area where the existing measurements fall relative to the curve for the boron absorption. The larger attenuation values at frequencies below about 100 Hz have been attributed (27) to scattering by large inhomogeneities within the deep sound channel, in which such measurements must necessarily be made. Theory shows the attenuation due to such scattering to be independent of frequency. We should note particularly the extremely small magnitude of this coefficient, averaging 0.003 dB/kyd or 3 dB in 500 miles of travel.

A recent thorough review of existing data and theory made by Fisher and Simmons (28) shows somewhat different results from those of Schulkin and Marsh cited above (12). The Fisher-Simmons expression is of the form

$$\alpha = A_1 P_1 \left(\frac{f^2}{f_1^2 + f^2}\right) f_1 + A_2 P_2 \left(\frac{f^2}{f_2^2 + f^2}\right) f_2 + A_3 P_3 f^2$$

The terms A_1, A_2, A_3, f_1, f_2 are complicated functions of temperature, and P_1, P_2, and P_3 are functions of pressure. The three terms represent the effects of boric acid, magnesium sulfate, and viscosity, respectively. The coefficient P_1 is

108 / *principles of underwater sound*

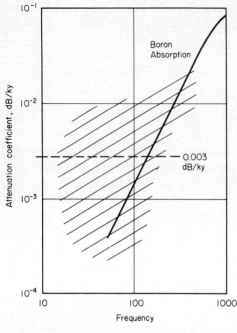

fig. 5.4 *Attenuation measurements at low frequencies, compiled in Ref. 26, fell in the crosshatched area. The great scatter in the data is the result of measurement uncertainties and the failure of the ocean to be uniform over the great distances required to obtain the data.*

arbitrarily taken to be a constant, since the pressure effect on the boron absorption, if any, has yet to be determined.

Figure 5.5 is a plot of the Fisher-Simmons expressions for three temperatures at zero depth.

A simpler and more specific expression due to Thorp (15) is

$$\alpha = \frac{0.1 f^2}{1 + f^2} + \frac{40 f^2}{4.100 + f^2} + 2.75 \times 10^{-4} f^2 + 0.003$$

where α is the attenuation coefficient in dB per kiloyard, and f is the frequency in kilohertz. The constant 0.003 has been added to take care of the attenuation at very low frequencies (Fig. 5.4). This expression applies for a temperature of 39°F (4°C) and a depth of about 3,000 ft, where most of the measurements on which it is based were made.

Variation with depth The effect of pressure on the absorption has been investigated theoretically and experimentally (30), using the resonator-decay method mentioned above. In the range of hydrostatic pressure found in the sea, the effect of pressure (12) is to reduce the absorption coefficient by the factor $(1 - 6.54 \times 10^{-4} P)$, where P is the pressure in atmospheres. Taking 1 atm as the pressure equivalent of 33.9 ft of water at 39°F, the absorption coefficient at depth d ft in the sea becomes

$$\alpha_d = \alpha_0 (1 - 1.93 \times 10^{-5} d)$$

in terms of its value α_0 at zero depth ($d = 0$). The absorption of sound in seawater accordingly decreases by about 2 percent for every increase of 1,000 ft in depth; at a depth of 15,000 ft the absorption coefficient decreases to 71 percent of its value at the surface.

A much larger depth dependence, by a factor of about two, has been observed in at-sea measurements. At a Pacific Ocean location, Bezdek (31) measured α at great depths by transmitting 75 kHz pulses over both vertical and horizontal paths between a deep source and a deep receiver. The coefficient was found to change from 20 dB/km at 910 meters to 13 dB/km at 3,350 meters, a decrease by a factor of 0.65 in only 2,440 meters (8,005 ft) of depth. If these measurements are typical of other oceanic locations and if they can be extrapolated downward to lower frequencies, the attenuation of sound over

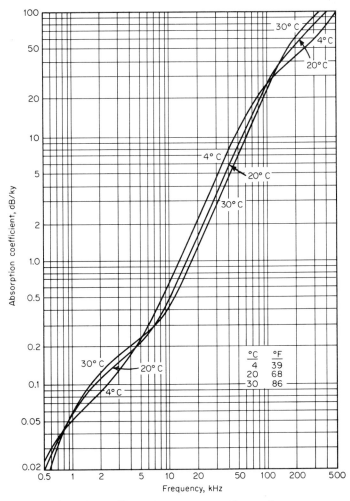

fig. 5.5 *Absorption coefficient in seawater according to the expressions of Fisher and Simmons (Ref. 28) for zero depth, salinity 35 ppt, pH = 8, and three temperatures.*

deep paths in the sea is apparently far lower than had been previously expected.

Figure 5.6 shows the effect of depth on the $MgSO_4$ absorption as calculated from the expressions of Fisher and Simmons. The effect is expressed as a factor of the coefficient at zero depth. We may note in passing that kilohertz frequencies traveling over deep refracted paths, such as to a convergence zone, suffer a much smaller loss due to absorption than would be predicted if the effect of pressure were neglected.

Summary of attenuation processes Figure 5.7 is a summary of the processes of attenuation and absorption in the sea as they are understood at the present time. At the lower left is a portion of the curve due to sound channel diffraction, caused by the fact that the deep sea is no longer an effective duct for such long wavelengths, while at the right is a dashed line showing the absorption due to viscosity.

Spherical spreading and absorption When propagation measurements are made at sea, it is found that spherical spreading, together with absorption, provides a reasonable fit to the measured data under a wide variety of conditions. That is to say, within the usually wide spread of measured data, the assumption of spherical spreading plus absorption yields reasonable agreement with observations. Indeed, spherical spreading often occurs when it has no right to occur—often under trapping conditions in sound channels where the gain due to trapping is apparently compensated for by the leakage of sound out of the channel.

When specific propagation conditions are of no interest and only a rough

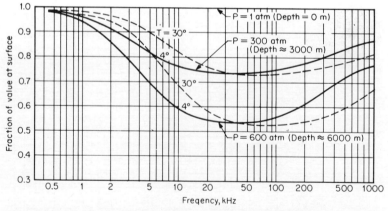

fig. 5.6 *Depth factor of the absorption coefficient as a function of frequency, according to Fisher and Simmons (Ref. 28). The solid curves are for a temperature of 4°C for two depths, approximately 3,000 and 6,000 meters. The dashed curves apply for a temperature of 30°C for the same depths. Salinity 35 ppt, pH = 8.*

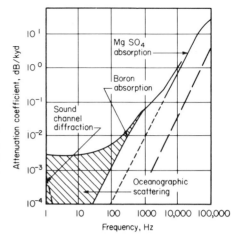

fig. 5.7 *The causes of attenuation and absorption in the sea. The absorption due to viscosity is the dashed line at the right. (From Ref. 29.)*

approximation of the transmission loss is adequate, the ubiquitous spherical-spreading law, plus an added loss due to absorption, is a handy working rule. It may be expressed as

$$\text{TL} = 20 \log r + \alpha r \times 10^{-3}$$

where the first term represents spherical spreading and the second term absorption, and where 10^{-3} is required to take care of the fact that r is in yards and α is, by custom, in decibels per kiloyard. Figure 5.8 is a nomogram to facilitate computations using this relationship. The dashed lines in the figure indicate how the nomogram is to be used.

Although spherical spreading often sensibly occurs in transmission measurements, free-field conditions seldom actually exist in the sea, except at very short ranges, because of refraction, scattering, and the presence of the ocean boundaries. The following pages will center about departures from spherical spreading caused by these and other effects occurring in the natural ocean medium.

5.4 Velocity of Sound in the Sea

Methods of measurement The velocity of sound* in the sea is an oceanographic variable that determines many of the peculiarities of sound transmission in the medium. It varies with depth, the seasons, geographic location, and time at a fixed location.

The history of attempts to measure sound velocity in a natural body of water dates back to 1827, when, as quoted by Wood (32, p. 261), Colladon and Sturm struck a submerged bell in Lake Geneva and simultaneously set off a charge of powder in the air. By timing the interval between the two signals

---

* Historically, and improperly, this scalar quantity has been called the "velocity" of sound. Sound "speed," instead of sound "velocity," is coming into common usage.

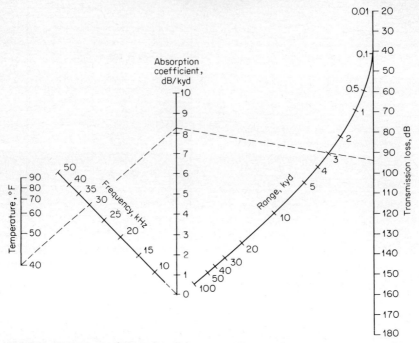

fig. 5.8 Nomogram for computing the transmission loss due to spherical spreading and absorption for a salinity of 35 ppt and zero depth. The dashed line indicates the method of use of the nomogram: at 40°F and 30 kHz, the absorption coefficient is 8.3 dB/kyd and the transmission loss to 3 kyd is 94 dB.

across the lake, they obtained the velocity of sound in water. The result was 1,435 m/s at 8.1°C, a value surprisingly close to its modern value. Subsequent investigators not only have timed the arrival of sound in the sea to derive the sound velocity (33–35), but also have attempted to relate it to the standard and more readily measured oceanographic parameters, such as temperature, salinity, and depth. A relationship of this kind can be established either by means of theory, using certain basic properties of water like its specific volume and ratio of specific heats, or by laboratory measurements of sound velocity over a range of temperatures, salinities, and pressures. By the former, or theoretical, method tables of sound velocity were long ago prepared by Heck and Service (34), Matthews (36), and Kuwahara (37), the last-mentioned serving as the standard tables of sound velocity for nearly 20 years. The more modern experimental method yields direct measurements of sound velocity by the use of laboratory techniques of one kind or another under carefully controlled conditions. Examples are the measurements of Weissler and Del Grosso (38), Del Grosso (39), and Wilson (40).

Variation with temperature, salinity, and pressure Both methods yield expressions for the velocity of sound in terms of the three basic quantities:

temperature, salinity, and pressure. Strangely enough, no other physical properties have been found to affect the velocity of sound in seawater, with the exception of contaminants such as air bubbles and biological organisms. Although these are the only three physical variables, the dependence of velocity on them is by no means a simple one—largely because of the precision to which velocity can be determined—and an empirical relationship giving velocity as a function of temperature, salinity, and pressure (or depth) is extremely complicated. Del Grosso (41) gives an equation containing 19 terms each to 12 significant figures in the powers and cross-products of the three variables, while a less complicated, though still cumbersome, expression has been given by Lovett (42). While these are suitable for computer programming, often a simpler expression will suffice for practical work when an error of a few parts in ten thousand or about 0.5 m/s is unimportant. A number of such simpler expressions have appeared in the literature, and are listed in Table 5.2, along with the range of temperature, salinity, and depth over which they are stated to be applicable. The errors in their use depend upon T, S, and D; the literature references should be consulted. An extremely precise knowledge of sound velocity is needed in applications such as precision depth sounding, active sonar range-finding for fire control, location by triangulation, or generally whenever an observed travel time needs to be accurately converted into a distance.

table 5.2 *Expressions for Sound Velocity in Meters per Second, in Terms of Temperature, Salinity, and Depth***

Expression	Limits	Reference
$c = 1492.9 + 3(T - 10) - 6 \times 10^{-3}(T - 10)^2$ $- 4 \times 10^{-2}(T - 18)^2 + 1.2(S - 35)$ $- 10^{-2}(T - 18)(S - 35) + D/61$	$-2 \leq T \leq 24.5°$ $30 \leq S \leq 42$ $0 \leq D \leq 1{,}000$	1
$c = 1449.2 + 4.6T - 5.5 \times 10^{-2}T^2$ $+ 2.9 \times 10^{-4}T^3 + (1.34 - 10^{-2}T)(S - 35)$ $+ 1.6 \times 10^{-2}D$	$0 \leq T \leq 35°$ $0 \leq S \leq 45$ $0 \leq D \leq 1{,}000$	2
$c = 1448.96 + 4.591T - 5.304 \times 10^{-2}T^2$ $+ 2.374 \times 10^{-4}T^3 + 1.340(S - 35)$ $+ 1.630 \times 10^{-2}D + 1.675 \times 10^{-7}D^2$ $- 1.025 \times 10^{-2}T(S - 35) - 7.139 \times 10^{-13}TD^3$	$0 \leq T \leq 30°$ $30 \leq S \leq 40$ $0 \leq D \leq 8{,}000$	3

* D = depth, in meters.
  S = salinity, in parts per thousand.
  T = temperature, in degrees Celsius.

REFERENCES FOR TABLE 5.2

1. Leroy, C. C.: Development of Simple Equations for Accurate and More Realistic Calculation of the Speed of Sound in Sea Water, *J. Acoust. Soc. Am.*, **46:**216 (1969).
2. Medwin, H.: Speed of Sound in Water For Realistic Parameters, *J. Acoust. Soc. Am.*, **58:**1318 (1975).
3. Mackenzie, K. V.: Nine-term Equation for Sound Speed in the Oceans, *J. Acoust. Soc. Am.*, **70:**807 (1981).

The velocity of sound in the sea *increases* with temperature, salinity, and depth. Approximate coefficients for the rate of change with these quantities are given in Table 5.3. In open deep water, salinity normally has only a small effect on the velocity.

table 5.3 *Approximate Coefficients of Sound Velocity*

Variation with	Coefficient	Coefficient
Temperature (near 70°F)	$\dfrac{\Delta c/c}{\Delta T} = +0.001/°F$	$\dfrac{\Delta c}{\Delta T} = +5$ ft/(s)(°F)
Salinity	$\dfrac{\Delta c/c}{\Delta S} = +0.0008/\text{ppt}$	$\dfrac{\Delta c}{\Delta S} = +4$ ft/(s)(ppt)
Depth	$\dfrac{\Delta c/c}{\Delta D} = +3.4 \times 10^{-6}/\text{ft}$	$\dfrac{\Delta c}{\Delta D} = +0.016$ ft/(s)(ft)

c = velocity, ft/s
T = temperature, °F
S = salinity, parts per thousand (ppt)
D = depth, ft

Field observations of sound velocity Two devices are commonly used for finding the velocity of sound as a function of depth in the sea. One, called the *bathythermograph*, measures temperature as a function of depth as it is lowered into the sea; the other, called the *velocimeter*, measures sound velocity directly in terms of the travel time of sound over a constant fixed path. Of the two, the bathythermograph has long been in use, for it is simple, inexpensive and nowadays, expendable. Moreover, in deep water far from coasts, where salinity gradients are relatively weak, it is generally an adequate indicator of the velocity-depth function.

The expendable bathythermograph (XBT) is a device for obtaining the temperature profile without having to retrieve the sensing unit. A cutaway view of an XBT is shown in Fig. 5.9. It consists basically of a thermistor probe that is ejected from the moving vessel and sinks at a known constant rate after launching. The probe is connected to measurement electronics on board ship by a fine wire contained on two reels, one remaining on board the ship under way, the other inside the streamlined probe housing. As the probe sinks at a fixed, known speed, both reels unwind simultaneously, leaving the wire at rest in the water with little or no tension placed on it. The thermistor bead changes its resistance with changing temperature and a trace is obtained of resistance, or temperature, against time, or depth, by means of a special recorder. This system has almost completely replaced the older, retrievable mechanical bathythermograph on board U.S. naval vessels. A variant of the XBT has been developed for use from aircraft; it incorporates a radio link instead of a connecting wire, as with sonobuoys. The aircraft version is called the Airborne Bathythermograph AN/SSQ-36.

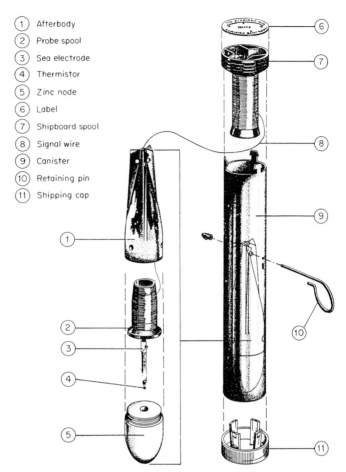

fig. 5.9 *An expanded view of an expendable bathythermograph. At left is shown the sinking streamlined probe whose most vital component is the thermistor bead 4. At right is the shipboard unit connected to a special recorder (not shown) which produces a paper trace of temperature versus depth. (Courtesy, Sippican Corp., Marion, Mass.)*

The sound *velocimeter* is an acoustic device, originally developed by Greenspan and Tschiegg (43) and now made by several manufacturers. It measures the travel time of short pulses between a projector and a receiver. It operates on the "sing-around," or "howler," principle, in which the arrival of a pulse at the receiver triggers the succeeding pulse from the projector. Since nearly all the time delay between pulses occurs as the acoustic delay in the water between projector and receiver, the repetition frequency of the pulses is determined by the sound velocity of the water between projector and receiver. Figure 5.10 is a close-up of the acoustic path in the velocimeter between the sound projector and receiver.

When the bathythermograph temperature-depth trace is converted to

fig. 5.10 *Transducers and acoustic path of a velocimeter. (Photograph by permission, Weston Instruments, Inc.)*

sound velocity, it gives a velocity-depth trace comparable to that of the velocimeter. This was demonstrated during World War II (44), when perhaps the first seagoing velocimeter, a variable-frequency fixed-path interferometer, was lowered into the sea side by side with a bathythermograph. An example of the traces obtained is given in Fig. 5.11. The similarity of the velocity-depth curve obtained with the interferometer to that obtained by converting the bathythermograph trace into sound velocity indicates that temperature and depth are the dominant determinants of sound velocity under normal conditions in the open sea. This same conclusion was reached subsequently by means of comparative measurements (45) with a more modern velocimeter. Under certain conditions, however, as in nearshore waters and under a melting ice cap, salinity changes and suspended material, rather than temperature alone, may play an important part in determining the sound velocity.

5.5 Velocity Structure of the Sea

Layers of the deep-sea velocity profile   By "velocity profile" is meant the variation of sound velocity with depth, or the velocity-depth function. In the deep sea, the velocity profile is obtained by the instruments described in the preceding section or, alternatively, by hydrographic observations of temperature, salinity, and depth.

A typical deep-sea profile is shown in Fig. 5.12. The profile may be divided into several layers having different characteristics and occurrence. Just below the sea surface is the *surface layer,* in which the velocity of sound is susceptible to daily and local changes of heating, cooling, and wind action. The surface layer may contain a mixed layer of isothermal water that is formed by the action of wind as it blows across the surface above. Sound tends to be trapped or channeled in this mixed layer. Under prolonged calm and sunny conditions the mixed layer disappears, and is replaced by water in which the temperature decreases with depth. Below the surface layer lies the *seasonal thermocline*—the word "thermocline" denoting a layer in which the temperature changes with depth. The seasonal thermocline is characterized by a *negative* thermal or velocity gradient (temperature or velocity *decreasing* with depth) that varies with the seasons. During the summer and fall, when the near-surface waters of the sea are warm, the seasonal thermocline is strong and well defined; during the winter and spring and in the Arctic, it tends to merge with, and be indistinguishable from, the surface layer. Underlying the seasonal thermocline is the *main thermocline,* which is affected only slightly by seasonal changes. The major increase of temperature over that of the deep cold depths of the sea occurs in the main thermocline. Below the main thermocline and extending to the sea bottom is the *deep isothermal layer* having a nearly constant temperature near 39°F, in which the velocity of sound in-

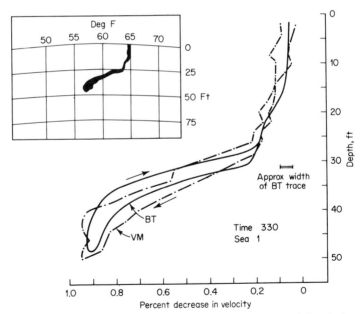

fig. 5.11 *Comparison of a bathythermogram (upper left) and the relative velocity recordings of an early velocimeter (dashed curves, right). The solid curves on the right were obtained by converting the bathythermogram to relative sound velocity and adjusting to agree with the velocimeter data at zero depth. (From Ref. 44.)*

118 / *principles of underwater sound*

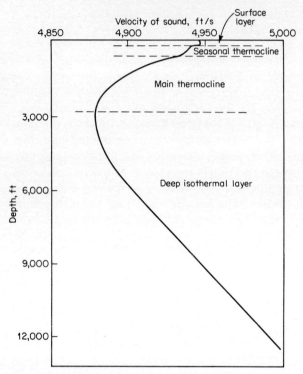

fig. 5.12 *Typical deep-sea velocity profile divided into layers.*

creases with depth because of the effect of pressure on sound velocity. Between the negative velocity gradient of the main thermocline and the positive gradient of the deep layer, there is a velocity minimum toward which sound traveling at great depths tends to be bent or focused by refraction. At high latitudes, the deep isothermal layer extends nearly to the sea surface.

Variation of the profile with latitude, season, and time of day The occurrence and thicknesses of these layers vary with latitude, season, time of day, and meteorological conditions. Some examples of this variability are shown in the following figures. Figure 5.13a illustrates the diurnal behavior of the surface layer. It shows a series of bathythermograms taken at different times of day to illustrate how the surface waters of the sea warm up during the daytime hours on a sunny day and cool off at night. These diurnal changes have a profound effect on sound transmission from a surface-ship sonar.* Figure 5.13b shows a series of bathythermograms taken in the Bermuda area and illustrates the development of the seasonal thermocline during the summer and autumn months. Figure 5.14 shows a number of velocity profiles for the four seasons

---

* Surface-ship echo ranging has long been known to be poorest in the afternoon—a phenomenon called the "afternoon effect."

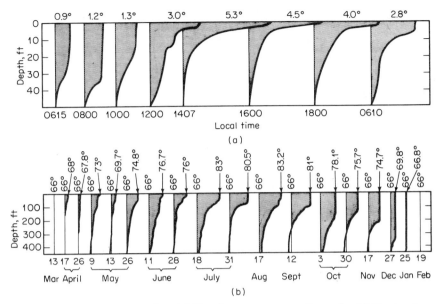

fig. 5.13 *Diurnal and seasonal variability of temperature near Bermuda:* (a) *Temperature profiles at various times of day showing the increase of surface temperature over the temperature at 50 ft.* (b) *Temperature profiles in different months of the year.* (From Ref. 46.)

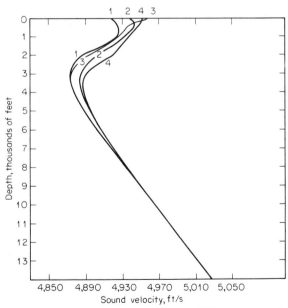

fig. 5.14 *Average velocity profiles in different seasons in an area halfway between Newfoundland and Great Britain. Latitudes 43 to 55°N, longitudes 20 to 40°W. (1) Winter. (2) Spring. (3) Summer. (4) Autumn. (Ref. 47.)*

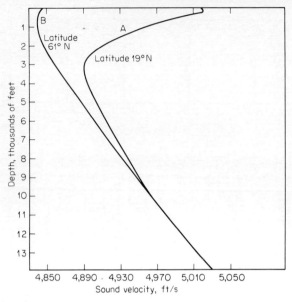

fig. 5.15 *Velocity profiles in different latitudes. A: 18°50′N, 30°01′W, spring. B: 61°02′N, 34°01′W, spring. (Ref. 47, profiles 41 and 143.)*

in a single area. It illustrates how the seasonal effect becomes less at the deeper depths. The effect of latitude on the sound-velocity profile in the deep sea is illustrated by Fig. 5.15, which shows the profile for two locations in the North Atlantic at the same season of the year. At low latitudes, the velocity minimum lies at a depth of about 4,000 ft. In high latitudes, the velocity minimum lies near the sea surface, and the main and seasonal thermoclines tend to disappear from the profile.

Some additional velocity profiles typical of different ocean areas of the world are given in Fig. 5.16.

Velocity profile in shallow water In the shallow waters of coastal regions and on the continental shelves, the velocity profile tends to be irregular and unpredictable, and is greatly influenced by surface heating and cooling, salinity changes, and water currents. The shallow-water profile is complicated by the effects of salinity changes caused by nearby sources of fresh water and contains numerous gradient layers of little temporal or spatial stability. Figure 5.17 shows some examples of bathythermograms selected at random for various seasons and locations off the East and Gulf Coasts of the United States.

5.6 Propagation Theory and Ray Tracing

Wave and ray theory The propagation of sound in an elastic medium can be described mathematically by solutions of the *wave equation* using the appro-

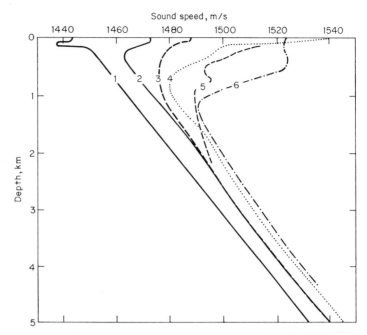

fig. 5.16 Characteristic velocity-depth profiles for the deep-ocean areas of the world. 1. Antarctic Ocean (60°S). 2. North Pacific, high latitudes (45 to 55°). 3. Southern oceans, high latitudes (45 to 55°). 4. Pacific and South Atlantic, low latitudes (40°). 5. Indian Ocean under influence of Red Sea outflow. 6. North Atlantic under influence of Mediterranean Sea outflow. (Ref. 48).

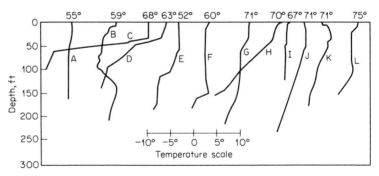

fig. 5.17 Temperature profiles in shallow water at various locations on the United States East and Gulf Coasts. Top figures are surface temperatures. (Ref. 49.)

priate boundary and medium conditions for a particular problem. The wave equation is a partial differential equation relating the acoustic pressure p to the coordinates x, y, z and the time t, and may be written as

$$\frac{\partial^2 p}{\partial t^2} = c^2 \left(\frac{\partial^2 p}{\partial x^2} + \frac{\partial^2 p}{\partial y^2} + \frac{\partial^2 p}{\partial z^2} \right)$$

where c is a quantity that has the general significance of sound velocity and may vary with the coordinates.

There are two theoretical approaches to a solution of the wave equation. One is called *normal-mode theory*, in which the propagation is described in terms of characteristic functions called *normal modes*, each of which is a solution of the equation. The normal modes are combined additively to satisfy the boundary and source conditions of interest. The result is a complicated mathematical function which, though adequate for computations on a digital computer, gives little insight, compared to ray theory, on the distribution of the energy of the source in space and time. However, normal-mode theory is particularly suited for a description of sound propagation in shallow water, and will be discussed more fully in that connection.

The other form of solution of the wave equation is *ray theory*, and the body of results and conclusions therefrom is called *ray acoustics*. The essence of ray theory is (1) the postulate of *wavefronts*, along which the phase or time function of the solution is constant, and (2) the existence of *rays* that describe where in space the sound emanating from the source is being sent. Like its analog in optics, ray acoustics has considerable intuitive appeal and presents a picture of the propagation in the form of the *ray diagram*. Ray theory has, however, certain shortcomings and does not provide a good solution under conditions where (1) the radius of curvature of the rays or (2) the pressure amplitude changes appreciably over the distance of one wavelength. Practically speaking, ray theory is therefore restricted to high frequencies or short wavelengths; it is useless for predicting the intensity of sound in shadow zones or caustics. The reader interested in the theory of sound propagation in the sea is referred to one of the NDRC Summary Technical Reports (50) and to books by Officer (51) and Brekhovskikh (52).

Table 5.4 presents a comparison between these two theoretical ways of describing the propagation of sound in the sea.

table 5.4   Comparison of Wave and Ray Theories

Wave theory	Ray theory
Gives a formally complete solution.	Does not handle diffraction problems, e.g., sound in a shadow.
Solution is difficult to interpret.	Rays are easily drawn. Sound distribution is easily visualized.
Cannot easily handle real boundary conditions.	Real boundary conditions are inserted easily, e.g., a sloping bottom.
Source function easily inserted.	Is independent of the source.
Requires a computer program, except in limiting cases when analytic answers exist, and presents computational difficulties in all but simplest boundary conditions.	Rays can be drawn by hand using Snell's law. However, a ray-trace computer program is normally used.
Valid at *all* frequencies, but practically is useful for *low* frequencies (few modes).	Valid only at *high* frequencies if (1) radius of ray curvature is $> \lambda$ or (2) sound velocity does not change much in a λ, i.e., $(\Delta V/V)/\Delta z \ll 1/\lambda$.

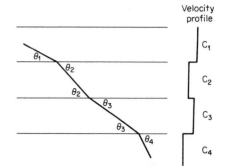

fig. 5.18 *Refraction in a layered medium.*

Snell's law One of the most important practical results of ray theory is *Snell's law*, which describes the refraction of sound rays in a medium of variable velocity. Snell's law states that in a medium consisting of constant velocity layers (Fig. 5.18), the grazing angles $\theta_1, \theta_2, \ldots$ of a ray at the layer boundaries are related to the sound velocity $c_1, c_2, \ldots$ of the layers by

$$\frac{\cos \theta_1}{c_1} = \frac{\cos \theta_2}{c_2} = \frac{\cos \theta_3}{c_3} = \cdots = \text{a constant for any one ray}$$

In this expression the ray constant is the reciprocal of the sound velocity in the layer in which the ray becomes horizontal, that is, where $\cos \theta = 1$. This expression is the basis of ray computation used by most analog and digital computers, since it enables a particular ray to be "traced out" by following it through the successive layers into which the velocity profile may have been divided. In a layered medium having layers of constant velocity, the rays consist of a series of straight-line segments joined together, in effect, by Snell's law.

A better understanding of Snell's law may be had from Fig. 5.19. In this figure the sloping lines show wavefronts, or surfaces of constant phase, in two fluids of sound velocities c_1 and c_2 separated by a plane boundary. The

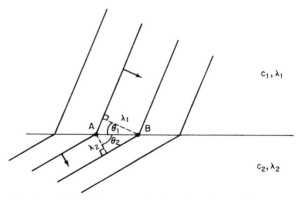

fig. 5.19 *Refraction across a plane boundary. The sloping lines are wavefronts with wavelengths λ_1 and λ_2 in the two mediums.*

distance AB along the boundary between any two wavefronts is related to the wavelengths λ_1 and λ_2 in the two mediums by

$$AB = \frac{\lambda_1}{\cos\theta_1} = \frac{\lambda_2}{\cos\theta_2}$$

If we remember that $\lambda_1 = c_1/f$ and $\lambda_2 = c_2/f$, we obtain Snell's law at once as

$$\frac{c_1}{\cos\theta_1} = \frac{c_2}{\cos\theta_2}$$

Linear variation of velocity with depth In a medium in which the velocity of sound changes linearly with depth, the sound rays can be shown to be arcs of circles, that is, to have a constant radius of curvature. A simple and heuristic demonstration of the circularity of rays in a medium with a linear gradient is given by Kinsler and Frey (53). In Fig. 5.20a, consider an arc of a circle of radius R connecting two points P_1 and P_2, where the velocity of sound is c_1 and c_2. This circle is horizontal at the depth where the velocity is c_0. Referring to the figure, we observe that

$$d_2 - d_1 = R\cos\theta_1 - R\cos\theta_2 \tag{1}$$

Since the velocity gradient is linear,

$$c_1 = c_0 + gd_1$$
$$c_2 = c_0 + gd_2$$

and

$$d_2 - d_1 = \frac{c_2 - c_1}{g} \tag{2}$$

The circle between P_1 and P_2 will be a *ray* if Snell's law is satisfied. This requires that

$$\cos\theta_1 = \frac{c_1}{c_0} \tag{3}$$

$$\cos\theta_2 = \frac{c_2}{c_0} \tag{4}$$

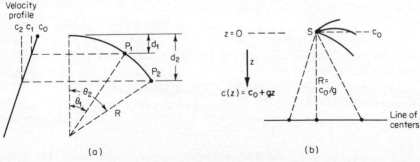

fig. 5.20 *Arcs of circles in a medium in which the velocity is a linear function of depth.*

On eliminating $d_2 - d_1$, $\cos \theta_1$, and $\cos \theta_2$ from Eqs. (1) to (4), one finds that the circle drawn between P_1 and P_2 will be a ray if, and only if,

$$R = -\frac{c_0}{g}$$

Since P_2 and P_1 were arbitrarily selected, it follows that *all* rays in the medium will be arcs of circles.*

As illustrated in Fig. 5.20b, the centers of curvature of the rays leaving a source S, at which the sound velocity is c_0, lie along a horizontal line at a distance c_0/g above or below the source, depending on the direction in which the velocity decreases with distance. In the sea, the distance c_0/g is normally very large. For isothermal water, such as in the mixed layer or in the deep isothermal layer, in which the velocity increases linearly with depth, the line of centers lies at a distance above or below the source equal to

$$R = \frac{c_0}{g} = \frac{5{,}000 \text{ ft/s}}{0.017 \text{ s}^{-1}} = 294{,}000 \text{ ft} = 48.4 \text{ nmi}$$

Nonlinear gradients   When the velocity happens to be some mathematically describable function of the depth, Snell's law permits the rays to be determined analytically. Various ray properties, such as the range to a vertex or the travel time, can be obtained mathematically. For example, if the profile has a minimum at a certain depth, and if the velocity at a distance z above or below the depth of minimum velocity is given by the function $c = c_0 \cosh Az$, where A is a constant, then it can be shown mathematically that all the rays leaving a source at $z = 0$ come to a point, or become focused, at a range equal to π/A and have a travel time equal to $\pi/(c_0 A)$. A summary of such mathematical relationships for a variety of velocity-depth functions has appeared in the seismic prospecting literature (54). Unfortunately the actual velocity profile in the sea can seldom be represented by a simple analytic function, and except for rough approximations, recourse must be made to ray tracing by a computer.

Ray tracing   The principle of circularity of rays in linear gradients is commonly employed in ray-tracing computers. The initial step in programming is to divide the velocity profile into layers of constant linear gradient and to program the computer to follow, by means of Snell's law, the arcs of rays leaving the source at different angles. An analog computer was developed (55) for this purpose as early as 1943, and consisted of a circular slide rule for finding the grazing angles of a ray at the various layers and the corresponding

---

* A more formal proof of the circularity relationship may be found in a book by Officer (51, pp. 59–60).

increments of range and depth. Other analog computers may also be used for drawing rays (56).

Ray tracing using Snell's law is easily adapted for high-speed digital computation.

Modern computer programs are extremely complex, and are able to handle such complications as sloping bottoms and arbitrary velocity profiles that change with range. Many programs compute the intensity of the sound field by finding the rays reaching a given point (called *eigenrays*) and summing the pressure contribution of each ray with regard to phase. Examples of two published ray-trace programs are CONGRATS (57) and GRASS (58). Some examples of computer-produced ray diagrams will be seen in the figures to follow. The discontinuities of velocity gradient that result from dividing up the profile into linear segments may give rise to spurious caustics in the ray diagram, but may be eliminated, as Pedersen showed (59), by using curved-line segments to approximate the velocity profile.

In a recent review (60), some 23 propagation models were identified, and were in some degree of use by different laboratories and contractors in the United States. Of these models, 13 were ray models and 10 were wave models; they differed in computer time and in their ability to handle sophisticated environmental parameters, such a range-variable bathythmetry and/or a range-variable velocity profile. Generally, they agree with one another for the same input data. Their ability to make valid predictions for the real ocean is, however, uncertain, because of uncertainties in the input parameters, especially when the ocean bottom is strongly involved in the propagation.

Another kind of modeling, besides the *mathematical* modeling just described, may be called *physical* modeling, in which experiments are carried out in laboratory tanks, and in lakes and reservoirs. This latter kind of modeling is often more revealing from a scientific standpoint than the former, but depends on the extent—often unknown—to which the model represents the real ocean. A fine summary of physical propagation modeling appears in a translation from Russian (61).

Transmission loss from ray diagrams The transmission loss between the source and any point in a ray diagram may be readily found in terms of the vertical spacing between rays that are adjacent at the source and pass above and below the distant point. Figure 5.21 shows a pair of rays separated by a small vertical angle $\Delta\theta$ leaving the source O at angles $\theta_1 + \Delta\theta/2$ and $\theta_1 - \Delta\theta/2$ with the horizontal. After traveling through the sea by any pair of refracted paths, however complex, as shown schematically by the dashed curves, they appear in the neighborhood of the field point P, at an angle θ_2 with the horizontal, with a vertical separation Δh.

Within the angle $\Delta\theta$, the source at O radiates a certain amount of power ΔP; by a fundamental property of ray acoustics, this same amount of power must appear within these same rays at P. This is equivalent to saying that energy or power does not "leak out"—as by scattering or diffraction—of the tube

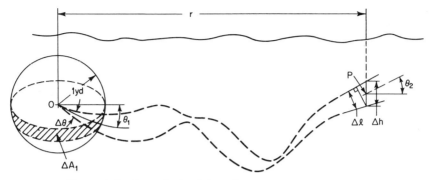

fig. 5.21 Transmission loss calculated from the spacing of rays in a ray diagram.

formed by the pair of rays. The intensity at 1 yd from the source in the θ_1 direction is

$$I_1 = \frac{\Delta P}{\Delta A_1}$$

where ΔA_1 is the area on a 1-yd sphere subtended by the pair of rays. At P, the intensity is similarly

$$I_2 = \frac{\Delta P}{\Delta A_2}$$

where ΔA_2 is the area included by the ray pair in the vicinity of P, taken normal to the rays. The transmission loss to P is therefore

$$\text{TL} = 10 \log \frac{I_1}{I_2} = 10 \log \frac{\Delta A_2}{\Delta A_1}$$

But

$$\Delta A_1 = 2\pi \cos \theta_1 \, \Delta \theta$$

and

$$\Delta A_2 = 2\pi r \, \Delta l = 2\pi r \, \Delta h \cos \theta_2$$

where r = horizontal distance from O to P
Δl = perpendicular distance between rays at P

The transmission loss is accordingly

$$\text{TL} = 10 \log \frac{r \, \Delta h}{\Delta \theta} \frac{\cos \theta_2}{\cos \theta_1}$$

But by Snell's law in a horizontally stratified medium,

$$\frac{\cos \theta_2}{\cos \theta_1} = \frac{c_2}{c_1}$$

where c_2 = velocity of sound at P
c_1 = velocity at O

Hence we have

$$\boxed{\mathrm{TL} = 10 \log \frac{r\,\Delta h}{\Delta \theta}\frac{c_2}{c_1}}$$

For practical use, r and Δh must be taken in yards, and $\Delta \theta$ must be in radians. In the sea, c_2/c_1 is so nearly unity that it may be neglected. As an example, let the vertical spacing of rays 1° apart at the source be 500 yd at a range of 50,000 yd. The transmission loss then becomes

$$\mathrm{TL} = 10 \log \frac{50{,}000 \times 500}{1/57.4} = 92 \text{ dB}$$

For straight-line paths, the expression above can be shown to reduce to TL = $10 \log r^2$, equivalent to spherical spreading.

The accuracy of this method of finding TL from a ray diagram—aside from the limitations of ray acoustics—depends on the density or number of rays drawn on the ray diagram, that is, on the size of the ray interval $\Delta \theta$. The method, along with ray acoustics generally, fails in the immediate vicinity of caustics, where wave theory (52, pp. 483–492) is required for making an estimate of the intensity. More elaborate treatments of the spreading loss under more general conditions are available (62, 63).

Yet the method for the simple conditions outlined gives useful predictions of transmission loss by the above simple expressions. Its validity rests on the basic assumption of ray theory: *there is no crossing of acoustic energy between rays*. All the energy emitted from the source within a pair of rays in the ray diagram remains within that pair of rays; there is no scattering by inhomogeneities and no diffraction as dictated by wave theory. For the refracted rays of the deep ocean sound channel, the ray diagram has been found valid down to the lowest frequency so far examined in propagation measurements.

5.7 The Sea Surface

Reflection and scattering The surface of the sea is both a reflector and a scatterer of sound and has a profound effect on propagation in most applications of underwater sound where source or receiver lie at shallow depth.

If the sea surface were perfectly smooth, it would form an almost perfect reflector of sound. The intensity of sound reflected from the smooth sea surface would be very nearly equal to that incident upon it. The *reflection loss*, equal to $10 \log I_r/I_i$, where I_r and I_i are the reflected and incident intensities of an incident plane wave, would be closely equal to zero decibels.

When the sea is rough, as it is to some extent at all times, the loss on reflection is found to be no longer zero. At 25 kHz, Urick and Saxton (64) measured an average sea-surface reflection loss of about 3 dB under conditions of 1-ft waves and grazing angles between 3 and 18°; using another method, Liebermann (65) found a median loss of 3 dB at 30 kHz for wave heights of 0.2 to 0.8 ft and a grazing angle of 8°. At lower frequencies, a smaller loss can be expected because of the fact that the sea is becoming smoother relative to a wavelength. This is borne out by the work of Addlington (66), who used explosive pulses reflected from the sea surface. In all octave bands between 400 and 6,400 Hz, over a range of wind speed 5 to 20 knots and grazing angle 10 to 55°, a zero-decibel median reflection loss was measured. A similar result may be inferred from the transmission measurements of Pedersen (67) at 530 and 1,030 Hz.

A criterion for the roughness or smoothness of a surface is given by the *Rayleigh parameter*, defined as $R = kH \sin \theta$, where k is the wave number $2\pi/\lambda$, H is the rms "wave height" (crest to trough), and θ is the grazing angle. When $R \ll 1$, the surface is primarily a *reflector* and produces a coherent *reflection* at the specular angle equal to the angle of incidence. When $R \gg 1$, the surface acts as a *scatterer*, sending incoherent energy in all directions. When certain theoretical assumptions are made, the amplitude reflection coefficient μ of an irregular surface (defined as the ratio of the reflected or coherent amplitude of the return to the incident amplitude) can be shown (68) to be given by the simple expression $\mu = \exp(-R)$. When $R \gg 1$, the return from the surface is incoherent scattering, instead of coherent reflection, with a distribution throughout space depending upon the nature of the surface roughness.

When the surface is in motion, as is always true of the surface of the sea, the vertical motion of the waves superposes itself, so to speak, upon the frequency of the sound incident upon it. The moving surface produces upper and lower sidebands in the spectrum of the reflected sound that are the duplicates of the spectrum of the surface motion. These extraneous dopplerlike frequencies have been demonstrated by the laboratory experiments of Roderick and Cron (69). Thus, the mobile sea surface produces a frequency-smearing effect on a constant-frequency signal. Although it is small, it is an effect having significance for narrow-band underwater acoustic communications.

Diagrammatically this is illustrated in Fig. 5.22 idealized from the work of Roderick and Cron. In (a) a sinusoidal pulse is incident on the sea surface and is reflected and scattered as a pulse of variable amplitude and frequency. Its spectrum is shown in (b) for a low and for a high sea state, and we note that energy appears to be extracted from the incident frequency centered at zero Hz and fed into upper and lower sidebands. The motional spectrum is shown in (c). Thus, the spectrum of the surface motion appears in, and widens, the spectrum of the incident sound. Surface sidebands have even been observed in RSR (Refracted Surface-Reflected) propagation, where the sea surface has been encountered many times in propagation out to hundreds of miles (70).

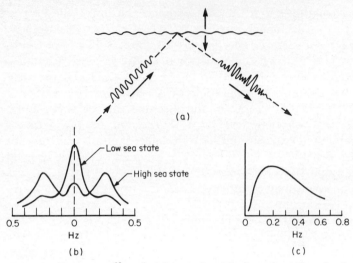

fig. 5.22 *Frequency effect of surface motion. The irregular reflected pulse in (a) acquires upper and lower sideband frequencies that are duplicates of the spectrum of the surface motion. (a) shows a reflected sinusoidal pulse, (b) its spectrum, and (c) the spectrum of the surface motion.*

Fluctuation of sea surface sound Large and rapid fluctuation in amplitude or intensity is produced by reflection at the sea surface. In the above-cited works of Urick and Saxton and of Liebermann, about 10 percent of the measured reflection losses of short pulses were greater by 10 dB or more above the average, and 10 percent were smaller by 3 dB or less. Similarly large fluctuations in surface-reflected amplitudes were found by Pollak (71) in Chesapeake Bay. Such fluctuations in amplitude may be attributed to reflections from wave facets on the sea surface that present a constantly changing size and inclination to the incident and reflected sound. Surface-reflection fluctuations occur even at low frequencies if the grazing angle is large, as observed, for example, by Brown and Ricard (72) at 168 Hz between a source and receiver deep in the sea.

The magnitude of the fluctuation of the amplitude of surface-reflected sound is related to the Rayleigh coefficient R mentioned above. Measurements by Gulin and Malyshev (73) at 108 Hz and by Brown (74) at four frequencies between 160 and 1,360 Hz tend to show that the coefficient of variation V, or the normalized standard deviation (Sec. 6.7), of the amplitudes of successive samples of a surface-reflected signal is given approximately by

$$V = 0.5R \quad R \ll 1$$
$$V = 0.5 \quad R \gg 1$$

Thus the fluctuation, as defined by the quantity V, is linearly proportional to R for small R, and is constant for large R at a value equal to about 0.5. It is noteworthy that the Rayleigh distribution, which one would expect to find for scattering from a very rough surface, has a coefficient of variation V equal to 0.52.

Image interference When the sea surface is not too rough, it creates an interference pattern in the underwater sound field. This pattern is caused by constructive and destructive interference between the direct and surface-reflected sound and is called the *Lloyd mirror*, or *image-interference*, effect.

Referring to Fig. 5.23, let a nondirectional CW sinusoidal point source be located at S in a refraction-free absorptionless sea. Let its pressure at unit distance be given by

$$P = P_0 \sin \omega t$$

At the receiver R (Fig. 5.23), this pressure becomes

$$P_1 = \frac{P_0}{l_1} \sin \omega(t + \tau_1)$$

where l_1 is the distance SR and τ_1 is the travel time from S to R, assuming straight-line paths. Via the surface reflected path, the pressure at R is

$$P_2 = \frac{\mu P_0}{l_2} \sin \omega(t + \tau_2)$$

where τ_1 has been taken to be zero and τ_2 has been replaced by τ for conven- travel time, respectively, for the surface reflection.

The total pressure at R is then

$$P_T = P_1 + P_2 = \frac{P_0}{l_1} \sin \omega t + \frac{\mu P_0}{l_2} \sin \omega(t + \tau)$$

where τ_1 has been taken to be zero and τ_2 has been replaced by τ for convenience.

The intensity at R is

$$I = \frac{\overline{P_T^2}}{\rho c} = \frac{P_0^2}{\rho c} \overline{\left[\frac{1}{l_1} \sin \omega t + \frac{\mu}{l_2} \sin \omega(t + \tau)\right]^2} \tag{1}$$

where the bar represents a time average over many periods of ω.

In what follows, we will divide the sound field whose intensity is given by Eq. (1) into three parts: (1) the *near field*, where the effect of the surface

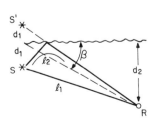

fig. 5.23 Interference at R between sound originating at the source S and at the surface image S'.

reflection is negligible and spherical spreading exists; (2) the *interference field*, where constructive and destructive interference occurs between the two paths just mentioned; and (3) the *far field*, where the source and its image act as a dipole and produce inverse fourth-power spreading.

The near field At ranges close to the source, where $l_2 \gg l_1$, the second term in Eq. (1) is insignificant and the intensity falls off inversely as the square of the direct distance l_1. With increasing range, the second or interfering term becomes increasingly important.

Let us take, as the limit of the near field and the beginning of the interference field, the distance at which the intensity of the second term is one-half, or 3 dB below, that of the first. That is, let us define the outer limit of the near field as the distance where

$$\frac{l_2^2}{l_1^2} = 2$$

Then it can be shown by geometry that

$$l_1 = 2(d_1 d_2)^{1/2}$$

This distance is the range to the beginning of the interference field according to the 3 dB-down criterion.

The interference field In this region the two paths may be taken to be of equal length, so that we can write $l_1 = l_2 = l$. Also, let us take for simplicity $\mu = 1$. The intensity now becomes, from Eq. (1)

$$I = \frac{P_0^2}{\rho c} \cdot \frac{1}{l^2} \overline{[\sin \omega t + \sin \omega(t + \tau)]^2}$$

which becomes after time averaging

$$I = \frac{I_0}{l^2} \cdot 2(1 - \cos \omega \tau) \tag{2}$$

where I_0 is free-field source intensity at unit distance ($l = 1$). To evaluate the time difference between the two paths, we note that

$$l_2^2 - l_1^2 = (l_2 + l_1)(l_2 - l_1) = 4 d_1 d_2$$

or

$$\Delta l = \frac{2 d_1 d_2}{l} = c\tau$$

where we have replaced $l_2 - l_1$ by Δl and $l_2 + l_1$ by $2l$ for brevity, and where c is the velocity of sound. Hence, Eq. (2) becomes

$$I = \frac{I_0}{l^2} \cdot 2\left(1 - \cos \frac{\omega}{c} \Delta l\right) = \frac{I_0}{l^2} \cdot 2\left(1 - \cos \frac{4\pi d_1 d_2}{\lambda l}\right) \tag{3}$$

where ω/c has been replaced by $2\pi/\lambda$. If now we define a reference distance l_0 to be

$$l_0 = \frac{4d_1 d_2}{\lambda}$$

the result is

$$I = \frac{I_0}{l^2} \cdot 2\left(1 - \cos \pi \frac{l_0}{l}\right) \tag{4}$$

and we will find that I will be zero when $l/l_0 = \frac{1}{2}, \frac{1}{4}, \frac{1}{6} \ldots$ and that it will be 4 times greater than I_0, or 6 dB, when $l/l_0 = 1, \frac{1}{3}, \frac{1}{5}, \frac{1}{7} \ldots$, inverse square spreading ($1/l^2$) neglected. Thus the interference field consists of a series of maxima and mimima extending out to the last maximum where $l = l_0$. The minima extend to zero for a perfectly smooth surface ($\mu = 1$) or for an infinitely narrow bandwidth, but fill up with increasing surface roughness and finite bandwidth.

The far field In Eq. (4), the argument of the cosine becomes small when l becomes large. If the cosine is expanded in terms of the powers of its argument, we obtain

$$I = \frac{I_0}{l^2}\left(\frac{4\pi d_1 d_2}{\lambda l}\right)^2 = \frac{I_0}{l^2}\left(\frac{\pi l_0}{l}\right)^2 \tag{5}$$

and we note that the intensity falls off as the fourth power of the range.

Figure 5.24 is a plot of the interference and far field of a source in terms of the *transmission anomaly*—so called during World War II to refer to unexplained deviations from spherical spreading—but used here simply as a way to eliminate the l^2 term in the preceding expressions.

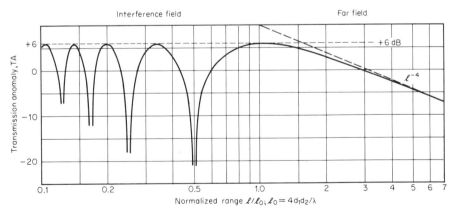

fig. 5.24 *The surface-image or "Lloyd Mirror" interference pattern. The vertical scale of transmission anomaly is the difference between the transmission loss and inverse-square spreading; the horizontal scale is a normalized range in terms of the quantity l_0. In the interference field the nulls occur at values of $\frac{1}{8}, \frac{1}{6}, \frac{1}{4}$, and $\frac{1}{2}$; the peaks at $\frac{1}{9}, \frac{1}{7}, \frac{1}{5}, \frac{1}{3}$ and 1.*

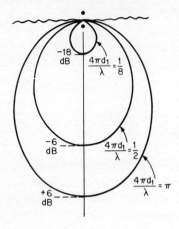

fig. 5.25 The single-lobe, or dipole, radiation pattern occurring when d_1/λ is equal to or less than $1/4$. The dB values refer to the radiated levels in the free field in the absence of the surface.

The dipole radiation pattern In Fig. 5.23 let us define an angle β such that $\sin\beta = d_2/l_1 = d_2/l_2$. Then Eq. (3) becomes, for small values of the argument of the cosine

$$I = \frac{I_0}{l^2} \cdot 2 \left[1 - \cos\left(\frac{4\pi d_1}{\lambda} \sin\beta\right) \right]$$

and the source, which is nondirectional in the free field, has acquired a directional pattern of loops and troughs as β is varied. For small arguments of the cosine, as, for example, for low-frequency sources near the surface, the preceding equation becomes

$$I = \frac{I_0}{l^2} \left(4\pi \frac{d_1}{\lambda} \sin\beta \right)^2 \tag{6}$$

and the pattern degenerates into a single lobe pointing downward, as shown in Fig. 5.25. The maximum radiation occurs when $d_1/\lambda = 1/4$, the quarter-wave depth; it decreases with decreasing d_1/λ and finally disappears altogether when $d_1 = 0$.

Effect of refraction When refraction exists, the two interfering paths are no longer straight lines and the interference pattern is perturbed. The effect is greatest at high frequencies and large depths, where the interference field limit l_0 lies at ranges of thousands of yards. The effect is small for ordinary velocity gradients. It was investigated by Young (75) during World War II and is reported in one of the earliest papers on underwater sound published in the Journal of the Acoustical Society of America.

Implications for sonar Most sonar measurements and observations are made in the far field where the irregularities of the interference field are absent. One effect is to require, by Eq. (5), a correction equal to $20 \log (\pi l_0/l)$ in the reduction of far-field measurements to free-field source levels (in addition to the spherical spreading term $20 \log l$). Another effect is to require in range calculations a corrected source level as given by Eq. (6), involving the source depth, wavelength, and, in ray-trace calculations, the angle β of the ray leaving the source. Finally, at a distant point, and at a shallow depth where a surface reflection may be involved, another correction of $\sin^2 \beta$ will be necessary to account for the image interference effect on reception. Corrections of these kinds needed in ray theory are unnecessary in wave-theory computer models in which the effects are, or should be, already included.

Shadow zones When a negative gradient exists just beneath the sea surface, a shadow is cast by the surface in the sound field of a shallow source. A *shadow zone* is produced, in which the intensity from the source is very low. An example of a computer-produced ray diagram showing a shadow zone is seen in Fig. 5.26. The velocity profile from which the ray diagram was computed is given on the right. Although no rays enter the shadow, some sound exists inside the shadow zone, and the shadow is not completely acoustically dark. At-sea measurements have shown (50, p. 125) that the intensity of 24-kHz sound inside the shadow is between 40 and 60 dB below what it would be in the free field; moreover, severe distortion of sinusoidal and explosive pulses occurs in transmission to a hydrophone located inside the shadow. The shadow boundary is very sharp at high frequencies and causes an attenuation in the vicinity of the boundary amounting to between 20 and 70 dB/kyd at 24 kHz.

Several causes of the sound inside a sea-surface shadow zone may be postulated. One is the diffraction of sound into the shadow zone as accounted for by wave theory. This is an effect likely to be of some consequence near the boundary of the shadow zone. Another is the likelihood that the gradient of temperature causing the shadow seldom extends all the way to the sea surface but approaches zero, or isothermal water, just underneath the surface; this effect was considered theoretically by Morse (76) and was found to account for the observed sound levels in shadow zones at 24 kHz. Finally, volume

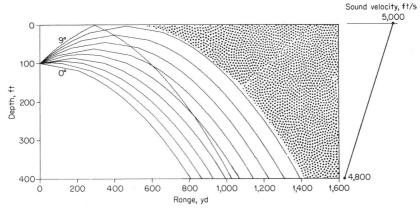

fig. 5.26 *Ray diagram for a source at a depth of 100 ft in the linear gradient shown at the right. Stippled area is the surface shadow zone.*

scatterers—especially those in the deep scattering layer—located in the insonified sound field short of the shadow will scatter sound forward into the shadow zone and will tend to "fill it up" with forward-scattered sound analogous to volume reverberation. This is probably the most likely cause of shadow-zone sound. If it is, the propagation is essentially identical with the tropospheric-scatter propagation of radio waves beyond the horizon.

5.8 The Sea Bottom

Comparison with the sea surface The sea bottom is a reflecting and scattering boundary of the sea having a number of characteristics similar to the sea surface. However, its effects are more complicated because of its diverse and multilayered composition. An example of this similar behavior is the fact that the sea bottom casts a shadow, or produces a shadow zone, in the upward-refracting water above it in the depths of the deep sea. In Fig. 5.27 the source is taken to be located 100 ft above the bottom; the distance to the beginning of the shadow zone at the bottom is 3,500 yd. Also, a Lloyd mirror effect may be expected in the neighborhood of the seabed, and an interference pattern may be expected to exist in the radiation of a nearby source. Examples of image-interference directivity patterns caused by bottom reflection have been computed by Mackenzie (77) for a rock, sand, and silt bottom.

For two reasons, the reflection of sound from the seabed is vastly more complex than that from the sea surface. First, the bottom is more variable in its acoustic properties because it may vary in composition from hard rock to soft mud. Second, it is often layered, with a density and a sound velocity that change gradually or abruptly with depth. For these reasons, the reflection loss of the seabed is less easily predicted than that of the sea surface.

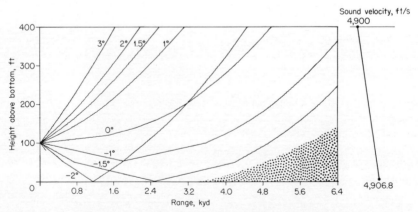

fig. 5.27 *Ray diagram for a source at a height of 100 ft above the bottom in isothermal water. The sound-velocity profile is shown at the right. Stippled area is the bottom shadow zone.*

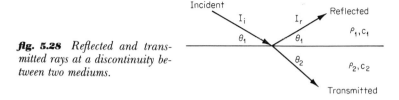

fig. 5.28 Reflected and transmitted rays at a discontinuity between two mediums.

Bottom-reflection loss The reflection loss of sound incident at an angle to a plane boundary between two fluids was worked out originally by Rayleigh, (8, p. 78). If a plane wave is incident at grazing angle θ_1 upon the boundary between fluids of density ρ_1 and ρ_2 and of sound velocity c_1 and c_2, as indicated in Fig. 5.28, then by the Rayleigh formula the intensity of the reflected wave I_r is related to the intensity of the incident wave I_i by

$$\frac{I_r}{I_i} = \left[\frac{m \sin \theta_1 - n \sin \theta_2}{m \sin \theta_1 + n \sin \theta_2}\right]^2 = \left[\frac{m \sin \theta_1 - (n^2 - \cos^2 \theta_1)^{1/2}}{m \sin \theta_1 + (n^2 - \cos^2 \theta_1)^{1/2}}\right]^2$$

where, following the notation of Brekhovskikh (52, pp. 16–20),

$$m = \frac{\rho_2}{\rho_1} \quad \text{and} \quad n = \frac{c_1}{c_2}$$

The reflection loss is the logarithmic expression of the above ratio, or $10 \log I_r/I_i$ dB. As a function of grazing angle, the reflection loss is therefore dependent on the ratios m and n. Figure 5.29, adapted from Brekhovskikh, shows the behavior of the loss with grazing angle for four different conditions on m and n. Of these four, the most common condition for natural bottoms is probably that of Fig. 5.29c, in which a *critical angle* θ_0 exists such that complete or total reflection occurs (zero loss) at grazing angles less than critical. In many soft mud bottoms, the sound velocity is *less* than that in the water above, and an *angle of intromission* θ_B may exist, as in Fig. 5.29a.

So far, absorption has not been brought into the picture. All bottom materials are to some extent absorptive, and the effect of absorption is to smooth out the variation of loss with angle so as to eliminate, or obscure, the sharp changes occurring at the critical angle θ_0 and the angle of intromission θ_B. An example of the effect of absorption is the dashed curve in Fig. 5.29c. The necessary theory was worked out by Mackenzie (77), who modified the theory stated above to include attenuation in the bottom materials.

Many measurements of sound attenuation in sediments have been made (78). They show that the attenuation coefficient of compressional waves in marine sediments is related to frequency by

$$\alpha = kf^n$$

where α is in decibels per meter, f is the frequency in kilohertz, and k and n are empirical constants. In the equation, n appears to be essentially unity for

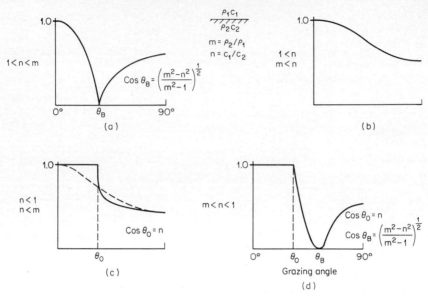

fig. 5.29 Ratio of reflected to incident intensities for four combinations of conditions of sound velocities and densities in lossless mediums separated by a plane interface. The dashed curve in (c) shows the effect of an attenuating lower medium. [After Brekhovskikh (Ref. 52, fig. 7).]

many measurements on sands, silts, clays, and the like; k depends upon porosity, being approximately equal to 0.5 over the range of porosity 35 to 60 percent.

It should be mentioned in passing that the acoustics of marine sediments has a vast literature, to which reference can be made to two books (79, 80), to the many papers of E. L. Hamilton, such as (81), and to review papers by Anderson and Hampton (82).

If the bottom in the most simple model is taken to be a homogeneous absorptive fluid with a plane interface, then the three bottom parameters that determine the reflection loss are its *density, sound velocity,* and *attenuation coefficient.* If the bottom happens to be a sedimentary material, these quantities are related to, and determined by, the *porosity* of the sediment. For example, Fig. 5.30 shows the results of many measurements made by Schreiber (83) of the sound velocity in core samples that had been taken during the extensive Marine Geophysical Survey (84) conducted during the years 1965–1968. Here the clear relationship between velocity and porosity of the sample is apparent. An even closer relationship exists between density and porosity (85), and a less exact one exists between attenuation coefficient and porosity (86). As a result, we can generate curves of reflection loss versus angle with porosity as a parameter that could be used for bottom loss prediction if the porosity is known or can be estimated.

However, a number of complications to this simple model occur in the real world. First of all, the ocean floor is not a perfectly plane interface, so that scattering as well as reflection takes place. In a very rough area, such as on the Mid-Atlantic Ridge, scattered sound dominates the bottom return. As a result, some sound is sent by the bottom in all directions, and the "beam pattern" of the bottom return shows no appreciable lobe or peak in the specular direction. An example of such a beam pattern is seen in Fig. 5.31, taken from the work of Hurdle, Flowers, and Thompson (87) at 19.5 kHz at a location on the Bermuda Rise southeast of Bermuda. Thus, it is a combination of density and velocity contrast (m, n) of the bottom and its roughness that determines the "beam pattern" of the bottom return. This shown in a highly diagrammatic way by Fig. 5.32 where the cross-hatching shows the amount and direction of sound that is returned back to the water when an impinging sound wave strikes the bottom obliquely. In A and B there is a marked discontinuity in the ρc ratio, or the ratio of *acoustic impedances* between water and bottom. When the interface is smooth, as in A, there is a strong reflected return, with only a small amount of sound going in other directions; in B, the reflected component is absent and the returned sound is entirely scattered. In both these cases

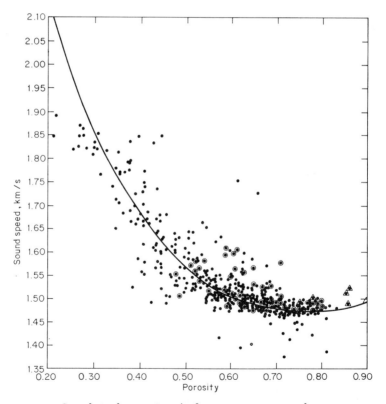

fig. 5.30 *Sound speed versus porosity from measurements made on core samples. The curve shows Wood's equation (Ref. 32) for a mixture of noninteracting solid particles in water. (Ref. 83.)*

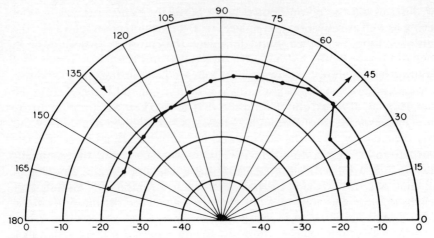

fig. 5.31 *Distribution of sound by the deep-sea bottom, showing the relative intensity in different directions. The arrow at the left shows the direction of incidence; that at the right the direction of specular reflection. Only a broad lobe exists in the specular direction. (Ref. 87.)*

only a small amount of sound enters the bottom itself. In C and D there is only a slight contrast in ρc between water and bottom. Here the directional patterns are similar in shape to A and B but are smaller since now more of the incident sound is able to penetrate into the bottom. For real bottoms, B and C are probably more common than A and D; B represents a rough hard sand

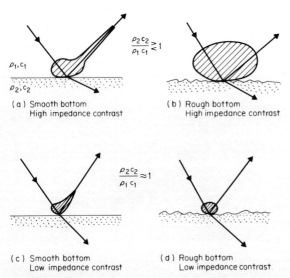

fig. 5.32 *Directional patterns of the return of sound from the bottom for different conditions of roughness and impedance (ρc) contrast.*

where current scour had produced the roughness, while C is a soft mud occurring in quiescent water.

Another complication is the effect of layering. Core samples show that most sedimentary bottoms have a layered structure, with alternating layers of material of different properties. Such bottoms will produce an enhanced return at some angles and frequencies, and a diminished return at others, according to the manner in which the reflections from the various layers interfere with one another. However, when the reflected contributions are summed with careful attention to phase, good agreement between theoretical and measured losses has been found by Cole (88) from field data and by Barnard, Bardin, and Hempkins (89) from laboratory measurements.

Still another complication is produced by the velocity gradient in the bottom. Because of compaction, the velocity of sound in sedimentary materials increases rapidly with depth (90). This causes upward refraction in the bottom and a return of sound back into the sea above. This refractive effect, as well as reflection from deep subbottom layers, can cause low-frequency sound to have a very small, or even a negative, reflection loss at the low grazing angles associated with long-range propagation in deep water.

Measured bottom losses In the late 1960s an extensive program, called the Marine Geophysical Survey, was carried out by contractors under the management of the U.S. Naval Oceanographic Office. Its purpose was to obtain data on the bottom loss in various areas for use with the then new bottom-bounce echo-ranging sonars, which obtain echoes by reflection from the deep ocean bottom. Figure 5.33 shows two average curves obtained in this survey for two broad areas in the North Atlantic in the frequency band 0.5–8.0 kHz. Individual measurements deviate greatly from these averages. For practical purposes the loss-angle relationship can be characterized by one of a series of

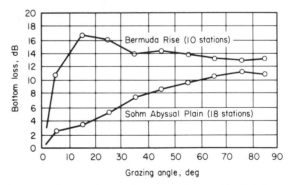

fig. 5.33 *Average bottom loss curves for two North Atlantic Ocean areas. One shows a suggestion of an angle of intromission (Fig. 5.29a), the other a critical angle masked by absorption (Fig. 5.29c).*

curves. Such a series is shown in Fig. 5.34 which, in practical use, are accompanied by charts that identify the curve best representing the bottom at that spot where the echo-ranging vessel is located.

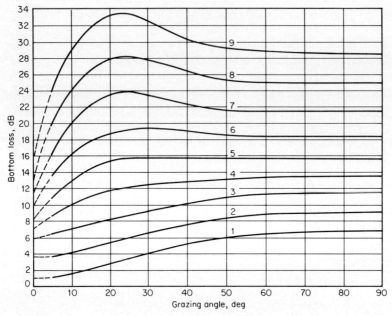

fig. 5.34 *Canonical loss curves, based on data, theory, and conjecture, to represent different loss versus angle categories in the frequency band 1–4 kHz.*

Although frequency measurements at kilohertz frequencies are highly variable from place to place, the bottom loss at lower frequencies shows much less variability. Figure 5.35 is based on measured data in many areas at frequencies near 100 Hz, where the measurements had the scatter shown by the dotted curves of one standard deviation from the mean. The negative loss values at angles less than 10° are the result of sound entering the bottom and coming out again by refraction, as mentioned above.

Older measurements exist at higher frequencies in coastal waters. Two sets

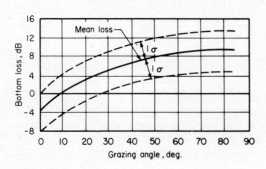

fig. 5.35 *Near 100 Hz, all measured data have been found to fall within the indicated range of 1 standard deviation. The negative losses at low angles are due to a failure to include bottom-refracted sound in the reduction of the data.*

of data are listed in Table 5.5. The meaning of numbers such as these, other than showing the general magnitude of loss values, is uncertain, being affected by the method employed and the experimental parameters. In these references, and others, the original reports should be examined to estimate the validity of the results for a particular purpose.

table 5.5 Measured Reflection Losses for Different Bottom Types

Type of bottom	Measured reflection loss, in decibels			
	24 kHz	16 kHz	7.5 kHz	4 kHz
10° grazing angle, 17 stations (Ref. 91)				
Mud		16		
Mud-sand		10		
Sand-mud		6		
Sand		4		
Stony		4		
Normal incidence, 7 stations (Ref. 77)				
Sandy silt (3 stations)		13	14	14
Fine sand (1 station)		6	3	7
Coarse sand (1 station)		8	8	7
Medium sand with rock (1 station)		10	6	8
Rock with some sand (1 station)		10	4	5

REFERENCES

1. Stephenson, E. B.: Transmission of Sound in Sea Water: Absorption and Reflection Coefficients and Temperature Gradients, *U.S. Nav. Res. Lab. Rep.* S-1204, 1935.
2. Stephenson, E. B.: Absorption Coefficients of Supersonic Sound in Open Sea Water, *U.S. Nav. Res. Lab. Rep.* S-1549, 1939.
3. Principles of Underwater Sound, *Nat. Def. Res. Comm. Sum. Tech. Rep.* 7, sec. 3.4, National Research Council, 1946. Also, Physics of Sound in the Sea, *Nat. Def. Res. Comm. Sum. Tech. Rep.* 8, sec. 5.2.2, National Research Council, 1946.
4. Wilson, O. B., and R. W. Leonard: Measurements of Sound Absorption in Aqueous Salt Solutions by a Resonator Method, *J. Acoust. Soc. Am.*, **26:**223 (1954).
5. Hansen, P. G.: Measurements of Low Frequency Sound in Small Samples of Sea Water, *U.S. Navy Electron. Lab. Rep.* 1135, 1962.
6. Glotov, V. P.: Reverberation Tank Method for the Study of Sound Absorption in the Sea, *Sov. Phys. Acoust.*, **4:**243 (1958).
7. Liebermann, L. N.: Sound Propagation in Chemically Active Media, *Phys. Rev.*, **76:**1520 (1949).
8. Rayleigh, Lord: "The Theory of Sound," vol. II, Dover Publications, Inc., New York, 1945.
9. Mason, W. P. (ed.): "Physical Acoustics," vol. II, pt. A, pp. 293–295, Academic Press Inc., New York, 1965.
10. Leonard, R. W., P. C. Combs, and L. R. Skidmore: Attenuation of Sound in Synthetic Sea Water, *J. Acoust. Soc. Am.*, **21:**63 (1949).
11. Liebermann, L. N.: Origin of Sound Absorption in Water and in Sea Water, *J. Acoust. Soc. Am.*, **20:**868 (1948).
12. Schulkin, M., and H. W. Marsh: Absorption of Sound in Sea Water, *J. Brit. IRE*, **25:**493 (1963). Also, Sound Absorption in Sea Water, *J. Acoust. Soc. Am.*, **34:**864 (1962).

13. Throp, W. H.: Deep Ocean Sound Attenuation in the Sub- and Low-Kilocycle per Second Region, *J. Acoust. Soc. Am.*, **38**:648 (1965).
14. Marsh, H. W.: Attenuation of Explosive Sounds in Sea Water, *J. Acoust. Soc. Am.*, **35**:1837 (1963).
15. Thorp, W. H.: Analytic Description of the Low Frequency Attenuation Coefficient, *J. Acoust. Soc. Am.*, **42**:270 (1967).
16. Urick, R. J.: Long Range Deep Sea Attenuation Measurement, *J. Acoust. Soc. Am.*, **39**:904 (1966).
17. Brown, C. B., and J. J. Raff: Theoretical Treatment of Low Frequency Sound Attenuation in the Deep Ocean, *J. Acoust. Soc. Am.*, **35**:2007 (1963).
18. Urick, R. J.: Low-Frequency Sound Attenuation in the Deep Ocean, *J. Acoust. Soc. Am.*, **35**:1413 (1963).
19. Schulkin, M.: Eddy Viscosity as a Possible Acoustic Absorption Mechanism in the Ocean, *J. Acoust. Soc. Am.*, **35**:253 (1963).
20. Duykers, L. R.: Sound Attenuation in Liquid-Solid Mixtures, *J. Acoust. Soc. Am.*, **41**:1330 (1967).
21. Yeager, E., F. H. Fisher, J. Miceli, and R. Bressel: Origin of the Low-Frequency Sound Absorption in Sea Water, *J. Acoust. Soc. Am.*, **53**:1705 (1973).
22. Browning, D. G., and W. H. Thorp: Attenuation of Low Frequency Sound in the Ocean, *U.S. Nav. Underwater Syst. Center Tech. Rep.* 4581 (1974).
23. Mellen, R. H., D. G. Browning, and J. M. Ross: Attenuation in Randomly Inhomogeneous Sound Channels, *J. Acoust. Soc. Am.*, **56**:80 (1974).
24. Mellen, R. H., D. G. Browning, and V. P. Simmons: Investigation of Chemical Sound Absorption in Seawater by the Resonator Method, Part I, *J. Acoust. Soc. Am.*, **68**:248 (1980). Part II: *J. Acoust. Soc. Am.*, **69**:1660 (1981).
25. Lovett, J. R.: Geographic Variation of Low Frequency Sound Absorption in the Atlantic, Indian and Pacific Oceans, *J. Acoust. Soc. Am.*, **67**:338 (1980).
26. Kibblewhite, A., and L. Hampton: Review of Deep Ocean Sound Attenuation Data at Very Low Frequencies, *J. Acoust. Soc. Am.*, **67**:147 (1980).
27. Mellen, R. H., D. G. Browning, and L. Goodman: Diffusion Loss in a Stratified Sound Channel, *J. Acoust. Soc. Am.*, **60**:1053 (1976).
28. Fisher, F. H., and V. P. Simmons: Sound Absorption in Sea Water, *J. Acoust. Soc. Am.*, **62**:558 (1977).
29. Browning, D., M. Fechner, and R. Mellen: Regional Dependence of Very Low Frequency Sound Attenuation in the Deep Sound Channel, *Nav. Underwater Systems Center Tech. Doc.* 6561, 1981.
30. Fisher, R. H.: Effect of High Pressure on Sound Absorption and Chemical Equilibrium, *J. Acoust. Soc. Am.*, **30**:442 (1958).
31. Bezdek, H. F.: Pressure Dependence of Sound Attenuation in the Pacific Ocean, *J. Acoust. Soc. Am.*, **53**:782 (1973).
32. Wood, A. B.: "A Textbook of Sound," The Macmillan Company, New York, 1941.
33. Stephenson, E. B.: Velocity of Sound in Sea Water, *Phys. Rev.*, **21**:181 (1923).
34. Heck, N. H., and J. H. Service: Velocity of Sound in Sea Water, *U.S. Coast Geod. Surv. Spec. Publ.* 108, 1924.
35. Garrison, G. R., P. C. Kirkland, and S. R., Murphy: Long Range Measurements of Sound Speed in Sea Water, *J. Acoust. Soc. Am.*, **33**:360 (1961).
36. Matthews, D. J.: Tables of the Velocity of Sound in Pure Water and Sea Water for Use in Echo Sounding and Sound Ranging, *Admiralty Hydrog. Dep. Publ.* 282, 1939.
37. Kuwahara, S.: Velocity of Sound in Sea Water and Calculation of the Velocity for Use in Sonic Sounding, *Hydrogr. Rev.*, **16**:123 (1939).
38. Weissler, A., and V. A. Del Grosso: The Velocity of Sound in Sea Water, *J. Acoust. Soc. Am.*, **23**:219 (1951).
39. Del Grosso, V. A.: Velocity of Sound in Sea Water at Zero Depth, *U.S. Nav. Res. Lab. Rep.* 4002, 1952.

40. Wilson, W. D.: Speed of Sound in Sea Water as a Function of Temperature, Pressure and Salinity, *J. Acoust. Soc. Am.*, **32**:641 (1960). Also, see extensions and revised formulas in *J. Acoust. Soc. Am.*, **32**:1357 (1960) and **34**:866 (1962).
41. Del Grosso, V. A.: New Equations for the Speed of Sound in Natural Waters with Comparisons to Other Equations, *J. Acoust. Soc. Am.*, **56**:1084 (1974).
42. Lovett, J. R.: Merged Seawater Sound Speed Equations, *J. Acoust. Soc. Am.*, **63**:1713 (1978).
43. Greenspan, M., and C. E. Tschiegg: Sing-around Ultrasonic Velocimeter for Liquids, *Rev. Sci. Instrum.*, **28**:897 (1957). Also, *J. Acoust. Soc. Am.*, **31**:1038 (1959).
44. Urick, R. J.: An Acoustic Interferometer for the Measurement of Sound Velocity in the Ocean, *U.S. Navy Radio Sound Lab. Rep.* S-18, 1944.
45. Brown, R. K.: Measurement of Sound Velocity in the Ocean, *J. Acoust. Soc. Am.*, **26**:64 (1954).
46. Application of Oceanography to Subsurface Warfare, *Nat. Def. Res. Comm. Div. 6 Sum. Tech. Rep.*, vol. 6A, figs, 17, 32, 1946.
47. An Interim Report on the Sound Velocity Distribution in the North Atlantic Ocean, *U.S. Navy Oceanogr. Office Tech. Rep.* 171, 1965.
48. "Ocean Science Program of the U.S. Navy," Office of the Oceanographer of the Navy, Alexandria, Va., 1970.
49. Woollard, G. P.: "Sound Transmission Measurements at 12 kc and 24 kc in Shallow Water," Woods Hole Oceanographic Institution unpublished memorandum, Contract NObs-2083, May 1, 1946.
50. Physics of Sound in the Sea, *Nat. Def. Res. Comm. Nat. Res. Counc. Div. 6 Sum. Tech. Rep.* 8, chaps. 2–4, 1946.
51. Officer, C. B.: "Introduction to the Theory of Sound Transmission," McGraw-Hill Book Company, New York, 1958.
52. Brekhovskikh, L. M.: "Waves in Layered Media," Academic Press Inc., New York, 1960.
53. Kinsler, L. E., and A. R. Frey: "Fundamentals of Acoustics," 2d ed., p. 466, John Wiley & Sons, Inc., New York, 1962.
54. Kaufman, H.: Velocity Functions in Seismic Prospecting, *Geophys.*, **18**:289 (1953).
55. "Calculation of Ray Paths Using the Refraction Slide Rule," U.S. Navy Department, Bureau of Ships, May 1943.
56. Galkin, O. P., and V. S. Grigorev: Instruments for Plotting Refracted Rays, *Sov. Phys. Acoust.*, **6**:20 (1960).
57. Cohen, J. S., and L. J. Einstein: Continuous Gradient Ray Tracing System (CONGRATS) II, *U.S. Nav. Underwater Syst. Center Rep.* 1069, 1970.
58. Cornyn, J. J.: GRASS: A Digital Ray-Tracing and Transmission Loss Prediction System: vol. I, Overall Description, *U.S. Nav. Res. Lab. Rep.* 7621, 1973; vol. II, User's Manual, *Rep.* 7642, 1973.
59. Pedersen, M. A.: Acoustic Intensity Anomalies Introduced by Constant Velocity Gradients, *J. Acoust. Soc. Am.*, **33**:465 (1961).
60. Etter, P. C. and R. S. Flum: "An Overview of the State of the Art in Naval Underwater Acoustic Modeling," MAR Inc., Rockville, Md., 1979. Also: Urick, R. J.: "Sound Propagation in the Sea," Peninsula Press, Los Altos, Calif., 1982.
61. Barkhatov, A. N.: "Modeling of Sound Propagation in the Sea," Consultants Bureau, Plenum Publishing Co., New York, 1971.
62. Anderson, G., R. Gocht, and D. Sirota: Spreading Loss in an Inhomogeneous Medium, *J. Acoust. Soc. Am.*, **36**:140 (1964).
63. Eby, E. S., and L. T. Einstein: General Spreading Loss Expression, *J. Acoust. Soc. Am.*, **37**:933 (1965).
64. Urick, R. J., and H. L. Saxton: Surface Reflection of Short Supersonic Pulses in the Ocean, *J. Acoust. Soc. Am.*, **19**:8 (1947).
65. Liebermann, L. N.: Reflection of Underwater Sound from the Sea Surface, *J. Acoust. Soc. Am.*, **20**:498 (1948).

66. Addlington, R. H.: Acoustic Reflection Losses at the Sea Surface Measured with Explosive Sources, *J. Acoust. Soc. Am.*, **35:**1834 (1963).
67. Pedersen, M. A.: Comparison of Experimental and Theoretical Image Interference in Deep Water Acoustics, *J. Acoust. Soc. Am.*, **34:**1197 (1962).
68. Beckmann, P., and A. Spizzichino: "Scattering of Electromagnetic Waves from Rough Surfaces," p. 93, The Macmillan Company, New York, 1963.
69. Roderick, W. I., and B. F. Cron: Frequency Spectra of Forward Scattered Sound from the Ocean Surface, *J. Acoust. Soc. Am.*, **48:**759 (1970).
70. Shooter, J. A., and S. K. Mitchell: Acoustic Sidebands in CW Tones Received at Long Ranges, *J. Acoust. Soc. Am.*, **60:**829 (1976).
71. Pollak, M. J.: Surface Reflections of Sound at 100 kc, *J. Acoust. Soc. Am.*, **30:**343 (1958).
72. Brown, M. V., and J. Ricard: Fluctuations in Surface Reflected Pulsed CW Arrivals, *J. Acoust. Soc. Am.*, **32:**1551 (1960).
73. Gulin, E. P., and K. I. Malyshev: Statistical Characteristics of Sound Signals Reflected from the Undulating Sea Surface, *Sov. Phys. Acoust.*, **8:**228 (1963).
74. Brown, M. V.: Intensity Fluctuations in Reflections from the Ocean Surface, *J. Acoust. Soc. Am.*, **46:**196 (1969).
75. Young, R. W.: Image Interference in the Presence of Refraction, *J. Acoust. Soc. Am.*, **19:**286(A) (1947).
76. Morse, R. W.: Dependence of Shadow Zone Sound on the Surface Sound Velocity Gradient, *J. Acoust. Soc. Am.*, **22:**857 (1950).
77. Mackenzie, K. V.: Reflection of Sound from Coastal Bottoms, *J. Acoust. Soc. Am.*, **32:**221 (1960).
78. Hamilton, E. L.: Compressional Wave Attenuation in Marine Sediments, *Geophys.*, **37:**620 (1972).
79. Hampton, L. (ed.): "Physics of Sound in Marine Sediments," Plenum Press, New York, 1974.
80. Kuperman, W. A., and F. B. Jensen: "Bottom-Interacting Ocean Acoustics," Plenum Press, New York, 1980.
81. Hamilton, E. L.: Geoacoustic Modeling of the Sea Floor, *J. Acoust. Soc. Am.*, **68:**1313 (1980).
82. Anderson, A. L., and L. D. Hampton: Acoustics of Gas-Bearing Sediments, I: Background, II: Measurements and Models, *J. Acoust. Soc. Am.*, **67:**1865 (1980).
83. Schreiber, B. C.: Sound Velocity in Deep Sea Sediments, *J. Mar. Res.*, **73:**1259 (1968).
84. Marine Geophysical Survey, Station Data Listing and Report Catalog, *Nav. Oceanogr. Office Spec. Publ.* 142, 1974.
85. Nafe, J. E., and C. L. Drake: Physical Properties of Marine Sediments, *Lamont Geol. Observ. Columbia Univ. Tech. Rep.* 2, 1961.
86. Shumway, G.: Sound Speed and Absorption Studies of Marine Sediments by a Resonance Method, pt. I, *Geophys.*, **25:**451 (1960); pt. II, *Geophys.*, **25:**659 (1960).
87. Hurdle, B. G., K. D. Flowers, and K. P. Thompson: Bistatic Acoustic Scattering from the Ocean Botton, *U.S. Nav. Res. Lab. Rep.* 7285, 1971.
88. Cole, B. F.: Marine Sediment Attenuation and Ocean-Bottom Reflected Sound, *J. Acoust. Soc. Am.*, **36:**1993(A) (1964).
89. Barnard, G. R., J. L. Bardin, and W. B. Hempkins: Underwater Sound Reflection from Layered Media, *J. Acoust. Soc. Am.*, **36:**2119 (1964).
90. Faust, L. Y.: Seismic Velocity as a Function of Depth and Geologic Time, *Geophys.*, **16:**192 (1951).
91. Liebermann, L. N.: Reflection of Sound from Coastal Bottoms, *J. Acoust. Soc. Am.*, **20:**305 (1948).

six

propagation of sound in the sea: transmission loss, II

Sound always travels to long distances in the sea by some form of ducted propagation. When sound travels in a *duct*, or *sound channel*, it is prevented from spreading in all directions, and remains confined between the boundaries of the sound channel. A number of kinds of ducts of common occurrence in the sea are the *mixed-layer sound channel*, the *deep sound channel*, and, in shallow water, the *shallow-water sound channel*.

6.1 The Mixed-Layer Sound Channel

Sound trapping in the mixed layer The fact that the upper portions of the sea often form a trap, or duct, in which relatively good propagation takes place was recognized long ago. In 1937 Steinberger (1) made propagation tests off Guantanamo, Cuba, when a mixed layer was present and correctly accounted for the good transmission as the result of ducting caused by the upward refraction of sound in the layer. The effect was all but forgotten during the World War II years and was essentially rediscovered shortly thereafter during propagation measurements at the lower frequencies then becoming of interest. The mixed-layer sound channel, like its radio analog, the ground-based duct, is now recognized as a regular characteristic of the medium in which transmission takes place.

In the cloudy, windy ocean areas of the world, the temperature profile regularly shows the presence of an isothermal layer just beneath the sea surface. This layer of isothermal water is created and maintained by

148 / *principles of underwater sound*

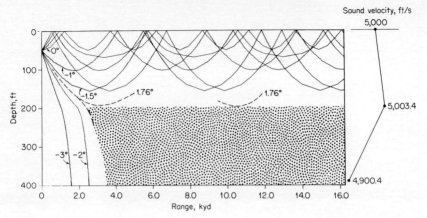

fig. 6.1 *Ray diagram for sound transmission from a 50-ft source in a 200-ft mixed layer. Rays are drawn at 1° intervals, with 1.5° and 1.76° added, for the profile at the right.*

turbulent wind mixing of the near-surface water of the sea. Within the layer—called simply the *mixed layer* from its manner of origin—the velocity of sound increases with depth because of the pressure effect on sound velocity. The upward refraction that results keeps a portion of the acoustic energy emitted by a shallow source close to the sea surface. The sound trapped in the layer propagates to long ranges by successive reflections from the sea surface along ray paths that are long arcs of circles between their encounters with the sea surface.

A computer-produced ray diagram for a source in a typical mixed layer is shown in Fig. 6.1.* Under the conditions for which the diagram was drawn, the ray leaving the source at an angle of 1.76° becomes horizontal at the base of the layer; rays leaving the source at smaller angles remain in the layer; rays leaving the source at greater angles are sent downward into the abyssal depths of the sea. A shadow zone is produced beneath the layer at ranges beyond the direct or close-in sound field. This shadow is not complete, but is insonified by scattered sound from the sea surface and by diffusion of sound out of the channel, caused by the nature of the lower boundary. By these two processes, sound "leaks out" of the channel at a rate given by a "leakage coefficient" that expresses the attenuation, in decibels per kiloyard, of sound trapped in the channel. The leakage coefficient varies with surface roughness, duct thickness, gradient below the layer, and frequency.

An early observation of trapping by the mixed layer (2) may be seen in Fig. 6.2. In the experiment one ship transmitted short pulses at 9.5 kHz to another ship 13.7 mi away. For each pulse sent out, two were received on the receiving ship, the first traveling via the mixed-layer duct, the second by bottom reflection. Sensitive thermal profiles were taken at the same time. As seen by Fig.

---

* The discontinuous form of the rays in this and other ray diagrams is caused by the way the velocity profile was subdivided in the computer program.

6.2, the first arrival began to disappear at about the time that the 80-ft mixed layer disappeared, leaving the second, or bottom-reflected, pulse unaffected by the near-surface changes.

Variation of transmission loss with depth With a source in the mixed layer, the intensity of sound tends to fall off with depth within the layer and to fall off faster with depth as the lower boundary of the mixed layer is crossed and as the shadow is entered.

Figure 6.3 shows measured transmission data for a source depth of 20 ft and a mixed layer 100 ft thick (3). In tests at sea, sound levels at 8 and 16 kHz were measured at five depths and at various ranges from a CW source. Measured levels have been plotted relative to those that would occur with spherical spreading and absorption, and are indicated by the shaded areas to the left (lower levels) or to the right (higher levels) of a vertical line at the measurement range. Of particular note are the higher levels obtained when the receiver is placed within the layer—indicating the existence of mixed-layer

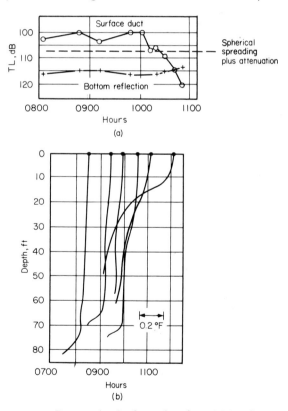

fig. 6.2 *Propagation in the surface duct. (a) Levels versus time during the morning hours of ducted and bottom-reflected pulses. (b) Sensitive thermal profiles at the receiving ship during the same period.*

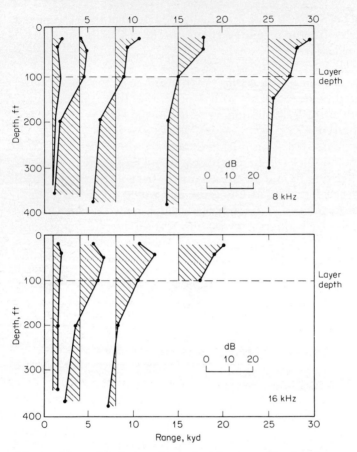

fig. 6.3 *Observed variation of 8 and 16 kHz sound level with depth in and below a mixed layer for a source at 20 ft. (Ref. 2.)*

sound channeling. At the same time, levels lower than those that would occur with spherical spreading and absorption exist when the receiver is in the shadow zone beneath the layer. At long ranges, the shadow tends to fill up with botton-reflected sound.

The quality of transmission in the layer and the acoustic darkness of the shadow beneath it vary greatly with the thickness of the layer, the conditions at its boundaries, and the frequency. Qualitatively, Fig. 6.4 shows two intensity-depth "profiles" to indicate the variation of transmitted sound level with depth for two contrasting sets of conditions. Profile *AA* is drawn for the conditions of high frequency, calm sea, strong negative velocity gradient beneath the layer, and moderate ranges. Profile *BB* is for low frequencies, rough seas, weak negative gradient, and long ranges.

That the high-frequency sound field below the layer is indeed surface-scattered sound has been clearly demonstrated by Schweitzer (4). Using the theory of Bechman and Spizzichino (5) for rough surface scattering, he was able to successfully account for field data obtained at 5 kHz below a 300-ft layer. The scattering was visualized as originating from a quasi-elliptical area located on the surface near the receiver in the direction toward the source of sound.

Low-frequency cutoff At very low frequencies, sound ceases to be trapped in the mixed layer or, indeed, in any sound channel. This occurs when the frequency approaches the cutoff frequency for the first mode of normal-mode theory, when, in a sense, the wavelength has become too large to "fit" in the duct. This maximum wavelength for duct transmission may be found from the theory of radio propagation in ground-based radio ducts to be (6)

$$\lambda_{max} = 8/3 \sqrt{2} \int_0^H \sqrt{N(z) - N(H)} \, dz$$

where $N(z)$ = index of refraction at any depth z in duct
$N(H)$ = index of refraction at base of duct

Using values of velocity and velocity gradient appropriate for sound transmission in the mixed layer, one obtains

$$\lambda_{max} = 4.7 \times 10^{-3} H^{3/2}$$

for the maximum wavelength, in feet, trapped in a mixed-layer duct H ft thick. For example, for a mixed layer 100 ft thick, the maximum trapped wavelength λ_{max} is 4.7 ft, corresponding to a frequency of 1,100 Hz. Although this does not represent a sharp cutoff, wavelengths much longer than this are strongly attenuated; wavelengths much shorter are attenuated by absorption and leakage. For a mixed layer of a given thickness, it follows that there is an optimum frequency of best transmission, at which the leakage of sound out of the layer is a minimum.

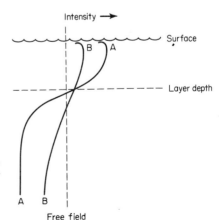

fig. 6.4 *Intensity versus depth in and below the mixed layer for two contrasting sets of conditions.*

Transmission-loss model An expression for the transmission loss in a mixed layer may be obtained through simple considerations. In Fig. 6.5, let a nondirectional source of sound be located at P in a mixed layer. Of all the rays leaving the source, only those within a certain limiting angle 2θ remain in the channel. At a distance of 1 yd, the power contained in this bundle of rays is distributed over a portion of a spherical surface A_1. At a long distance r, this same amount of power—in the absence of absorption and leakage—is distributed over a cylindrical surface A_2. Because the power crossing areas A_1 and A_2 is the same, the transmission loss to range r, averaged over the duct thickness H, is

$$\text{TL} = 10 \log \frac{A_2}{A_1}$$

By geometry

$$A_2 = 2\pi r H$$

and

$$A_1 = 2\pi \int_{-\theta}^{\theta} \cos\theta \, d\theta = 4\pi \sin\theta$$

Therefore, for the mixed-layer duct,

$$\text{TL} = 10 \log \frac{rH}{2 \sin\theta} = 10 \log r\, r_0 = 10 \log r_0^2 \frac{r}{r_0}$$

where $r_0 = H/2 \sin\theta$
θ = inclination at source of maximum trapped ray

The quantity $r_0^2 r/r_0$ indicates that the transmission to range r may be viewed as the result of spherical spreading out to a *transition range* r_0, followed by cylindrical spreading from r_0 to r. When the attenuation due to absorption

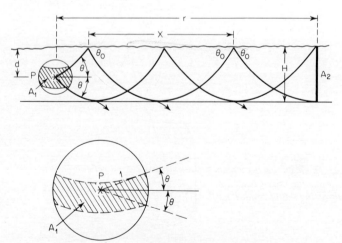

fig. 6.5 *Transmission loss in a mixed-layer channel.*

and leakage is added, the duct-transmission-loss expression becomes

$$\boxed{TL = 10 \log r_0 + 10 \log r + (\alpha + \alpha_L) r \times 10^{-3}}$$

where α = absorption coefficient, dB/kyd
α_L = leakage coefficient, dB/kyd, expressing rate at which acoustic energy leaks out of duct
r = range, yd

For a layer of constant velocity gradient, in which the rays are arcs of circles of radius of curvature $R = c_0/g$, the following relationships may be found by geometry for the conditions of large R ($R \gg H$) and small θ_0 ($\sin \theta_0 \ll 1$):

$$x = \sqrt{8RH} = \text{skip distance limiting ray}$$

$$\theta_0 = \sqrt{\frac{2H}{R}} = \text{maximum angle limiting ray, rad}$$

$$\theta = \sqrt{\frac{2(H-d)}{R}} = \text{angle limiting ray at source depth, rad}$$

$$r_0 = \sqrt{\frac{RH}{8}} \sqrt{\frac{H}{H-d}} = \text{transition range}$$

$$= \frac{x}{8} \sqrt{\frac{H}{H-d}}$$

As an example, for a typical mixed layer of thickness $H = 300$ ft and for a source at the surface ($d = 0$), we would have

$$R = \frac{5{,}000}{0.017} = 98{,}000 \text{ yd}$$

$x = 8{,}800$ yd, $\theta_0 = \theta = 2°$, and $r_0 = 1{,}100$ yd.

Leakage of sound from the layer The leakage coefficient for mixed-layer transmission may be conceived as being due to two causes. One is scattering of sound out of the layer by the rough sea surface, a process that is the source of the sound field below the mixed-layer duct, as we have seen. The other component of the leakage has been called "transverse diffusion" out of the mixed layer (7), and depends on the sharpness of the discontinuity between the layer and the thermocline below. The former causes the leakage to increase with frequency; the latter causes it to decrease with frequency. The net result is that, for a given layer, there is a frequency of minimum leakage coefficient where the transmission of sound by the layer is best. This optimum frequency lies between a few hundred and a few thousand hertz, depending upon sea state and layer depth.

An extensive series of measurements of the losses associated with the mixed-layer data were made under the AMOS program (8) during the years 1953–1954. Based on these data, Schulkin (9) found that the loss over the horizontal distance between surface bounces of the limiting ray (x in Fig. 6.5) amounts to $1.64(fh)^{1/2}$ decibels, where f is the frequency in kilohertz and h is the mean wave height in feet. This expression was said to be valid over the range $3 < fh < 70$ kHz · ft. Noting that $x = \sqrt{8RH}$ and relating h to the sea state S by the empirical expression $h = 0.4S^2$, we obtain

$$\alpha_L = 2S \left(\frac{f}{H}\right)^{1/2}$$

where α_L = leakage coefficient, dB/kyd
S = sea-state number
f = frequency, kHz
H = layer, depth, ft

A somewhat different empirical formula was obtained by Saxton, Baker, and Shear (10) from field data at 10 kHz. They fitted their data by the relationship $\alpha_L = 1,200(1.4)^S(RH)^{-1/2}$ dB/kyd, where the radius of ray curvature R and the layer depth H are in yards. If we take $R = 10^5$ yd and convert the layer depth to feet, we obtain

$$\alpha_L = 6.6(1.4)^S H^{-1/2}$$

for a frequency of 10 kHz. Although this is different in form from the expression given by Schulkin, the two expressions give substantially identical values for α_L (near 1.9 dB/kyd) at 10 kHz for a 100-ft layer in sea state 3.

Typical transmission curves Analysis of the vast quantity of data gathered during the AMOS measurement program just mentioned led to a series of formulas from which the curves of Fig. 6.6 and Fig. 6.7 were computed (11). These give the transmission loss at 2 and 8 kHz for various combinations of source and receiver depths, with layer depth as a parameter. They apply for low sea states and for an average velocity gradient below the mixed layer. They show average transmission-loss values to be expected in the presence of a wind-mixed isothermal layer at the two frequencies 2 and 8 kHz.

More recent observations have resulted in the expressions of W. F. Baker (12) given in Table 6.1. These expressions follow the model outlined above in assuming spherical spreading out to a certain range and cylindrical spreading beyond. The expression for $10 \log r_0$ is 10 times the log of the transition range as given above for $d = 0$. The range of experimental and field conditions on which this model is based is given at the foot of the table.

However, of the various models for surface duct propagation available at the present time, none of them can satisfactorily predict measured losses in actual open-ocean ducts. This is the conclusion of a paper by Hall (13) which compared several models with 25 measurements of transmission loss to 6 km at 4, 8, and 16 kHz for a variety of source and receiver depths and duct

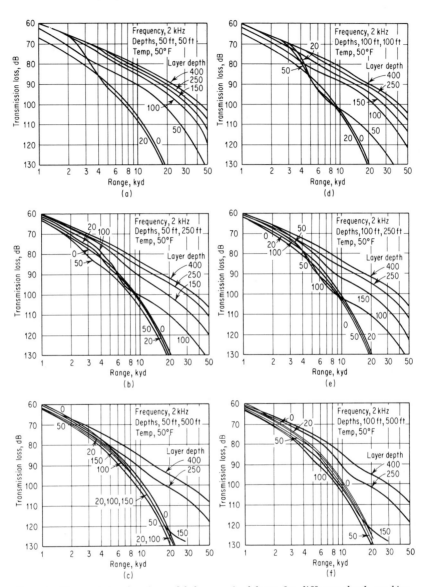

fig. 6.6 *Transmission loss in and below a mixed layer for different depth combinations, 2 kHz. Layer depths given in feet. (Ref. 11.)*

conditions. The poorest results were had with flat-surface normal mode theory; the best with the AMOS model modified to include a surface roughness term. The unmodified AMOS model (Figs. 6.6, 6.7) and the Baker model of Table 6.1 fell in between. But, as stated above, none of the four models were able to predict the losses as measured at sea satisfactorily well.

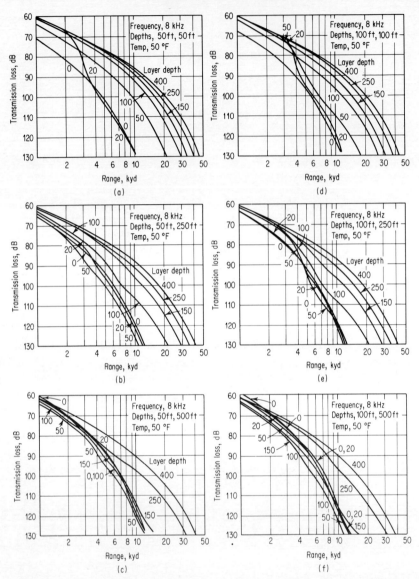

fig. 6.7 *Transmission loss in and below a mixed layer for different depth combinations, 8 kHz. Layer depths given in feet. (Ref. 11.).*

A comparison of measured transmission losses within and below the duct at a range of 6 km led to the following expression:

$$\Delta = -25 + 25 \log\left(\frac{x}{\lambda}\right) \pm 7 \text{ dB}$$

where Δ is the difference in level in and below the duct, λ is the wavelength, and x the "significant" wave height, defined by oceanographers as the average of the heights (crest-trough) of the one-third highest waves, and related to wind speed for a fully developed sea by $x = 0.0182\, U^2$ where x is in feet and U is the wind speed in knots (14). The data leading to the expression for Δ ranged in duct thickness up to 150 m, wind speeds from 0 to 15 m/s, the significant wave heights from 0.3 to 4.5 m.

table 6.1 Duct Expressions According to W. F. Baker (Ref. 12)

Short ranges
$(r < 350\, H^{1/2})$ $TL = 20 \log r + (\alpha + \alpha_L)r \times 10^{-3}$

Long ranges
$(r > 350\, H^{1/2})$ $TL = 10 \log r_0 + 10 \log r + (\alpha + \alpha_L)r \times 10^{-3}$

$10 \log r_0 = 20.9 + 5 \log H$

$$\alpha_L = \frac{26.6 f(1.4)^S}{[(1452 + 3.5t)H]^{1/2}} \quad \text{dB/kyd}$$

Limits of field data
f: 3.25 to 7.5
r: 1,000 to 51,000
H: 80 to 220
S: 2 to 5

f = frequency, in kilohertz.
H = layer depth, in feet.
r = range, in yards.
S = sea-state number.
t = temperature, in degrees Celsius.

Occurrence of the layer   Statistical data on the existence and thickness of the mixed layer in the North Atlantic are given in Fig. 6.8. These distribution curves, based on numerous bathythermograph observations, show the percentage of the time that the layer thickness exceeds the value shown as ordinate for two bands of latitude and for the four seasons of the year.

Internal waves   Over short distances, the thickness of the mixed layer may vary because of internal waves in the thermocline below. Internal waves are oscillations of the thermocline that are manifested by a changing temperature at a fixed point in the sea (15). In one series of observations (16) at a location just off the California shore, these oscillations were found to have periods of 4 to 8 min and to have heights of about 2 to 8 ft. Internal waves propagate along the density discontinuity between the mixed layer and the water below, with a velocity, as measured in the study just referred to, of about 0.3 knot and a wavelength of 50 yd. Longer wavelengths are observed in deep water. Internal waves complicate the propagation of underwater sound by causing variations in transmission loss over short distances, as has been demonstrated by means of ray diagrams (17). Fluctuations of echoes between a surface-ship source and a target below the mixed layer are doubtless attributable, in part at least, to the existence of such internal waves in the thermocline (18).

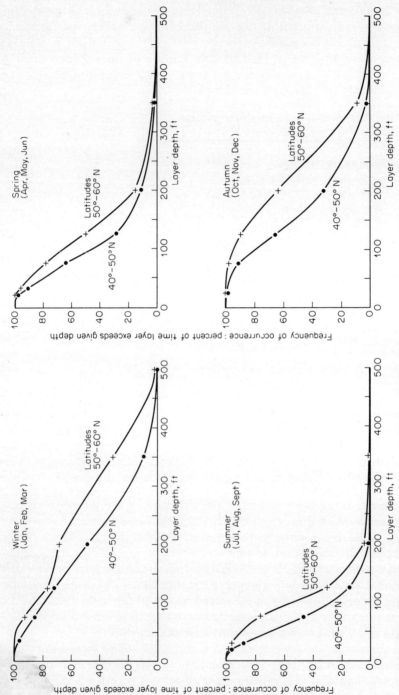

fig. 6.8 Distribution curves of layer depth in the North Atlantic. Ordinate is the percentage of occurrences of a layer depth greater than abscissa. Based on original data provided by the Navy Oceanographic Office.

6.2 The Deep Sound Channel

Sound channeling in the deep sea The deep sound channel, sometimes called the sofar channel, is a consequence of the characteristic velocity profile of the deep sea. It will be recalled that this profile has a minimum at a depth which varies from about 4,000 ft in midlatitudes to near the surface in the polar regions. This velocity minimum causes the sea to act like a kind of lens; above and below the minimum, the velocity gradient continually bends the sound rays toward the depth of minimum velocity. A portion of the power radiated by a source in the deep sound channel accordingly remains within the channel and encounters no acoustic losses by reflection from the surface and bottom. Because of the low transmission loss, very long ranges can be obtained from a source of moderate acoustic power output, especially when it is located near the depth of minimum velocity. This depth is called the *axis* of the sound channel.

The deep sound channel was originally investigated by Ewing and Worzel (19). Its remarkable transmission characteristics were utilized in the *sofar* system for the rescue of aviators downed at sea. In sofar (the letters denote sound *f*ixing *a*nd *r*anging) a small explosive charge is dropped at sea by a downed aviator and is received at shore stations thousands of miles away (20). The time of arrival at two or more stations gives a "fix" locating the point at which the detonation of the charge took place. More recently, the ability to measure accurately the arrival time of explosive signals traveling along the axis of the deep sound channel has been used for geodetic distance determinations (21) and for missile-impact location (22).

Transmission characteristics Figure 6.9 is a ray diagram, originally published by Ewing and Worzel, for a sound source on the channel axis at a depth of 4,000 ft. This ray diagram was drawn during World War II by hand methods. It clearly shows what has since become known as a *convergence zone* at a range of 32½ miles, where the rays reaching the surface are brought together, or converged, by the lenslike action of the sea. A more modern computer-drawn version is given in Fig. 6.10. Here the complexity of the deep-sea sound field is apparent. Evidently, the deep sound channel is full of convergence zones, caustics, foci, and shadow zones. Moreover, in addition to the refracted rays shown in Figs. 6.9 and 6.10 there are reflected rays leaving the source at higher angles and suffering one or more encounters from the ocean boundaries. These bottom-surface bounce rays tend to fill up the shadow zones and further complicate the sound field.

Between the source and a point a great distance away, a number of refracted propagation paths exist, each having a different travel time and crossing the channel axis at different intervals. The path with the greatest excursion from the axis has the shortest travel time; the shortest path, straight down the axis of the channel, has the longest travel time, since it lies entirely at the depth of minimum velocity. Of all these paths, the axial path carries the

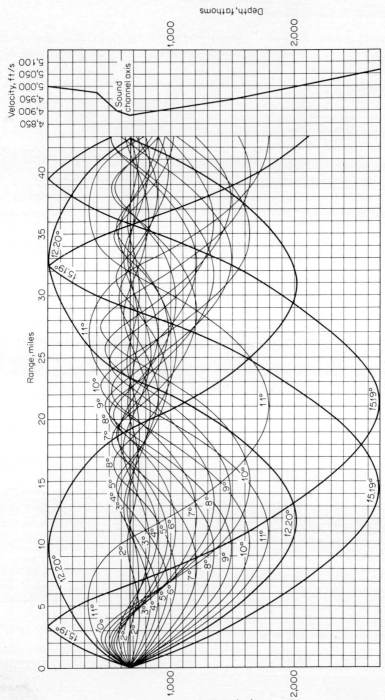

fig. 6.9 Ray diagram of transmission in the deep sound channel for a source on the axis. (Ref. 19.)

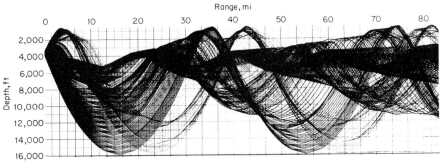

fig. 6.10 *Ray diagram of the deep-ocean sound channel for a source near the axis. Reflected rays are omitted.*

greatest amount of acoustic energy and is the path of minimum transmission loss.

Because of these travel-time and intensity characteristics, an explosion on the channel axis is heard at a distance as a long-drawn-out signal. Its amplitude rises slowly to a climax and has an abrupt cutoff at the instant of arrival of sound traveling along the channel axis. Examples of these characteristic sofar-type signatures observed when both source and receiver are on the channel axis are shown in Fig. 6.11. This is a sequence of recordings, made with a hydrophone on the axis at Bermuda, of the sound of 4-lb explosive sofar signals dropped from an aircraft and detonating on the axis at different distances in a northeasterly direction from Bermuda. The distances vary from 43 miles (Drop 6) to 1,632 miles (Drop 19). Of particular note is the severe distortion accompanying deep-sound-channel transmission. The time stretching of the initially short explosive pulse in this series of signals amounts to 9.4 seconds per 1,000 miles of travel. However, when both source and receiver are *not* located on the axis of the channel, the signature no longer shows the smooth buildup to its climax, but consists of a series of discrete short pulses that can be accounted for by accurate travel time and ray computation on a digital computer (23).

Transmission-loss model In a way similar to that already described for the mixed layer, the transmission loss in the deep sound channel can be conceived to be due to spherical spreading out to a transition range r_0, with cylindrical spreading beyond, plus a loss proportional to range. Because of the signal distortion in long-range propagation, the transmission loss is more properly described in terms of energy. This model gives

$$\text{TL} = E_0 - E_r = 10 \log r_0 + 10 \log r + \alpha r \times 10^{-3}$$

where E_r = energy-flux-density level at r yd
 E_0 = energy-flux-density level of source at 1 yd
 α = attenuation coefficient, dB/kyd

For explosives, values for E_0 are given in Figs. 4.19 and 4.21.

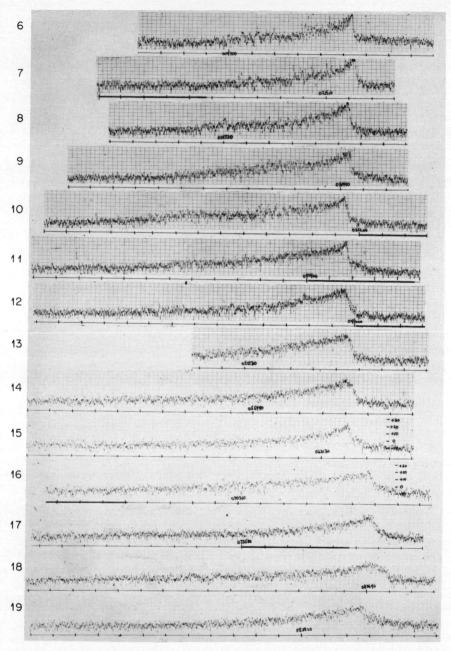

fig. 6.11 *Sofar signal envelopes at ranges about 100 miles apart. Horizontal scale has 1-second time ticks; vertical scale is in decibels; total width of each recording, 50 dB.*

In analyzing field measurements of loss versus range, the quantities r_0 and α can be readily obtained by plotting the quantity (TL $-$ 10 log r) against range on a linear range scale. The result is, or should be, a straight line having a slope equal to α and an intercept at zero range equal to $10 \log r_0$.

Measurements of α by various investigators for transmission in the deep sound channel are reasonably concordant. However, published determinations of r_0 are much less so. Examples of individual determinations of r_0 that may be cited are 1.6–4.0 kyd (24); 40–150 kyd (25); 3 kyd (26); and 90–250 kyd (27). The disparity stems from differences in the velocity profile between measurements, together with the fact that r_0 is particularly sensitive to the calibration of the measurement system employed.

Ray paths in the deep sound channel In terms of the velocity profile, the upper and lower limits of the channel are defined by the two depths of equal maximum velocity in the profile between which a velocity minimum exists. In Fig. 6.12 these limits of the deep sound channel are the depths AA' in the velocity profile. Different ray paths from a source in the channel exist, depending on whether or not the channel extends to the sea surface or bottom. In Fig. 6.12a the velocity at the surface and bottom is the same. All depths in

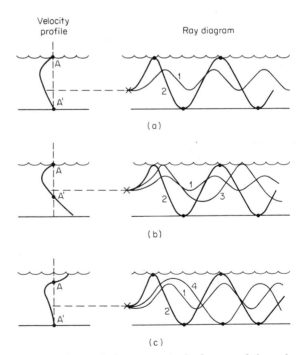

fig. 6.12 *Ray paths for a source in the deep sound channel. In (a) the channel extends between the sea surface and bottom; in (b) and (c) it is cut off by the sea surface and by the sea bottom, respectively.*

the sea lie within the channel, and sound is propagated via paths that are either *refracted* (path 1) or *reflected* (with consequent losses) at the sea surface and bottom (path 2). In Fig. 6.12b the upper bound of the deep sound channel lies at the sea surface. Here, in addition to the two types of paths 1 and 2, *refracted surface-reflected* (RSR) paths occur (path 3) involving losses intermediate between those of paths 1 and 2. In Fig. 6.12c the channel is cut off by the sea bottom, and *refracted bottom-reflected* (RBR) paths exist (path 4). The entirely refracted paths and the low transmission losses associated with these paths do not exist when the source or the receiver is outside the depth limits AA' of the channel.

6.3 Caustics and Convergence Zones

Within the channel, *caustics* and *convergence zones* are formed in which relatively high sound intensities occur. In a ray diagram, a *caustic* is the envelope formed by the intersection of adjacent rays. It is a region of partial focusing, in which wave theory is required to find the intensity—for example, the wave theory of Brekhovskikh (Chap. 5, Ref. 52, pp. 483–496). When a caustic intersects the sea surface, the region at or near the surface, within which high sound levels occur, is known as a *convergence zone*.

Shallow convergence zones The characteristics of convergence zones for a shallow source were described in a paper by Hale (28). At intervals of 30 to 35 miles in temperate and tropical latitudes and under conditions when the deep sound channel extends up to the surface (Fig. 6.12a and b), successive narrow zones of high intensity are encountered as the range from a shallow source to a shallow receiver is increased. An extreme example of a series of convergence zones, as reported by Hale from observed data, is given in Fig. 6.13. The peaks here extend about 25 dB above the level that would be found with spherical spreading and absorption; this increase has been called the *conver-*

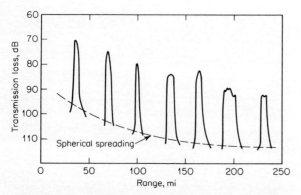

fig. 6.13 *Transmission loss versus range for a shallow source and receiver, showing convergence zones at 35-mile intervals.* (Ref. 28.)

gence gain. Convergence gains of 10 to 15 dB are more common. The width of the zones is of the order of 5 to 10 percent of the range, with the first zone about 3 miles in width. In between the zones, only the relatively low levels typical of bottom-reflected signals occur. Each convergence zone is bounded on its near side, or at its boundary closest to the source, by a caustic. This boundary is extremely sharp; observations show that the level of the sound from a distant source rises some 20 dB or more within a few tens of yards in this region. While the near side is sharp, the far side of the zone is less well defined, and the level falls off more slowly with range. The width of the zones increases with the *order* of the zone; the second zone (near 70 miles) is wider than the first, and so on, until eventually at ranges of several hundred miles the zones overlap and become indistinguishable. The width of a zone also increases with depth, being wider at, say, 500 feet than at the surface, and eventually splits into two half zones, as will be described below. When a mixed layer is present, the convergence zone tends to become wider and more diffuse at shallow depths.

In order for convergence to exist, both the source and receiver must lie within the sound channel. This means that, for a shallow source and receiver, the water depth must be sufficiently great for deep-going rays to be refracted and brought together without encountering the bottom. In water of insufficient depth the deep rays are reflected by the bottom and are prevented from converging. It follows that for a given temperature, and therefore sound velocity, at or near the surface, there is a *minimum* water depth for convergence to occur. This minimum water depth is plotted against surface temperature in Fig. 6.14. The water depth must be *greater* than the depth given by the curve for a convergence zone to exist.

We have already seen (Sec. 5.5) that the velocity profile of the deep oceans varies with latitude. In cold surface waters, the depth of the axis of the deep-ocean sound channel is shallow, and the range to the convergence zone, and therefore the range interval between zones, is small. Accordingly we would expect to find a relation between zone range and surface temperature. This relationship is shown in Fig. 6.14*b* for midlatitudes in the North Atlantic and North Pacific Oceans. Together, Fig. 6.14*a* and *b* demonstrate the overriding importance of surface temperature in determining both the existence and the range of convergence zones in deep water.

Deep convergence zones For a deep source or receiver, the single convergence zone found when both source and receiver are near the surface splits into two *half zones*. These convergence half zones are regions near the surface where deep-going rays are brought by refraction to shallow depths. The ranges to the half zones vary with the depth of the source or receiver. With increasing depth, the *inner* half zone moves *inward* in range and the *outer* half zone moves *outward*. This migration with depth may be seen by examination of the ray diagrams of Fig. 6.15 for source depths of 300, 3,000, 7,500, and 12,000 ft for a velocity profile having nearly equal sound velocity at the surface and at 16,000 ft. This zone splitting and migration with depth was verified in field trials using explosive sound signals and a deep hydrophone (29); half-zone widths of 2 to 5 miles were found, with convergence gains to 5 to 10 dB in the half zones.

This behavior is a result of a changing pattern of caustics in the deep sound channel with changing source depth. Figure 6.16 shows diagrammatically the pattern of caustics for a sound source above, on, and below the axis. The arrows indicate the movement of the various caustics with increasing depth of the source.

The variation with range of the level from a source as it passes through the half zones, as it was observed in a field experiment (30), is illustrated in Fig.

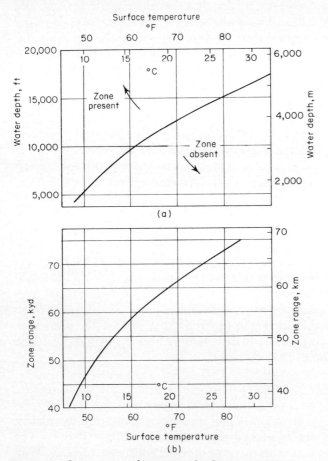

fig. 6.14 *Occurrence and range to the first convergence zone for a shallow source and receiver as determined by the surface temperature: (a) Water depth required for convergence. (b) Range to the convergence zone. For midlatitudes in the North Atlantic and North Pacific Oceans.*

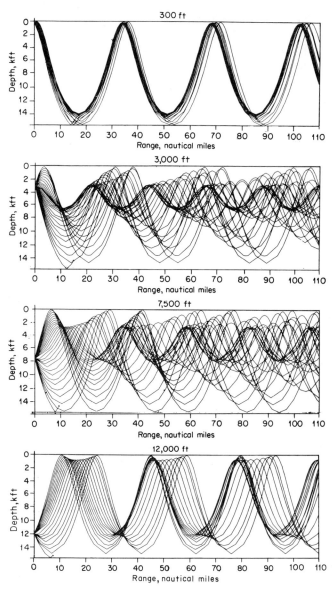

fig. 6.15 Ray diagrams for a source at four source depths in the deep sea.

6.17. In this figure the upper portion is a trace of the level from a 1,120-Hz CW source as it was towed outward in range at a depth of 150 feet between 28 and 43 miles from a receiving hydrophone at a depth of 3,800 feet. The lower portion shows the ray diagram for rays drawn at $1/10$-degree intervals. If the receiver were at a shallow depth, the two half zones would coalesce and a sensibly single convergence zone would be observed.

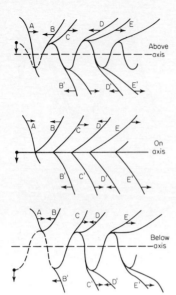

fig. 6.16 *Caustics in the deep sound channel. These should be compared with the ray diagrams of Fig. 6.15.*

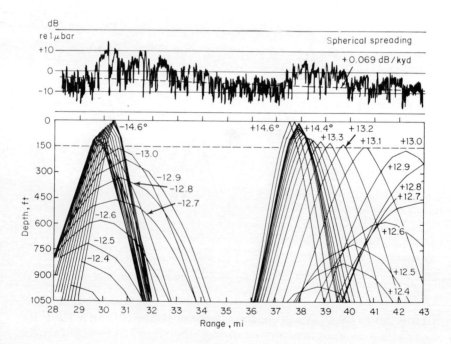

fig. 6.17 *Level from a shallow 1,120-Hz CW source received at a deep hydrophone. Upper trace shows the received level along with spherical spreading plus attenuation as a reference. Lower trace is the corresponding ray diagram; the numbers show the angle at which each ray leaves the source; upward negative, downward positive. (Ref. 30.)*

6.4 Internal Sound Channels

Intermediate in depth between the deep-ocean sound channel, with its axis at a depth between 3,000 and 4,000 feet (about 1,000 meters), and the ice-covered Arctic where it lies at or just below the surface, there are sound channels of a more restricted area and of a seasonal occurrence. Such channels have their axes at a shallow depth, and, although they are of a more local and transitory nature, they often have a strong effect on sonar operations.

Two examples may be mentioned. One occurs during the summer between Long Island and Bermuda where the peculiar conditions of the Gulf Stream cause a reversal of the normal velocity gradient and a velocity minimum or channel axis near 300 feet (100 meters). Examples of velocity profiles in this area are given in Fig. 6.18a.

Another shallow summertime channel exists in the Mediterranean Sea. In this region the heating by the sun of the upper layers of water, together with an absence of mixing by the wind, causes a strong near-surface negative gradient to develop during the spring and summer months. This thermocline overlies isothermal water at greater depths. The result is a strong sound channel with its axial depth near 300 feet (100 meters). During the autumn, the profile gradually returns to its wintertime conditions, wherein isothermal water and a positive velocity gradient extend throughout the water column all the way up to the surface (Fig. 6.18b). During the summer season the near-surface negative gradient and the resulting strong downward refraction greatly limit the range of surface ship sonars. By way of compensation, the summertime channel in the Mediterranean produces convergence zones for a near-surface source in the same way as does the deep-ocean sound channel, although the zone range is much less (typically, 20 mi or 37 km) because of the smaller vertical extent of the channel.

6.5 Arctic Propagation

In the Arctic regions of the world, the axis of the deep sound channel lies at, or near, the ice-covered surface. Sound reaches long ranges by repeated reflections from the undersurface of the ice, to which it is repeatedly returned by upward refraction. This is shown in the ray diagram of Fig. 6.19 for a velocity profile typical of the Arctic.

Under an ice cover, the combination of upward refraction and downward reflection from the rough under-ice surface creates a number of unique propagation effects. One peculiarity of Arctic transmission is its similarity to a bandpass filter. Both high and low frequencies are rapidly attenuated, the former by reflection losses from the ice cover, the latter by the fact that very low frequencies are not effectively trapped in the channel. In the Arctic, the best propagation has been found to occur in the octave 15 to 30 Hz, approximately. Measured data show rapid attenuation with increasing frequency above 30 Hz, as well as a higher loss at 10 Hz (the lower limit of measurement

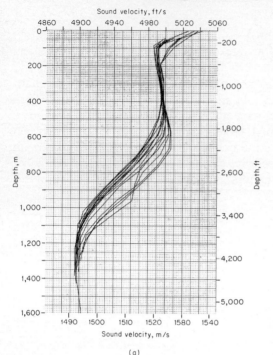

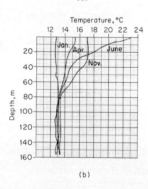

fig. 6.18 Examples of shallow sound channels: (a) Velocity profiles at latitudes 34 to 35°N, longitudes at 65 to 66°W northwest of Bermuda, month of June. (Ref. 31.) (b) Temperature profiles during different months in the Mediterranean Sea south of France. (Ref. 32.)

thus far) than at 20 Hz. Within the passband, dispersion occurs. Wave packets of different frequency from an explosive shot are observed to have different group velocities, with low frequencies traveling faster than high ones. In addition, the time stretching that is a dominant characteristic of sofar signals occurs as well, causing, in one experiment (33), an increase of duration of 15 seconds per 1,000 miles of travel. Because of these two effects—dispersion in both frequency and time—a distant explosion in Arctic waters is received as a long-drawn-out pulse of several seconds' duration, with frequencies near the lower end of the passband appearing at the beginning of the signal and higher frequencies appearing at the end.

Some success has been had by Diachok (34) in predicting the under-ice scattering losses by means of a model which represents the ice ridges as infinitely long randomly distributed half-cylinders. The number and dimensions of the under-ice ridges were inferred from laser measurements taken from an aircraft. The theory for the model yields theoretical predictions that are in surprisingly good agreement with measured losses out to a range of 300 km at 40, 50, and 200 Hz.

A number of measurements have been made of the transmission loss in the ice-covered upward-refractive Arctic, principally during the years 1960–1965. All were made by means of explosive sound signals detonated at ranges out to several hundred miles from a stationary receiving hydrophone. Examples are measurements of Marsh and Mellen (33), Buck and Greene (35), and Milne (36). These and others have been summarized by Buck (37), with the result shown in Fig. 6.20. These curves show the average measured transmission loss in the Arctic at a number of frequencies, based on the available data; the standard deviation of the various data points from the smooth curve is stated for each frequency. The dashed line shows spherical spreading (TL = 20 log r). It is evident that the transmission in the Arctic degrades rapidly with increasing frequency above 20 Hz. It is better than it would be in the free field (i.e., with spherical spreading) out to some range, and is poorer beyond. This peculiarity is the result of opposing influences on the propagation. At short and moderate ranges, ducting improves the transmission; at long ranges the repeated encounters with the under-ice surface degrade it. These same effects occur in shallow-water ducts as well. Subsequent data confirming the validity of the curves of Fig. 6.20 have been reported by Bradley (38) from work done in the marginal ice zone east of Greenland.

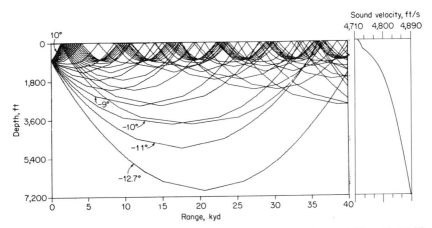

fig. 6.19 *Ray diagram for transmission in the Arctic. Ray interval 1°, with 12.7° added. The velocity profile is shown at the right.*

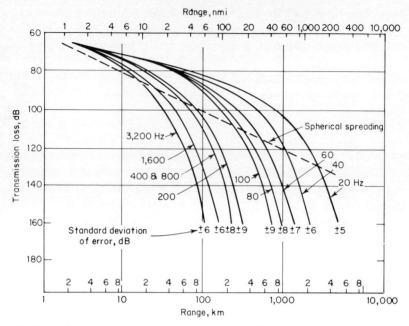

fig. 6.20 *Average transmission loss in the Arctic.* (Ref. 37.)

6.6 The Shallow-Water Channel

Sound channeling by shallow water In the present context, "shallow" means a water depth in which sound is propagated to a distance by repeated reflections from both surface and bottom. In water that is thus acoustically shallow, the acoustic characteristics of both surface and bottom are important determinants of the sound field. Although no precise definition can be given, shallow water, acoustically speaking, may be said to mean propagation to distances at least several times the water depth, under conditions where both boundaries have an effect on transmission. In a geographic sense, shallow water refers to the inland waters of bays and harbors and to coastal waters less than 100 fathoms deep, which often extend outward to the edge of the continental shelf. Water that is "shallow" forms a sound channel between the surface and bottom, in which, as in other kinds of channels, sound is trapped between the upper and lower channel boundaries.

Of the many aspects of sound propagation in the sea, the propagation of sound in shallow water has received perhaps the greatest amount of attention. The subject is characterized by both theoretical complexity and difficulty in formulating a mathematical description of the acoustic conditions existing at the boundaries. Unlike the propagation of sound in deep water, the propagation of sound in shallow water is the subject of an abundant literature.

The initial studies of the subject—both theoretical and with field observations—were made by Ide, Post, and Fry (39), Roe (40), and Pekeris (41). Good theoretical summaries may be found in books by Brekhovskikh (Chap. 5, Ref. 52) and Officer (Chap. 5, Ref. 51).

Two theoretical approaches exist for describing the shallow-water sound field. Both involve functions that are solutions of the wave equation and are added up with the correct coefficients to satisfy the acoustic conditions at the boundaries and at the source.

Ray theory Ray theory represents the sound field as a sum of *ray* contributions, each ray emanating from the source or its *image* in the surface and the bottom. In Fig. 6.21 a source is located at a point O in a shallow-water channel, and a receiver is at P. To satisfy the boundary condition at the surface (zero pressure at the pressure-release surface), an image O_{10} of the source in the surface is added, with a source strength and phase shift (180°) to account for the reflection coefficient of the sea surface. To match the reflectivity of the sea bottom, a pair of images O_{01} and O_{11} are added in the bottom. But now the boundary condition at the sea surface is no longer satisfied, and to correct for the presence of O_{01} and O_{11}, their images O_{11} and O_{21} in the sea surface must be added. But now the bottom condition is upset, and it is necessary to add still another pair of images in the bottom. By this back-and-forth process an infinite string of images is built up, with higher-order images tending to become insignificant because of weakening by repeated reflections and by

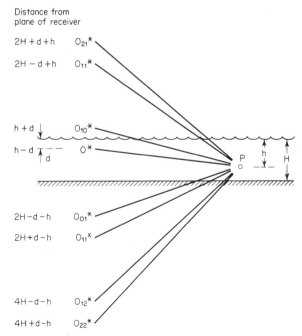

fig. 6.21 *Images in shallow-water transmission.*

greater distance. At a distant point P the sound pressure is the vector sum of the pressure contributions of the images, and may be written as

$$P = p_0 \sum_{m=0}^{\infty} \frac{R_m e^{ikr_m}}{r_m}$$

where p_0 = source pressure (time factor omitted)
r_m = distance of mth image from P
R_m = amplitude reflection coefficient appropriate for mth image
k = wave number equal to $2\pi/\lambda$

The term e^{ikr_m} is the phase factor for waves of wavelength λ propagating over the distance r_m. The reflection coefficient R_m varies with angle and will, in general, be complex, with a real and imaginary part, because of phase shifts on reflection.

Normal-mode theory An alternative expression of the shallow-water sound field can be given in terms of *normal modes*. These are complicated functions, each representing a wave traveling outward from the source with an amplitude that is a function of the source and receiver depths. Various expressions for the normal-mode solution of the wave equation are given by different authors. Brekhovskikh (Chap. 5, Ref. 52, eq. 26.21) gives the solution for a pressure-release surface and a perfectly rigid bottom as

$$\Psi = \frac{2\pi i}{h} \sum_{l=0}^{\infty} \cosh b_l z_0 \cosh b_l z H_0^{(1)}(x_l r)$$

$$b_l = \frac{i(l + \frac{1}{2})\pi}{h}$$

$$hx_l = h(b_l^2 + k^2)^{1/2} = [(kh)^2 - (l + \frac{1}{2})^2 \pi^2]^{1/2}$$

$$l = 0, 1, 2, \ldots$$

$$k = \frac{2\pi}{\lambda}$$

where h = water depth
z_0 = source depth
z = receiver depth
$H_0^{(1)}$ = Hankel function of the first kind

Modes propagating without attenuation are those for which the argument $(x_l r)$ of the Hankel function is real; these are the modes for which $kh > \pi/2$ or $h > \lambda/4$, that is, those for which the water depth is greater than one-quarter wavelength. The frequency corresponding to $h = \lambda/4$ is termed the cutoff frequency; frequencies lower than the cutoff frequency are propagated in the channel only with attenuation and are not effectively trapped in the duct.

fig. 6.22 *Sound pressure versus depth for the first four modes for a pressure-release (soft) surface and a rigid (hard) bottom.*

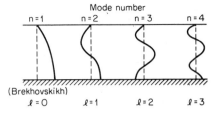

When the lower boundary is not rigid but is the boundary of a fluid medium of sound velocity c_2, the cutoff frequency becomes

$$f_0 = \frac{c_1}{4h}\sqrt{\frac{1}{1-(c_1/c_2)^2}}$$

where c_1 is the velocity in the upper water layer of thickness h. This reduces to the former condition for $c_2 \gg c_1$. The variation of pressure with depth of the first four modes, for the conditions of a free (pressure-release) surface and a rigid bottom, is shown in Fig. 6.22.

Each mode can be conceived to correspond to a pair of plane waves incident upon the boundaries at an angle β and propagating in a zigzag fashion by successive reflections. For the first two modes these equivalent plane waves are shown in Fig. 6.23, where the lines denote the pressure nodes (zero pressure), and the + and − signs give the polarity of the pressure between the pressure nodes. The relation between the grazing angle β and the mode number n for a pressure-release surface and a rigid bottom is

$$\sin\beta = \left(n - \frac{1}{2}\right)\frac{\lambda}{d}$$

The higher modes thus correspond to steeper angles. Modes for which the grazing angle of the equivalent plane wave is less than the critical angle at the bottom will be propagated with relatively low attenuation. Because the angle

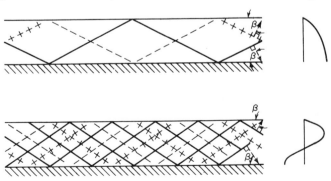

fig. 6.23 *Plane-wave equivalents of the first and second modes in a duct with a pressure-release surface and a rigid bottom.*

varies with the mode number, different normal modes will propagate with different phase velocities.

Comparison of ray and mode theories Ray theory is more convenient to use at *short* ranges, where the higher-order images rapidly die out because of reflection losses and an increasingly great distance from the field point. Accordingly, only a few images need ordinarily to be summed at short ranges. Normal-mode theory is more appropriate at *long* ranges because of the greater attenuation with distance of the higher-order modes; only a few modes need to be taken to describe the transmission. A "crossover" range between the regions of convenient usefulness of the two theories is given by the expression (42)

$$r = \frac{H^2}{\lambda}$$

where H is water depth and λ is wavelength. With modern digital-computer techniques, however, the number of summations required is no longer a serious restriction, and both theories can be successfully used to describe the sound field at long ranges. Examples are the work of Mackenzie (43) on ray summation (with phase neglected) to provide a basis of comparison between theory and field data at frequencies of 200, 600, and 1,000 Hz and to ranges of about 30 miles, and of Bucker and Morris (44) on normal-mode theory, where some 152 modes were summed to account for the transmission of 1½-kHz sound out to 25 miles.

Theoretical models The ray and normal mode theories in practical use require the summation of rays or modes on a digital computer. If a computer is not available, or if the task at hand does not justify programming a computer when one is available, it is convenient to have a theoretical model that provides tractable answers based on simplying theoretical assumptions. A number of such models are available. Brekhovskikh (Chap. 5, Ref. 52, eqs. 33.21–33.25) derived some simple expressions for the intensity from a shallow-water source under limited conditions of range, water depth, and frequency. Urick (45, 46) and MacPherson and Daintith (47) summed modes and rays without regard to phase and obtained closed-form answers for the shallow-water sound field. These models employ theoretical approximations compatible with the approximations necessary to fit exact theories to real-world conditions. Still and all, although the models are sometimes useful as first approximations, it must be said that they usually fall short of predicting shallow-water transmission losses with any degree of accuracy.

For both shallow and deep water, a large number of computer models, based upon wave theory, ray theory, or some combination of both, are available (Chap. 5, Ref. 60). For the same input data, the various models generally agree with one another, and the choice between them rests upon computer details such as running time, storage requirements, ease of implementation, complexity of program execution, etc. Their usefulness for predicting the transmission loss in the real world, however, depends upon the accuracy and validity of the input data, such as the velocity profile, bathymetry, bottom loss, etc. In shallow water, where the bottom plays a large part in determining the transmission loss, this usefulness for prediction is limited, since bottom interaction effects on transmitted sound are largely unknown.

Modifications to the simple theories Complications arise when the simple two-fluid theories are modified to allow for more realistic conditions of the surface and bottom and of the water medium. Normal-mode theory for a multilayered lossless fluid bottom has been presented in a series of papers by Tolstoy (48–50) and Clay (51). Ray theory for a multilayered lossless fluid bottom has been given by McLeroy (52). The effect of attenuation in the bottom is considered by Kornhauser and Raney (53) and by Eby, Williams, Ryan, and Tamarkin (54), and the effect of refraction in the fluid layer on the normal modes is described by Williams (55). Propagation over an elastic bottom supporting both shear and compressional waves is considered by Spitznogle and McLeroy (56) and Victor, Spitznogle, and McLeroy (57), and the effect of a variable water depth over the length of the propagation path—such as a sloping bottom—has been dealt with by Weston (58). Measurements in shallow water beneath an ice-covered surface have been reported by Milne (59), and the effect of surface roughness on shallow-water propagation has been dealt with by Clay (60). In most of the papers just cited, agreement is claimed between theory and data measured in the field or in laboratory tanks. All in all, the transmission of sound in a liquid layer overlying an elastic, absorbing layered bottom is a remarkably complex subject, as was dramatically demonstrated by the sound-field patterns obtained by Wood (61) in tanks of water.

A further complication concerns the dispersion of sound in shallow water. Because different modes propagate with different phase velocities, peculiar interference patterns are observed from a broadband source. The signal from an explosive source in shallow water is an extended pulse containing different frequency components at different instants during its duration interval. The unique dispersive effects peculiar to shallow water were pointed out long ago in the classic paper by Pekeris (41) and have been more recently observed under a shallow-water ice cover in the Arctic (62).

Transmission loss in shallow water The transmission loss for sound propagation in shallow water depends upon many natural variables of the sea surface, water medium, and bottom. Because of its sensitivity to these variables, the transmission loss in shallow water is only approximately predictable in the absence of specific knowledge of variables—especially the sound velocity and density structure of the bottom. For rough prediction purposes, the semiempirical expressions published by Marsh and Schulkin (63) are particularly useful. These are based on some 100,000 measurements in shallow water in the frequency range 0.1 to 10 kHz. Three equations are used in different ranges of a parameter H defined by $H = [\frac{1}{8}(D + L)]^{1/2}$, where D is the water depth in feet, L is the layer depth in feet, and H is in kiloyards. At short ranges, such that the range r is less than H, the transmission loss is in decibels,

$$\text{TL} = 20 \log r + \alpha r + 60 - k_L$$

where r is in kiloyards, α is the absorption coefficient of seawater in decibels per kiloyard (Sec. 5.3), and k_L is a "near-field anomaly" dependent on sea state and bottom type, as given in Table 6.2 a. At intermediate ranges, such that $H \leq r \leq 8H$, the transmission loss is

$$\text{TL} = 15 \log r + \alpha r + a_T \left(\frac{r}{H} - 1\right) + 5 \log H + 60 - k_L$$

where r and H are in kiloyards, and a_T is a shallow-water attenuation coefficient listed in Table 6.2 b. For long ranges, such that $r > 8H$,

$$\text{TL} = 10 \log r + \alpha r + a_T \left(\frac{r}{H} - 1\right) + 10 \log H + 64.5 - k_L$$

These equations verify the expected transition between spherical spreading near the source and cylindrical spreading at great distances. Table 6.2 c gives the probable errors of the transmission loss computed by these expressions relative to the individual observations on which they are based.

table 6.2   Factors and Error Estimate for Shallow-Water Transmission Loss (Ref. 63)

Sea state f, kHz	0		1		2		3		4		5	
	Sand	Mud	Sand	Mud	Sand	Mud	Sand	Mud	Sand	Mud	Sand	Mud
(a) Near-field anomaly k_L, dB												
0.1	7.0	6.2	7.0	6.2	7.0	6.2	7.0	6.2	7.0	6.2	7.0	6.2
0.2	6.2	6.1	6.2	6.1	6.2	6.1	6.2	6.1	6.2	6.0	6.2	6.0
0.4	6.1	5.8	6.1	5.8	6.1	5.8	6.1	5.8	6.1	5.8	4.7	4.5
0.8	6.0	5.7	6.0	5.6	5.9	5.6	5.3	5.0	4.3	3.9	3.9	3.6
1.0	6.0	5.6	5.9	5.5	5.7	5.3	4.6	4.2	4.1	3.7	3.8	3.4
2.0	5.8	5.4	5.3	4.9	4.2	3.8	3.8	3.4	3.5	3.1	3.1	2.8
4.0	5.7	5.1	3.9	3.5	3.6	3.1	3.2	2.8	2.9	2.4	2.6	2.2
8.0	4.3	3.8	3.3	2.8	2.9	2.5	2.6	2.2	2.3	1.9	2.1	1.7
10.0	3.9	3.4	3.1	2.6	2.7	2.2	2.4	2.0	2.2	1.7	2.0	1.6
(b) Attenuation factor a_T, dB												
0.1	1.0	1.3	1.0	1.3	1.0	1.3	1.0	1.3	1.0	1.3	1.0	1.3
0.2	1.3	1.7	1.3	1.7	1.3	1.7	1.3	1.7	1.3	1.7	1.4	1.7
0.4	1.6	2.2	1.6	2.2	1.6	2.2	1.6	2.2	1.7	2.4	2.2	3.0
0.8	1.8	2.5	1.8	2.5	1.9	2.6	2.2	3.0	2.4	3.8	2.9	4.0
1.0	1.8	2.7	1.9	2.7	2.1	2.9	2.6	3.7	2.9	4.1	3.1	4.3
2.0	2.0	3.0	2.4	3.5	3.1	4.4	3.3	4.7	3.5	5.0	3.7	5.2
4.0	2.3	3.6	3.5	5.2	3.7	5.5	3.9	5.8	4.1	6.2	4.3	6.4

table 6.2 **Factors and Error Estimate for Shallow-Water Transmission Loss (Ref. 63)** *(Continued)*

Sea state f, kHz	0 Sand	0 Mud	1 Sand	1 Mud	2 Sand	2 Mud	3 Sand	3 Mud	4 Sand	4 Mud	5 Sand	5 Mud
8.0	3.6	5.3	4.3	6.3	4.5	6.7	4.7	6.9	5.0	7.3	5.1	7.5
10.0	4.0	5.9	4.5	6.8	4.8	7.2	5.0	7.5	5.2	7.8	5.3	8.0

(c) Probable error (semi-interquartile range) of computed loss, dB

Range, kyd	Frequency, Hz			
	112	446	1,120	2,820
3	2	4	4	4
9	2	4	5	6
30	4	9	11	11
60	5	9	11	12
90	6	9	11	12

At higher frequencies, an extensive series of transmission measurements were made during World War II (Chap. 5, Ref. 49). Some 617 runs were made at many locations off the East and Gulf Coasts of the United States. During these runs, the sound level from a CW source at 12 and 24 kHz was measured out to distances between 2 and 10 miles, depending on the quality of the propagation. Examples of good and poor transmission are shown in Fig. 6.24, giving the measured excess of loss over spherical spreading for some typical conditions. The "good" condition corresponds to well-mixed, nearly isothermal water overlying a sand bottom; the "poor" condition, to water of negative temperature gradient (downward refraction) over a mud bottom. At these frequencies it was found that the transmission improved with (1) decreasing frequency, (2) smaller negative temperature gradient, (3) increasing water depth (between 60 and 600 ft); and (4) increasing depth of the receiving hydrophone in a negative gradient; the transmission was best over sand bottoms and poorest over mud bottoms. More recently, the sensitivity of shallow-water transmission to temperature gradients has been noted by Macpherson and Fothergill (64) in measurements off the Scotian shelf at different seasons; good transmission was found in winter, when the temperature—and therefore the sound velocity—increased with depth; poor transmission occurred in summer when downward refraction existed.

The seasonal variability of shallow water transmission is well known. It was observed in a number of areas, and correctly accounted for, by German acousticians apparently before World War I (Chap. 1, Ref. 6), and was pointed out by Weston (65) as a feature of the transmission in the North Sea. Recent quantitative data on the attenuation coefficient (66) in the Baltic Sea during the summer and winter seasons are given in Fig. 6.25. In winter, the low attenuation in the Baltic Sea over much of the frequency range is due to

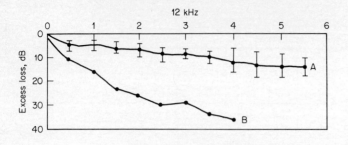

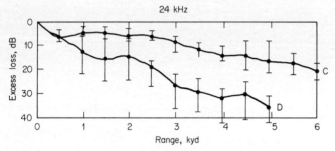

fig. 6.24 *Shallow-water transmission at 12 and 24 kHz. Vertical lines show ±1 standard deviation of individual measurements from the mean. (Chap. 5, Ref. 49.)*

Curve	Bottom	Temperature gradient	Hydrophone depth	Wind force (Beaufort scale)	No. runs	No. areas
A	Sand	None or small	Shallow	1–3	28	14
B	Mud	Negative	Middepth	0–2	3	2
C	Sand	None or small	Shallow	1	18	7
D	Mud	Negative	Shallow	0–2	15	4

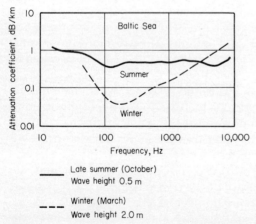

——— Late summer (October)
Wave height 0.5 m

- - - Winter (March)
Wave height 2.0 m

fig. 6.25 *Attenuation coefficient in the Baltic Sea in summer and in winter.*

isothermal water at this season and the resulting upward refraction, while, at the high end of the frequency scale, the high attenuation is likely to be the result of greater surface roughness during the stormy winter season.

Another example is seen in Fig. 6.26, showing the results of two field tests in 200 ft of water off the west coast of Florida using a receiver at a depth of 80 ft and explosive sound signals detonating at 60 ft. One test was made in October, when a strong thermocline existed below a mixed layer; the other in August, when a more gentle thermocline extended up to the surface. The low frequencies appear to have been unaffected by the change in the thermal gradient, but the high frequencies show better transmission in October than in August because of the presence of the mixed layer. All frequencies tend to show better transmission than spherical spreading (20 log r) at short ranges because of trapping, and to show poorer transmission at long ranges because of attenuation.

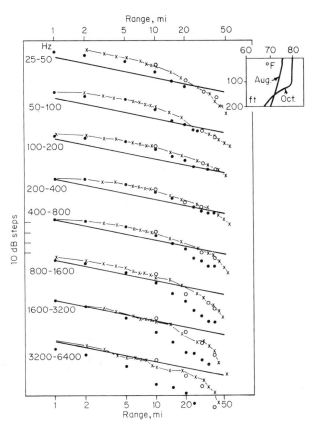

fig. 6.26 Transmission loss versus range in consecutive octave bands in water 200 ft deep off Florida. The light points connected by lines show measurements made in October; the heavy points in August. Explosive sound signals were used. Source depth 60 ft, receiver depth 80 ft. The straight sloping lines show spherical spreading and provide the scale for the transmission loss. (Ref. 67.)

In addition to such seasonal changes, an added complication in shallow water is the fact that frequencies too low to be well trapped within the water column travel to long ranges by way of the bottom. This was observed long ago by Worzel and Ewing (68) in wartime work motivated by the needs of the then newly discovered acoustic mine. It was recognized at that time that the acoustic properties of the bottom down to considerable depths govern the propagation of low-frequency sound in shallow water. Indeed, it has been suggested, and there is some experimental field evidence to show, that seismic sensors, such as those used in geophysical prospecting on land, and responding to bottom vibration, may be better sensors for low frequencies in shallow water than are conventional hydrophones responding to pressure in the water (69).

Because of these complexities, the transmission loss to be expected at a shallow-water location may be said to be, for many purposes, unpredictable (70). Recourse to direct measurements is necessary. For this purpose an aircraft, and the use of explosive sound signals (Sec. 4.4) and sonobuoys, has been found to provide a rapidly deployable platform for making on-site acoustic measurements. As an example, Fig. 6.27 shows some transmission-loss contours obtained in this way out to 30 miles in different directions about a location on the continental shelf off the East Coast of the United States.

6.7 *Fluctuation of Transmitted Sound*

When the noise from a distant sound in the sea is observed, one of its most obvious characteristics is that it fluctuates, or changes with time. Rapid and often violent changes of amplitude and phase are observed, especially with sinusoidal or narrow-band signals; very slow, large, long-period changes are encountered when observations are continued over a sufficiently long period of time.

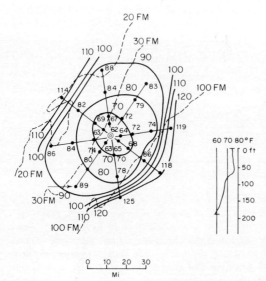

fig. 6.27 *Transmission loss contours about a shallow-water site. An aircraft dropped a sonobuoy at the center of the pattern and later, explosive sound signals at the dots along the radial arms out to a distance of 30 miles. The number alongside each dot is the measured transmission loss. The dashed lines are bottom contours. At the right is the temperature profile obtained with an airborne bathythermograph (Sec. 5.4). (Ref. 70.)*

Amplitude fluctuations cause erratic detections of sonar targets. Targets tend to be detected when the signal is strong, and lost when the signal is weak. The erratic, ephemeral behavior of sonar echoes was the motivation behind a thorough study of fluctuations during World War II. At that time many of its causes were identified (71). Since then, the subject has received continued, though often sporadic and ad hoc, attention. Yet it can be said that the magnitude, time scale, and causes of the fluctuation to be expected under a given set of circumstances—except for high frequencies via direct paths—remain unpredictable.

As a general rule, it may be said that fluctuations are caused by *moving inhomogeneities* in the sea. The *motion* may be that of the sonar platform and its target, or it may be associated with the sea itself, such as the motion of surface waves or turbulences in the body of the sea. If the sea became completely frozen over, along with all the objects in it, fluctuations would no longer exist.

We may discuss signal fluctuations under a number of categories, depending broadly on the circumstances and the properties and causes of the fluctuation.

Deterministic and stochastic fluctuations An important distinction is necessary between *deterministic* and *stochastic or statistical* fluctuations. Signals that fluctuate in a deterministic way can be predicted quantitatively; fades and surges of transmitted sound can, at least to some extent, be foretold from our knowledge of underwater sound and from models based on that knowledge. Many kinds of deterministic fluctuations have been described in Chapters 5 and 6. Examples are the seasonal and diurnal changes in transmission loss, the image interference fluctuations of a moving receiver, and the increase of sound from a moving source of noise as it transits a convergence zone. On the other hand, stochastic fluctuations can be predicted only statistically, such as the fraction of the levels of a large number of signal samples that will be larger or smaller by a given number of decibels than the mean level. In this category fall the fluctuations due to thermal microstructure and the fluctuations caused by multipath propagation. For these kinds of fluctuations, our prediction capability will depend on the time-stationarity of the sea and hence upon the time duration of the ensemble of signal samples.

Platform motion The motion of a sonar and its platform is an obvious cause of fluctuation. Motion causes a wandering in space of the directional sonar beam relative to the target, and results in amplitude changes of the target signal.

This particular cause of fluctuation was given much attention during World War II (71); it demonstrated the need for mechanical or electronic stabilization of active sonars on surface ships.

Thermal microstructure (high frequencies, short ranges) The body of the sea is not uniform, but—like the atmosphere—contains inhomogeneities of temperature—or more generally, of index of refraction—that are in a state of

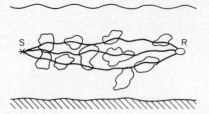

fig. 6.28 Rays between a source and a receiver in a medium containing blobs of inhomogeneity.

turbulent motion. Figure 6.28 shows a segment of an inhomogeneous ocean between a source S and a receiver R. Here the inhomogeneities are crudely visualized as irregular blobs of water of different sound velocity or of different index of refraction. The receiver at R will receive sound from S via a number of refracted and scattered paths, such as the three paths shown in the figure. Because of turbulence and water currents, a time-varying signal due to interference effects among the various propagation paths is observed at R. This fluctuation causes sonar pings and echoes to have a ping-to-ping and an echo-to-echo variability, even in the absence of interfering propagation paths reflected from the sea surface or bottom.

When a sensitive thermometer is towed through the sea at a constant depth, small temperature fluctuations amounting to a few thousandths to a few tenths of a degree are observed. This temperature microstructure has a thermal and spatial size that can be quantitatively described in terms of a mean-square temperature deviation and a "patch size." The latter quantity is the autocorrelation distance of the microstructure, within which the temperature variations maintain coherence. Two correlation functions are useful for theoretical investigations: one exponential, in which the autocorrelation function of temperature $\rho(d)$ is expressed in terms of distance d by

$$\rho(d) = e^{-d/a}$$

the other gaussian,

$$\rho(d) = e^{-(d/a)^2}$$

in which a is the patch size. Of the two, the latter is the more physically reasonable since the former involves discontinuous changes in the temperature fluctuations. Figure 6.29 shows sample records of the temperature fluctuations in the mixed layer and in the thermocline below recorded by Urick and Searfoss (72) by means of a sensitive submarine-mounted thermopile carried through the sea by the submarine's motion. The larger and longer fluctuations in the thermocline are probably associated with internal waves. Other observations of microstructure were made by Liebermann (73) using the same method, but with a recording thermometer of more rapid response.

Such data yield values for the patch size a and for a quantity μ describing the relative change in sound velocity, as measured by temperature, along the path on which measurements were made. The rms deviation μ of the relative sound velocity is defined by

$$\mu = \left[\overline{\left(\frac{\Delta c}{c}\right)^2}\right]^{1/2}$$

where $\overline{(\Delta c)^2}$ = mean square deviation of velocity of sound
c = mean value of velocity of sound

Urick and Searfoss found $\mu^2 = 8 \times 10^{-10}$ and $a = 500$ cm at a depth of about 20 ft in the mixed layer off Key West, Florida; Liebermann observed $\mu^2 = 5 \times 10^{-9}$ and $a = 60$ cm at a depth of 50 m off the California coast.

The fluctuation in transmitted sound produced by the microstructure is commonly expressed as the coefficient of variation of the amplitudes of a series of short pulses. If P is the absolute magnitude of the acoustic pressure of a transmitted pulse, then the coefficient of variation V is the fractional standard deviation of the pressure amplitude, defined as

$$V = \left[\frac{\overline{(P - \bar{P})^2}}{(\bar{P})^2}\right]^{1/2} = \left[\frac{\overline{P^2} - (\bar{P})^2}{(\bar{P})^2}\right]^{1/2}$$

where the bars represent averages of a large number of short pulses, or of a large number of readings of the envelope of a continuous signal.

Theoretical studies by Mintzer (74) and others (73, 75–77) have related the fluctuation described by the quantity V to the microstructure as described by the quantities a and μ. At short ranges and high frequencies, such that $r \ll ka^2$, where $k = 2\pi/\lambda = 2\pi f/c$ and r is the range, *ray theory* yields

$$V = \left(\frac{4}{15}\pi^{1/2}\frac{\mu^2 r^3}{a^3}\right)^{1/2}$$

At such short ranges, the dominant physical process producing the fluctuations may be considered to be the focusing and defocusing of sound by the

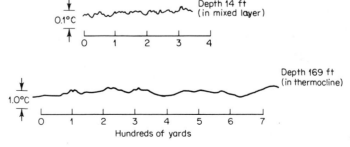

fig. 6.29 *Temperature fluctuations observed with a sensitive thermometer aboard a moving submarine at two different depths. (Ref. 72.)*

patches of inhomogeneity. On the other hand, at long ranges and low frequencies such that $r \gg ka^2$, *wave theory* gives the result

$$V = \left(\frac{\sqrt{\pi}}{2} \mu^2 k^2 ar\right)^{1/2}$$

At long ranges, foward scattering by the inhomogeneities is the principal cause of the fluctuation. Of particular note is the dependence of V on $r^{3/2}$ at short ranges and on $r^{1/2}$ at long ranges, a dependence that has been repeatedly verified experimentally, both for underwater sound (78–80) and for acoustic and electromagnetic propagation in the atmosphere.

Examples of the observed fluctuation as a function of range of short pulses transmitted through the sea are shown in Fig. 6.30. These are based on averages of the amplitude of a series of short pulses at different ranges. Figure 6.30a shows the average and the spread of the fluctuations found by Sheehy (81) at 24 kHz; Fig. 6.30b shows similar data reported by Whitmarsh, Skudryzk, and Urick (82) from a different series of observations at 25 kHz. In both cases the coefficient of variation V for transmission through the body of the sea was found to follow the $r^{1/2}$ relationship predicted by theory for long ranges.

All the above implies isotropicity; that is to say, μ is assumed to be independent of direction, and the "patches" are assumed to be spherical. While this condition doubtless exists in a well-mixed surface layer, the inhomogeneities below the layer are far from being isotropic. Recent work with sensitive temperature sensors of high resolution has demonstrated that the thermocline has a *laminar* microstructure. Near Bermuda, for example, the thermocline was found (83) to be composed of horizontal layers a few meters in thickness, differing in temperature from one another by about 0.4°C, and having a remarkably large horizontal extent of 400 to 1,000 meters. This kind of stepped laminar microstructure has been found to exist in many of the

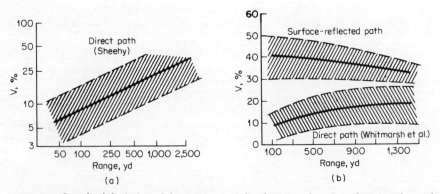

fig. 6.30 *Standard deviation of the pressure amplitude V as a function of range, observed in two different field experiments: [(a) Ref. 81. (b) Ref. 82.] Shaded areas show limits of determination from different groups of pulses.*

world's oceans. It produces maximum fluctuation when there is a ray between source and receiver that reaches a vertex and becomes horizontal (84). This fluctuation can be extremely large. Whitmarsh and Leiss (85), for example, found pulse-to-pulse amplitude fluctuations of up to 15 dB within a distance of 10 in. vertically, as a source and a separated receiver sank together slowly through a thermocline.

Surface reflection The fluctuation of pulses reflected from the sea surface obeys a different law with range, as can be seen from Fig. 6.30b. At short ranges, the fluctuation of the reflection from the rough sea surface is greater, but tends to decrease with increasing range as the grazing angle on the surface becomes less and as the sea surface begins to act more and more like a perfect mirror. At long ranges, the fluctuation of sound traveling along the two paths tends to become the same and tends to be determined by the microstructure in the body of the sea. The fluctuation of surface-reflected sound in the sea has been the subject of a number of more recent investigations (86–89) and has been briefly described above (Sec. 5.7) in connection with sound reflection from the sea surface.

Low frequencies—long ranges Because of changes that take place within the medium and on its boundaries, fluctuations also exist in the transmission of sound between a fixed bottomed source and a fixed bottomed receiver. In this connection, a long series of observations has been made by Steinberg and associates (90) across the Straits of Florida, a distance of 41 miles traversed by the Gulf Stream. Over this path, variations of 420-Hz CW sound in both amplitude and phase, with periods in the range from seconds to weeks, were observed. One peculiar and significant finding was that there were periods of up to two hours during which the phase, and to a lesser extent the amplitude, of the received signal was stable. In a different experiment, Kennedy (91) also found that, for pulses of 800-Hz sound transmitted over a single refracted path 25 miles long, the phase of the received signal was more stable than the amplitude. In this experiment most of the phase fluctuations had periods longer than one hour, while the amplitude fluctuations were much more rapid, becoming decorrelated over an interval of only 0.12 hour.

Over the longer 700-mile distance between Bermuda and Eleuthera in the Bahamas, Nichols and Young (92) transmitted 270-Hz sound and found two kinds of fluctuations: relatively fast fluctuations with periods near eight seconds, apparently caused by surface wave motion near the source, and slow fluctuations with periods longer than 10 minutes that were ascribed to internal waves.

In a multipath environment, where the transmission occurs by several different transmission paths, the motion in range of the source or the receiver, or both, acts to change the interference pattern of the multipaths and so produce amplitude fluctuations in a received signal.

This was believed to be the cause of surges and fades having periods between one minute and one hour observed in the transmission from a 142- and 275-Hz CW source towed outward in range at a speed of five knots from a receiver (93). In this experiment, the propagation was entirely by bottom-bounce multipaths. In addition to these slow fluctuations, faster fluctuations with a spectral peak near eight seconds occurred which were attributed to the effects of surface waves on the bottom-bounce paths, all but one of which encountered the sea surface.

Shallow water In shallow water, multipath transmission occurs even at short ranges. One would expect to encounter fluctuations in an extreme form in shallow water. Two examples may be mentioned. Mackenzie (94) observed variations of as much as 50 dB in the intensity of a CW tone transmitted over a range of a few kiloyards. In addition, frequency spreading of an initially pure tone, amounting to a few tenths of a hertz at 1 kHz, was found to occur, presumably because of repeated encounters with the changing sea surface. Another example is the finding of fluctuations of 10 to 20 dB in the sound from a 1,120-Hz source located ¾ mile from a receiver in 60 feet of water (95). These changes occurred regularly during the tidal cycle, with high signal levels at high tide and low levels at low tide. The cause was believed to be a changing-mode (or ray path) interference pattern brought about by the 3-ft change in water depth caused by the tides.

These and other causes of fluctuation have been repeatedly observed by Weston and coworkers (96) in the shallow waters of the British Isles. From work extending over a period of years at frequencies of 1, 2, and 3 kHz and at ranges out to 74 miles, a number of fluctuation-producing mechanisms were identified and described. These included (1) seasonal effects, such as a changing temperature gradient producing better transmission during winter, (2) storms, (3) tidal changes, such as the effect of tidal flow on the phase delay between source and receiver, (4) fish, producing better transmission by day than by night because of clustering in schools by day and dispersal at night, and (5) surface waves.

Other effects Other important aspects of the fluctuation of transmitted sound are the variation in apparent bearing of a distant target and the correlation of received signals between separated receivers (97, 98). The latter effect is important in determining the array gain (Sec. 3.1) of an array of hydrophones in the sea. These and other manifestations of transmission fluctuation are lucidly described in English translations of books by Chernov (99) and Tatarski (100) and in a nonmathematical survey paper by Brekhovskikh (101).

Signal fluctuation statistics A model for the statistics of the amplitude of signal fluctuations can be derived from simple considerations. At the outset of this section, it was noted that the fluctuations of the signal from a steady distant source in the sea may be said to be caused by propagation in an inhomogeneous moving medium. Some examples of inhomogeneities in the sea are the rough sea surface, the temperature and salinity microstructure, and the biological matter existing in the body of the sea. These various inhomogeneities are always in motion relative to each other and to the source or receiver, or both, because of currents, turbulences, and source or receiver motion relative to the medium. In these and other cases, multipath propagation is involved; the received signal has traveled from the source along closely adjacent or widely separated multipaths.

The result is that a received signal is likely to consist, in whole or in part, of a number of contributions of random and time-varying phase and amplitude. These random multipath contributions will be negligible near the source but, with increasing distance, will tend to overwhelm the steady, direct path component and produce, at a long enough range, a resultant consisting only of components of varying phase and amplitude. As an example, the propagation through random microstructure may be viewed conceptually as consisting of a steady, direct path component that decreases with range, together with scattered or diffracted components that increase with range and eventually dominate the received signal. At any range, the resultant is the sum of a steady and a random component, with an amplitude distribution that depends only on the fraction of the total average power in the random component.

In a classic paper published in 1945, S. O. Rice (102) derived the distribution function of the envelope of a sine wave plus narrow-band gaussian random noise. This is the same as the distribution of the sum of a constant vector and a random vector whose x and y coordinates are gaussian time-functions. In his honor, this function is called the *Rician Distribution*.

Let P be the magnitude of the constant vector, and let V be the magnitude of the sum of the constant vector and a random vector whose x and y coordinates have unit variance. Then Rice showed that the probability density of V is given by

$$p(V) = V \exp\left[-\left(\frac{V^2 + P^2}{2}\right)\right] \cdot I_0(PV)$$

where $I_0(PV)$ is the modified zero-order Bessel function of argument PV, for which tables are available (103). When the constant component vanishes, the result is the Rayleigh distribution

$$p(V) = V \exp\left(\frac{-V^2}{2}\right)$$

The signal is now made up entirely of random components. At the other extreme, when the constant component is very large, or more accurately, when $PV \gg 1$, the distribution is essentially gaussian with unit variance:

$$p(V) = \left(\frac{V}{2\pi P}\right)^{1/2} \exp\left[\frac{-(V-P)^2}{2}\right]$$

We introduce now a ratio T, which may be called the *randomicity* of the signal, defined as the fraction of the power of the received signal that is random, or

$$T = \frac{\text{random power}}{\text{total power}} = \frac{2}{2 + P^2}$$

where the mean power of the random vector is 1, and P^2 is the power of the constant vector. The randomicity T is small close to the source, or in a convergence zone, or generally when there is a single, strong, dominant propagation path; under these conditions the amplitude distribution tends to be gaussian. T approaches unity when the signal is composed principally of scattered contributions, as in a shadow zone at high frequencies in deep water, or generally when many multipaths exist; now the distribution will approach the Rayleigh limit.

Fig. 6.31 gives cumulative distribution curves of the ratio $10 \log I/\bar{I}$ where I is the power, or amplitude squared, of a given sample of signal and $\bar{I}$ is the mean power, or mean squared amplitude, of an ensemble of samples. When T can be estimated from the existing propagation conditions, the corresponding cumulative distribution curves give the fraction of signal samples having an intensity equal to or less than the mean intensity of the sample population. When T is greater than about 0.5, the distribution is insensitive to T. From a prediction standpoint, the fluctuation of a received signal about its mean intensity can often be predicted better than can the mean intensity itself.

The model has been validated empirically by comparison with field data observed under a wide variety of propagation conditions (104). Included in the comparison was transmission to a receiver in and below a surface duct, low frequency CW bottom-bounce transmission to ranges short of the first convergence zone, CW transmission from Eleuthera to Bermuda, and several other cases. Of course the model applies only to fluctuations that are stochastic, rather than deterministic, in origin.

6.8 Horizontal Variability

When sonar ranges were short it was assumed, and justly so, that the sea was horizontally uniform and therefore that sonar ranges would be the same in all horizontal directions. But for the longer ranges of modern sonars that are often measured in hundreds of miles, this is no longer the case. Moreover, in a single direction from a source of sound, great changes with range can, and often do, occur. Examples are the changes in the velocity profile and bathymetry with range from a source; they present challenging theoretical problems to the propagation modeler. The importance, and even the existence, of large-scale inhomogeneities in the sea, such as fronts, eddies, and different water masses, as well as bottom irregularities such as slopes and seamounts, have been recognized in recent years. In the following we can present but a brief mention of some of the research studies on these problems.

In one investigation, six ships ran east to west parallel to one another in the

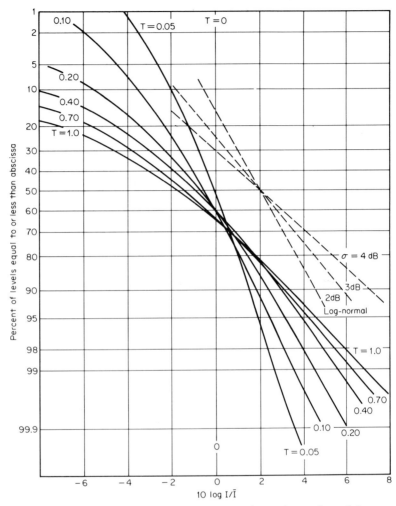

fig. 6.31 *Cumulative Rician distribution curves for various values of the randomicity T. They show the fraction of a large number of statistically stationary samples of the signal from a distant steady source that are greater or less than the mean intensity $\bar{I}$. For example, for T = 0.4, 15% of the signal levels will be equal to or less than 6 dB below the mean level 10 log $\bar{I}$. The dashed lines show for comparison log-normal distributions for three values of the standard deviation σ in dB.*

North Pacific making simultaneous oceanographic observations to obtain "mesoscale" variations (changes of the order of hundreds of kilometers) of the velocity profile (105). In another study (106) the attenuation coefficient at low frequencies in the South Pacific was found to have a regional dependence, apparently as a result of water-mass differences. In the Atlantic, propagation across the Gulf Stream from the colder water to the northwest to the warmer water of the Sargasso Sea results in lower transmission losses from a shallow

192 / **principles of underwater sound**

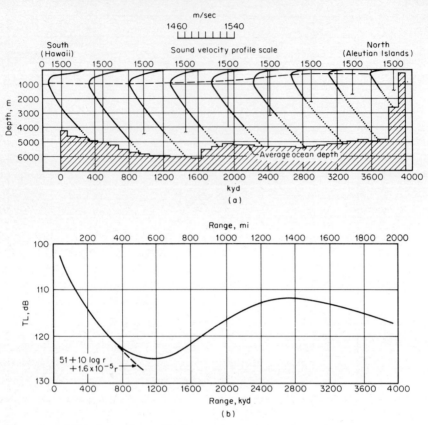

fig. 6.32 (a) *Velocity profiles and bathymetry between Hawaii and the Aleutian Islands in the North Pacific.* (b) *Transmission loss versus range from shallow explosive sound signals to a deep receiver at Hawaii near the sound channel axis.*

source to a deep receiver at Bermuda (107). An even more striking effect of a regularly changing velocity profile is seen in Fig. 6.32, showing transmission loss versus range from a series of shallow shots southward to a deep receiver located near the axis of the deep sound channel. Because the axis of the channel becomes more shallow toward the north (Sec. 5.5), shallow shots are located near or at the channel axis at the north end. Sound can travel easily along axial or near-axial paths to a deep receiver. The result is a much lower transmission loss than would otherwise occur, even becoming less by 10 dB at 1,400 miles than at 600 miles.

The effect of bottom slope on propagation from deep to shallow water, as in approaching a continental shelf, has been called the *slope enhancement effect*. Sound traveling over deep refracted paths strikes the bottom as the bottom becomes shallow, and reflection takes place up into shallow water. The result is lower transmission loss than would otherwise be observed—a kind of reverse megaphone effect. It was observed off of the East Coast of the United States (108) and also for sound traveling up the slope of the Mid-Atlantic Ridge (109). In the opposite direction, it is suspected (Sec. 7.13) to be the cause of higher ambient noise levels at a deep receiver.

6.9 Coherence of Transmitted Sound

By *coherence* is meant the similarity of signal waveform at spatially separated receivers. Coherence is measured by the crosscorrelation of the amplitude of the two waveforms, and, along with the coherence of the background of ambient noise (Sec. 7.10) or of reverberation, determines the gain of sonar arrays (Sec. 3.1).

The coherence of signals depends upon the separation and orientation of the two receivers, the frequency band employed, and integration time used. Especially important is the multipath structure of the sea between the source and the two receivers, that is, upon the manner of propagation. When there is a strong dominant propagation path, as inside a convergence zone, the coherence is high (110); when there are many multipaths in the transmission geometry the coherence is low (111).

The measurement literature is sparse and diverse, consisting of field data obtained under different and never-repeated conditions. It is impossible to present here meaningful numerical values for practical use. A few additional literature references can be cited (112–115). There has not been a thorough authoritative study of this somewhat esoteric subject.

From what has been done, however, it may be said qualitatively that the coherence between separated receivers varies with *separation* (improves with decreasing separation), *frequency* (improves with decreasing frequency), *bandwidth* (improves with decreasing bandwidth), *orientation* (horizontal better than vertical), *integration time* (improves with decreasing integration time) and, finally, the *multipath structure* of the medium (improves when a single dominant path exists and multipaths are effectively absent).

6.10 Multipaths in the Sea

Multipath propagation occurs whenever there is more than one propagation path between source and receiver. As we have seen, this is a common occurrence in long-range propagation. Multipaths commonly occur in ducted propagation, as in the surface duct (Fig. 6.1), in the deep ocean duct (Fig. 6.10), and in shallow water; they occur often in the absence of a duct, as when multiple bottom reflections take place. Sometimes the multipaths are widely separated in angle at the source or receiver or both, as when there is a direct path and a bottom reflected path in deep water. In this case, the effects of multipaths can be reduced or eliminated by transducer directivity. But under other circumstances, the multipaths are close together in angle—as in propa-

194 / principles of underwater sound

gation through microstructure or within the surface duct—and transducer directivity is to no avail. Multipaths that carry nearly the same amounts of energy—that is, have about the same transmission loss—vary among themselves in phase and amplitude when the source or receiver, or both, are in motion, and various interference effects are the result.

First of all, multipath propagation causes *fluctuations* in phase and amplitude at a single receiver. Amplitude fluctuations are beneficial for target detection at long ranges and low signal-to-noise ratios because they permit target detection during signal surges (see Sec. 12.4); at high signal-to-noise ratios they reduce the probability of detection during periods of signal fading. Second, multipaths cause signal *distortion* because the travel times along different paths are different; the most flagrant example is the propagation from an axial source to an axial receiver in the deep sound channel (Fig. 6.11). The practical importance of distortion is, among other effects, that it prohibits the use of replica correlation for signal detection. A third result of multipath propagation is *decorrelation* of phase and amplitude between separated receivers, causing a degradation of the gain of arrays (Sec. 3.1). Finally, multipaths cause *frequency broadening*, as when source/receiver motion causes different doppler shifts among multipaths or when there are one or more encounters with the sea surface (Sec. 5.7).

In summary, it may not be an exaggeration to say that the effects of multipaths in the sea are deleterious to the detection of long-range sonar targets as well as in many short-range applications. The sea, as is self-evident by now, is not a simple propagation medium for sonar, as is the atmosphere for radar, but is variable in time and space and in supporting propagation via multipaths.

6.11 Deep-Sea Paths and Losses: A Summary

Propagation paths in the deep sea Figure 6.33 illustrates the different types of propagation paths that exist between a source and a receiver in the deep sea. At short ranges A, there is a nearly straight-line path between the two, and the transmission loss is determined by spherical spreading (Sec. 5.2), plus a loss due to absorption if the frequency is high enough (Sec. 5.3), and modified by interference effects of the surface reflection for a near-surface source and receiver (Sec. 5.7).

At longer ranges, propagation occurs in the mixed-layer channel B, involving repeated surface reflections and leakage out of the channel to a receiver below (Sec. 6.1). Propagation may occur also via "bottom-bounce" reflected paths C, which suffer a reflection loss at the sea bottom and along which the transmission loss is determined by spherical spreading plus absorption and by the bottom-reflection loss (Sec. 5.8). At still greater ranges, convergence-zone transmission D takes place where, within a zone a few miles wide, convergence gains between 5 and 20 dB occur, depending on depth (Sec. 6.3). Beyond the first pair of convergence half zones, the transmission between a shallow

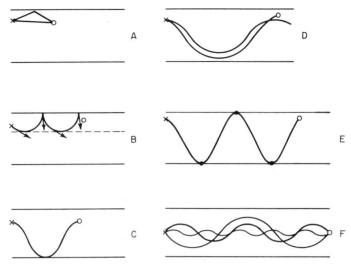

fig. 6.33 Propagation paths between a source and a receiver in deep water.

source and a shallow receiver involves multiple reflections from the bottom E. In the deep sound channel, long-range propagation occurs over internally refracted paths F and is especially good when both source and receiver lie on the channel axis (Sec. 6.2).

The vagaries of transmission between a shallow source and a shallow receiver may be avoided by placing the source or the receiver deep in the sea. Propagation to moderate ranges then can take place via the so-called *reliable acoustic paths,* a number of which are shown in Fig. 6.34. Such paths are "reliable" because they are sensitive neither to near-surface effects nor to the varying losses on reflection that characterize "bottom-bounce" propagation (C, Fig. 6.33).

Transmission loss as a function of range As an illustration of the transmission losses associated with some of these paths, Fig. 6.35 shows computed curves for conditions of (1) frequency 2 kHz, (2) source depth 50 ft, (3) receiver depth 500 ft, (4) mixed-layer thickness 100 ft, and (5) water depth 15,000 ft. Curves B and B' are taken from Fig. 6.7 for receiver depths of 50

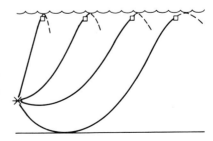

fig. 6.34 "Reliable" acoustic paths from a deep source to a shallow receiver in the deep sea.

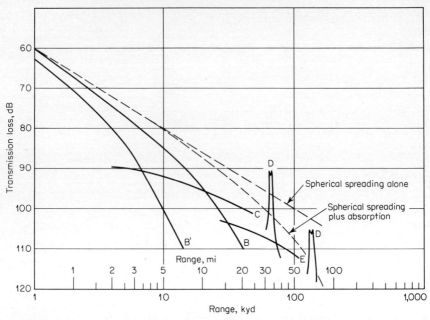

fig. 6.35 *Transmission loss versus range for different propagation paths. The letters on each curve correspond to the paths of Fig. 6.33*

and 500 ft, respectively. Curve C is the loss by bottom reflection obtained by adding (1) the loss due to spherical spreading over the slant range to the bottom, (2) a loss due to absorption and equal to 0.08 dB/kyd, and (3) the reflection loss for the appropriate grazing angle. The peaks marked D illustrate the expected losses to the first and second convergence zones at 65 and 130 kyd, respectively, and E is the transmission loss via the second bottom reflection. The dashed curves illustrate the degree to which spherical spreading and spherical spreading plus absorption approximate the computed losses over the various paths.

Concluding remarks The example just given illustrates the fact that, almost always, there is more than one propagation path between a source and receiver in the sea. Often one path will be dominant, and the transmission loss corresponding to it will be a minimum compared with other possible paths. But under some conditions, multipath propagation occurs and more than one acoustic path of comparable loss exists between the two points, depending on the source and receiver depths and on the acoustic conditions in the body of the sea and in its boundaries.

Because of the complex nature of sound propagation in the sea, the selection of a reasonably satisfactory value of transmission loss for use in a practical problem is often a vexatious problem in itself. Indeed, of all the sonar parameters, transmission loss presents the greatest difficulty from a quantitative standpoint, since the choice of a numerical value must rest on a clear understanding of the transmission characteristics of the medium. Furthermore, other characteristics, such as the distortion, fluctuation, and coherence of signals transmitted through the sea, are often of comparable importance to the transmission loss itself. In short, the propagation vagaries of sound in the sea are such that, in spite of the extent to which they have been studied, they still present formidable problems, both to the engineer interested in a sonar application and to the research scientist.

REFERENCES

1. Steinberger, R. L.: "Underwater Sound Investigation of Water Conditions, Guantanamo Bay Area, Feb. 1937," Navy Yard Sound Laboratory, Washington, D.C., May 3, 1937.
2. Urick, R. J.: Recent Results on Sound Propagation to Long Ranges in the Ocean, *U. S. Navy J. Underwater Acoust.*, **2**:57 (1951).
3. Urick, R. J.: Sound Transmission Measurements in the Long Island–Bermuda Region, Summer 1949, *U.S. Nav. Res. Lab. Rep.* 3630, 1950.
4. Schweitzer, B. J.: Sound Scattering Into the Shallow Zone below an Isothermal Layer, *J. Acoust. Soc. Am.*, **44**:525 (1968).
5. Beckmann, P., and A. Spizzichino: "Scattering of Electromagnetic Waves from Rough Surfaces," The Macmillan Company, New York, 1963.
6. Kerr, D. E. (ed.): "Propagation of Short Radio Waves," M.I.T. Radiation Laboratory Series, vol. 13, McGraw-Hill Book Company, New York, 1951.
7. Brekhovskikh, L. M., and I. D. Ivanov: Concerning One Type of Attenuation of Waves Propagating in Inhomogeneously Stratified Media, *Sov. Phys. Acoust.*, **1**:23 (1955).
8. Marsh, H. W., and M. Schulkin: Report on the Status of Project AMOS (Acoustic, Meterological and Oceanographic Survey), *U.S. Navy Underwater Sound Lab. Rep.* 255A, reprinted 1967.
9. Schulkin, M.: Surface Coupled Losses in Surface Sound Channels, *J. Acoust. Soc. Am.*, **44**:1152 (1968).
10. Saxton, H. L., H. Baker, and N. Shear: 10-Kilocycle Long Range Search Sonar, *U.S. Nav. Res. Lab. Rep.* 4515, 1955.
11. Condron, T. P., P. M. Onyx, and K. R. Dickson: Contours of Propagation Loss and Plots of Propagation Loss vs. Range for Standard Conditions at 2, 5, and 8 kc, *U.S. Navy Underwater Sound Lab. Tech. Memo* 1110-14-55, 1955.
12. Baker, W. F.: New Formula for Calculating Acoustic Propagation Loss in a Surface Duct, *J. Acoust. Soc. Am.*, **57**:1198 (1975).
13. Hall, M.: Surface-duct Propagation: An Evaluation of Models of the Effects of Surface Roughness, *J. Acoust. Am.*, **67**:803 (1980).
14. Neumann, G., and W. J. Pierson, Jr., "Principles of Physical Oceanography," p. 351, Prentice-Hall, Englewood Cliffs, N.J., 1966.
15. Hill, M. N. (ed.): Internal Waves, chap. 22, in E. C. Lafond, "The Sea," Interscience Publishers (division of John Wiley & Sons, Inc.), New York, 1962.
16. Lafond, E. C., and O. S. Lee: Internal Waves in the Ocean, *Navigation*, **9**:231 (1962).
17. Lee, O. S.: Effect of an Internal Wave on Sound in the Ocean, *J. Acoust. Soc. Am.*, **33**:677 (1961).
18. Barakos, P. A.: On the Theory of Acoustic Wave Scattering and Refraction in Internal Waves, *U.S. Navy Underwater Sound Lab. Rep.* 649, 1965.
19. Ewing, M., and J. L. Worzel: Long-Range Sound Transmission, *Geol. Soc. Am. Memo* 27, 1948.
20. Stifler, W. W., and W. F. Saars: Sofar, *Electron.*, **21**:98 (1948).

21. Bryan, G. M., M. Truchan, and J. I. Ewing: Long Range SOFAR Studies in the South Atlantic Ocean, *J. Acoust. Soc. Am.*, **35:**273 (1963).
22. Baker, H. H.: Missile Impact Locating System, *Bell Teleph. Lab. Rec.*, **39:**195 (1961).
23. Urick, R. J.: Ray Identification in Long Range Sound Transmission, *U.S. Nav. Ord. Lab. Tech. Rep.* 65-104, 1965.
24. Urick, R. J.: Long Range Deep Sea Attenuation Measurement, *J. Acoust. Soc. Am.*, **34:**904 (1966).
25. Thorp, W. H.: Deep-Ocean Sound Attenuation in the Sub and Low Kilocycle per Second Region, *J. Acoust. Soc. Am.*, **38:**648 (1965).
26. Webb, D. C., and M. J. Tucker: Transmission Characteristics of the SOFAR Channel, *J. Acoust. Soc. Am.*, **48:**767 (1970).
27. Sussman, B., and others: Low Frequency Attenuation Studies, *U.S. Navy Underwater Sound Lab. Tech. Memo* 911-16-64, 1964.
28. Hale, F. E.: Long Range Sound Propagation in the Deep Ocean, *J. Acoust. Soc. Am.*, **33:**456 (1961).
29. Urick, R. J.: Caustics and Convergence Zones in Deep-Water Sound Transmission, *J. Acoust. Soc. Am.*, **38:**348 (1965).
30. Urick, R. J., and G. R. Lund: Coherence of Convergence Zone Sound, *J. Acoust. Soc. Am.*, **43:**723 (1968).
31. "Velocity Profile Atlas of the North Atlantic," vol. II, U.S. Naval Air Development Center, 1966.
32. Leroy, C. C.: Sound Propagation in the Mediterranean Sea, in V. M. Albers (ed.), "Underwater Acoustics," vol. 2, p. 203, Plenum Press, New York, 1967.
33. Marsh, H. W., and R. H. Mellen: Underwater Sound Propagation in the Arctic Ocean, *J. Acoust. Soc. Am.*, **35:**552 (1963).
34. Diachok, O. I.: Effects of Sea-Ice Ridges on Sound Propagation in the Arctic Ocean, *J. Acoust. Soc. Am.*, **59:**110 (1976).
35. Buck, B. M., and C. R. Greene: Arctic Deep-Water Propagation Measurements, *J. Acoust. Soc. Am.*, **36:**1526 (1964).
36. Milne, A. R.: A 90-km Sound Transmission Test in the Arctic, *J. Acoust. Soc. Am.*, **35:**1459 (1963).
37. Buck, N. M.: Arctic Acoustic Transmission Loss and Ambient Noise, in J. E. Sater (ed.), "Arctic Drifting Stations," Arctic Institute of North America, 1968.
38. Bradley, D. L.: Long Range Acoustic Transmission Loss in the Marginal Ice Zone North of Iceland, *U.S. Nav. Ord. Lab. Rep.* 72-217, 1973.
39. Ide, J. M., R. F. Post, and W. J. Fry: Propagation of Underwater Sound at Low Frequencies as a Function of the Acoustic Properties of the Bottom, *U.S. Nav. Res. Lab. Rep.* S-2113, 1943.
40. Roe, G. M.: Propagation of Sound in Shallow Water, *U.S. Navy Bur. Ships Minesweeping Branch Rep.* 65, 1943. Also, Addendum, *Rep.* 88, 1946.
41. Pekeris, C. L.: Theory of Propagation of Explosive Sound in Shallow Water, *Geol. Soc. Am. Memo* 27, 1948.
42. Weston, D. E.: Propagation of Sound in Shallow Water, 1962 Sonar Systems Symposium, University of Birmingham, England, *J. Brit. IRE*, **26:**329 (1963).
43. Mackenzie, K. V.: Long-Range Shallow-Water Transmission. *J. Acoust. Soc. Am.*, **33:**1505 (1961).
44. Bucker, H. P., and H. E. Morris: Normal Mode Calculations for a Constant-Depth Shallow Water Channel, *J. Acoust. Soc. Am.*, **38:**1010 (1965).
45. Urick, R. J.: A Prediction Model for Shallow Water Sound Transmission, *U.S. Nav. Ord. Lab. Rep.* 67-12, 1967.
46. Urick, R. J.: Intensity Summation of Modes and Images in Shallow-Water Sound Transmission, *J. Acoust. Soc. Am.*, **46:**780 (1969).
47. MacPherson, J. D., and M. J. Daintith: Practical Model of Shallow Water Acoustic Propagation, *J. Acoust. Soc. Am.*, **41:**850 (1967).

48. Tolstoy, I.: Resonant Frequencies and High Modes in Layered Wave Guides, *J. Acoust. Soc. Am.*, **28**:1182 (1956).
49. Tolstoy, I.: Dispersion and Simple Harmonic Point Sources in Wave Ducts, *J. Acoust. Soc. Am.*, **27**:897 (1955).
50. Tolstoy, I.: Guided Waves in a Fluid with Continuously Variable Velocity Overlying an Elastic Solid: Theory and Experiment, *J. Acoust. Soc. Am.*, **32**:81 (1960).
51. Clay, C. S.: Propagation of Band-limited Noise in a Layered Wave Guide, *J. Acoust. Soc. Am.*, **31**:1473 (1959).
52. McLeroy, E. G.: Complex Image Theory of Low Frequency Sound Propagation in Shallow Water, *J. Acoust. Soc. Am.*, **33**:1120 (1961).
53. Kornhauser, E. T., and W. P. Raney: Attenuation in Shallow Water Propagation Due to an Absorbing Bottom, *J. Acoust. Soc. Am.*, **27**:689 (1955).
54. Eby, R. K., A. O. Williams, R. P. Ryan, and P. Tamarkin: Study of Acoustic Propagation in a Two-layered Model, *J. Acoust. Soc. Am.*, **32**:88 (1960).
55. Williams, A. O.: Some Effects of Velocity Structure on Low Frequency Propagation in Shallow Water, *J. Acoust. Soc. Am.*, **32**:363 (1960).
56. Spitznogle, F. R., and E. G. McLeroy: Propagation at Short Ranges of Elastic Waves from an Impulsive Source near a Fluid Two-Layer Solid Interface, *J. Acoust. Soc. Am.*, **35**:1808 (1963).
57. Victor, A. S., F. R. Spitznogle, and E. G. McLeroy: Propagation at Short Ranges of Elastic Waves from an Impulsive Source in a Shallow Field Overlying a Layered Elastic Solid, *J. Acoust. Soc. Am.*, **37**:894 (1965).
58. Weston, D. E.: Guided Propagation in a Slow-varying Medium, *Proc. Phys. Soc.* London, **73**:365 (1958).
59. Milne, A. R.: Shallow Water Under-Ice Acoustics in Barrow Strait, *J. Acoust. Soc. Am.*, **32**:1007 (1960).
60. Clay, C. S.: Effect of a Slightly Irregular Boundary on the Coherence of Waveguide Propagation, *J. Acoust. Soc. Am.*, **36**:833 (1964).
61. Wood, A. B.: Model Experiments on Sound Propagation in Shallow Seas, *J. Acoust. Soc. Am.*, **31**:1213 (1959).
62. Hunkins, K., and H. Kutschale: Shallow-water Propagation in the Arctic Ocean, *J. Acoust. Soc. Am.*, **35**:542 (1963).
63. Marsh, H. W., and M. Schulkin: Shallow Water Transmission, *J. Acoust. Soc. Am.*, **34**:863 (1962).
64. Macpherson, J. D., and N. O. Fothergill: Study of Low Frequency Sound Propagation in the Hartlen Point Region of the Scotian Shelf, *J. Acoust. Soc. Am.*, **34**:967 (1962).
65. Weston, D. E.: Propagation of Sound in Shallow Water, 1962 Sonar Systems Symposium, University of Birmingham, England, *J. Brit. IRE*, **26**:329 (1963).
66. Schellstede, G. S., and P. Wille: Measurements of Sound Attenuation in Standard Areas of the North Sea and Baltic, Proceedings of a SACLANTCEN Conference on Shallow Water Sound Propagation, September 1974.
67. Urick, R. J.: Shallow-Water Revisited: Further Acoustic Observations at a Site off the Coast of Florida, *U.S. Nav. Ord. Lab. Rep.* 69-234, 1970.
68. Worzel, J. L., and M. Ewing: Explosion Sounds in Shallow Water, *Geol. Soc. Am. Memo* 27, 1948.
69. Proceedings of a Workshop on Seismic Propagation in Shallow Water, Office of Naval Research, Arlington, Va., 6–7 July 1978.
70. Urick, R. J.: Some Perspectives and Recent Findings in Shallow Water Acoustics, *U.S. Nav. Ord. Lab. Rep.* 71-204 (1971).
71. Eckart, C., and R. R. Carhart: "Fluctuation of Sound in the Sea, Basic Problems of Underwater Acoustics," chap. 5, National Research Council, 1950.
72. Urick, R. J., and C. W. Searfoss: The Microthermal Structure of the Ocean near Key West, Florida: Part I—Description; Part II—Analysis, *U.S. Nav. Res. Lab. Reps.* S-3392, 1948, and S-3444, 1949.

73. Liebermann, L.: Effect of Temperature Inhomogeneities in the Ocean on the Propagation of Sound, *J. Acoust. Soc. Am.*, **23**:563 (1951).
74. Mintzer, D.: Wave Propagation in a Randomly Inhomogeneous Medium, I: *J. Acoust. Soc. Am.*, **25**:922 (1953); II: **25**:1107 (1953); III: **26**:186 (1954).
75. Skudryzk, E.: Scattering in an Inhomogeneous Medium, *J. Acoust. Soc. Am.*, **29**:50 (1957).
76. Potter, D. S., and S. R. Murphy: On Wave Propagation in a Random Inhomogeneous Medium, *J. Acoust. Soc. Am.*, **29**:197 (1957).
77. Knollman, G. C.: Wave Propagation in a Medium with Random Spheroidal Inhomogeneities, *J. Acoust. Soc. Am.*, **36**:681 (1964).
78. Stone, R. G., and D. Mintzer: Range Dependence of Acoustic Fluctuations in a Randomly Inhomogeneous Medium, *J. Acoust. Soc. Am.*, **34**:647 (1962).
79. Whitmarsh, D. C.: Underwater Acoustic Transmission Measurements, *J. Acoust. Soc. Am.*, **35**:2014 (1963).
80. Sagar, F. H.: Acoustic Intensity Fluctuations and Temperature Microstructure in the Sea, *J. Acoust. Soc. Am.*, **32**:112 (1960).
81. Sheehy, M. J.: Transmission of 24 kc Underwater Sound from a Deep Source, *J. Acoust. Soc. Am.*, **22**:24 (1950).
82. Whitmarsh, D. C., E. Skudryzk, and R. J. Urick: Forward Scattering of Sound in the Sea and Its Correlation with the Temperature Microstructure, *J. Acoust. Soc. Am.*, **29**:1124 (1957).
83. Cooper, J. W., and H. Stommel: Regularly Spaced Steps in the Main Thermocline near Bermuda, *J. Geophys. Res.*, **73**:5849 (1968).
84. Melberg, L. E., and O. M. Johannessen: Layered Oceanic Microstructure—Its Effect on Sound Propagation, *J. Acoust. Soc. Am.*, **53**:571 (1973).
85. Whitmarsh, D. C., and W. J. Leiss: Fluctuation in Horizontal Acoustic Propagation over Short Depth Increments, *J. Acoust. Soc. Am.*, **43**:1036 (1968).
86. Brown, M. V., and J. Ricard: Fluctuations in Surface Reflected Pulsed CW Arrivals, *J. Acoust. Soc. Am.*, **32**:1551 (1960).
87. Clay, C. S.: Fluctuations of Sound Reflected from the Sea Surface, *J. Acoust. Soc. Am.*, **32**:1547 (1960).
88. Scrimger, J. A.: Signal Amplitude and Phase Fluctuations Induced by Surface Waves in Ducted Sound Propagation, *J. Acoust. Soc. Am.*, **33**:239 (1961).
89. Beckerly, J. C.: Effects of Ocean Waves on Acoustic Signals to Very Deep Hydrophones, *J. Acoust. Soc. Am.*, **35**:267 (1963).
90. Steinberg, J. C., and others: Fixed System Studies of Underwater Acoustic Propagation, *J. Acoust. Soc. Am.*, **52**:1521 (1972).
91. Kennedy, R. M.: Phase and Amplitude Fluctuations Through a Layered Ocean, *J. Acoust. Soc. Am.*, **46**:737 (1969).
92. Nichols, R. H., and H. J. Young: Fluctuations in Low-Frequency Acoustic Propagation in the Ocean, *J. Acoust. Soc. Am.*, **43**:716 (1968).
93. Urick, R. J.: Amplitude Fluctuations of the Sound From a Low-Frequency Moving Source in the Deep Sea, *U.S. Nav. Ord. Lab. Tech. Rep.* 74-43, 1971.
94. Mackenzie, K. V.: Long Range Shallow Water Signal Level Fluctuations and Frequency Spreading, *J. Acoust. Soc. Am.*, **34**:67 (1962).
95. Urick, R. J., G. R. Lund, and D. L. Bradley: Observations of Fluctuation of Transmitted Sound in Shallow Water, *J. Acoust. Soc. Am.*, **45**:868 (1969).
96. Weston, D. E., and others: Studies of Sound Transmission Fluctuations in Shallow Coastal Waters, *Phil. Trans. Roy. Soc. London*, **265**:567 (1969).
97. Gershman, S. G., and Y. I. Tuzhilkin: Measurements of the Transverse Correlation Coefficient of a Continuous Sound Signal in the Sea, *Sov. Phys. Acoust.*, **6**:291 (1961).
98. Gulin, E. P., and K. I. Malyshev: Experiments in the Spatial Correlation of the Amplitude and Phase Fluctuations of Acoustic Signals Reflected from a Rough Ocean Surface, *Sov. Phys. Acoust.*, **10**:365 (1965).
99. Chernov, L. A.: "Wave Propagation in a Random Medium," translation, McGraw-Hill Book Company, New York, 1960.

100. Tatarski, V. I.: "Wave Propagation in a Turbulent Medium," translation, McGraw-Hill Book Company, New York, 1961.
101. Brekhovskikh, L. M.: Propagation of Sound in Inhomogeneous Media: A Survey, *Sov. Phys. Acoust.*, **2**:247 (1956).
102. Rice, S. O.: Mathematical Analysis of Random Noise, *Bell System Tech. J.*, **24**:46, Art. 3.10 (1945).
103. Abramowitz, M., and I. Stegun (eds.): "Handbook of Mathematical Functions," *U.S. Dept. of Commerce Math. Series*, **55**:Sec. 9 (1964).
104. Urick, R. J.: A Statistical Model for the Fluctuation of Sound Transmission in the Sea, *Tech. Rep.* 75-18, Naval Surface Weapons Center, White Oak, Md., 1975. Also, Models for the Amplitude Fluctuation of Narrow-Band Signals and Noise in the Sea, *J. Acoust. Soc. Am.*, **62**:878 (1977).
105. Emery, W. J., and others: Mesoscale Variations in the Deep Sound Channel and Effects on Low Frequency Propagation, *J. Acoust. Soc. Am.*, **66**:835 (1979).
106. Kibblewhite, A. C., N. R. Bedford, and S. K. Mitchell: Regional Dependence of Low-Frequency Attenuation in the North Pacific Ocean, *J. Acoust. Soc. Am.*, **61**:1169 (1977).
107. Levenson, C., and R. Doblar: Long Range Acoustic Propagation Through the Gulf Stream, *J. Acoust. Soc. Am.*, **59**:1134 (1936).
108. Urick, R. J.: Sound Transmission from Deep to Shallow Water, *U.S. Nav. Ord. Lab. Tech. Rep.* 72-1, 1972.
109. LaPlante, P., D. Koenigs, and D. G. Browning: Effect of the Mid-Atlantic Ridge on Long-Range Sound Propagation, *Nav. Underwater Systems Center Tech. Doc.* 6555, 1981.
110. Urick, R. J., and G. R. Lund: Coherence of Convergence Zone Sound, *J. Acoust. Soc. Am.*, **43**:723 (1968).
111. Urick, R. J.: Measurements of the Vertical Coherence of the Sound from a Near-Surface Source in the Sea, *J. Acoust. Soc. Am.*, **54**:115 (1973).
112. Gershman, S. G., and Y. I. Tuzhilkin: Measurement of the Transverse Correlation of a Continuous Signal in the Sea, *Sov. Phys. Acoust.*, **6**:291 (1961).
113. Gulin, E. P., and K. I. Malyshev: Spatial Correlation of Amplitude Fluctuations of a Continuous Tone Signal with Reflections from Ocean Surface Waves, *Sov. Phys. Acoust.*, **11**:428 (1966). See also, same authors, *Sov. Phys. Acoust.* **10**:365 (1964).
114. Parkins, B. E., and G. R. Fox: Measurement of the Coherence and Fading of Long Range Acoustic Signals, *IEEE Trans. Audio and Electr. Acoust.*, **19**:158 (1971).
115. Wille, P., and R. Thiele: Transverse Horizontal Coherence of Explosive Signals in Shallow Water, *J. Acoust. Soc. Am.*, **50**:348 (1971).

seven

the noise background of the sea: ambient-noise level

In a loose manner of speaking, ambient noise may be said to be the noise of the sea itself. It is that part of the total noise background observed with a nondirectional hydrophone which is not due to the hydrophone and its manner of mounting called "self-noise," or to some identifiable localized source of noise. It is what is "left over" after all identifiable noise sources are accounted for.

According to Webster's Collegiate Dictionary, the word *ambient* means "encompassing, surrounding on all sides." In the present context, ambient noise truly surrounds the hydrophone on all sides, though unequally and in a nonisotropic manner. In deep water, the ambient noise of the sea has a directionality of its own, which will be described later on.

The ambient-noise level, as a sonar parameter, is the intensity, in decibels, of the ambient background measured with a nondirectional hydrophone and referred to the intensity of a plane wave having an rms pressure of 1 μPa. Although they are measured in different frequency bands, ambient levels are always reduced to a 1-Hz frequency band and are then called *ambient-noise spectrum levels*.

The ambient background of the sea often presents difficult measurement problems. For a valid measurement of ambient noise, all possible sources of self-noise must be eliminated, or at least reduced to an insignificant contribution to the total noise level. Self-noise sources, such as cable strumming, splashes of waves against the hydrophone cable, 60-Hz-hum pickup, and sometimes even crabs crawling on the hydrophone, must be absent. Also, identifiable distant noise sources, such as individual ships, must not contribute to the noise background. Thus, ambient noise is the residual noise background in the absence of individual identifiable sources that may be considered the natural noise environment of the hydrophone. To the ear, ambient noise often sounds like a low rumble at low frequencies and a nontonal frying, crackling hiss at high frequencies. Sometimes peculiar sounds are heard as part of the ambient background, like the noise of marine animals or the noise made by cracks forming in the ice sheet over a hydrophone.

7.1 Sources of Ambient Noise in Deep Water

The studies of ambient noise made in recent years, largely with deep-bottomed hydrophones, have vastly improved and extended our knowledge about the characteristics and sources of deep-sea ambient noise. Measurements have been made over a frequency range extending from below 1 Hz up to about 100 kHz. Over this broad frequency range, the data show that ambient noise has different characteristics at different frequencies, with a different spectral slope and a different behavior with varying conditions, such as wind speed, in different parts of the spectrum. Because of this, it follows that the noise must be due to a variety of different sources. In any one region of the spectrum, one or more of these are apt to be dominant over the others. In the following sections, the principal sources of noise, as presently understood, will be briefly described. We will begin with those sources likely to be important at the low end of the spectrum and end with the source of noise that completely dominates all the others at high kilohertz frequencies. Some of the noise sources that we will consider are illustrated diagrammatically in Fig. 7.1.

Tides and the hydrostatic effects of waves Most underwater sound hydrophones are pressure-sensitive and will respond to changes in ambient pressure, whether of acoustic origin or not. Tides and waves cause hydrostatic pressure changes of relatively large amplitude at the low-frequency end of the

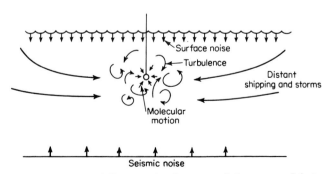

fig. 7.1 *Conceptual diagram showing some of the sources of deep-water ambient noise.*

spectrum. The tides, for example, may be imagined to create line components, or tones, in the pressure spectrum at frequencies corresponding to 1 or 2 cycles/day. The magnitude of these tidally produced pressure changes is, of course, enormous; the pressure equivalent of a 1-ft head of water, for example, amounts to 3×10^4 dyn/cm^2. Fortunately such pressure changes lie far below the frequencies of interest in underwater sound and are restricted to narrow regions of the spectrum.

An interesting effect of the tides results from the fact that tidal currents produce changes in water temperature. Since piezoelectric transducers are also pyroelectric, they will produce a changing voltage when the temperature changes. The pyroelectric effect is surprisingly large, and has been estimated (1) to amount to 12 volts per degree Celsius! Thus, thermal shields are always required around hydrophones at infrasonic frequencies; at the extremely low tidal frequencies, no amount of thermal shielding will be effective and thermal and pressure effects will be inseparable. In addition to this temperature effect, tidal currents may cause flow-induced vibrations of the transducer and its support (1). For these reasons and others, the response of a transducer to the changing tides is complex and difficult to predict.

Surface waves are also a source of hydrostatic pressure changes at a depth in the sea. However, they have a pressure amplitude which falls off rapidly with increasing depth and with decreasing wavelength of the surface waves. Figure 7.2 shows how the hydrostatic pressure produced by waves of a given wave period becomes attenuated with depth below the surface. This attenuation is extremely rapid with increasing depth. In shallow water, however, the depth may not be great enough to eliminate completely the pressure effects of waves passing over a bottomed hydrophone; in such instances, the rough surface can become the dominant source of the low-frequency background of a pressure-sensitive hydrophone.

Seismic disturbances   Because the earth is in a constant state of seismic activity, earth unrest is probably an important cause of low-frequency noise in the sea. One very strong and almost continuous form of seismicity is *microseisms* having a nearly regular periodicity of about ⅐ Hz and a vertical amplitude on land of the order of 10^{-4} cm. If such amplitudes exist in the deep seabed, and if the disturbance is assumed to be sinusoidal, the pressure that would be produced in the sea above can be found from the expression

$$p = 2\pi f \rho c a$$

where f = frequency
ρ = density of water
c = sound velocity of water
a = displacement amplitude

With f = ⅐ Hz and $a = 10^{-4}$ cm, the resulting pressure amplitude is 10 dyn/cm^2, or 120 dB above 1 μPa. This is roughly of the same level as that

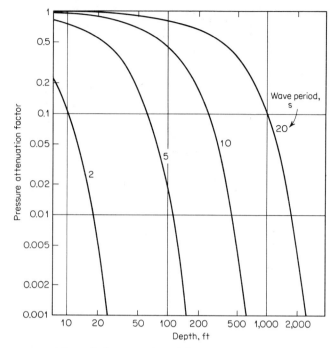

fig. 7.2 *Theoretical attenuation of the pressure changes produced by surface waves of a given period at a depth below the surface.*

measured as the ambient noise pressures at frequencies below 1 Hz. Above 1 Hz, measurements of seismic unrest with underwater seismometers suggest that the importance of this source extends well into the decade 10 to 100 Hz, at least in areas of low ship traffic (2). It may therefore be surmised that microseismic disturbances, and perhaps earth seismicity in general, are likely sources of sea noise at very low frequencies. In addition, intermittent seismic sources such as individual earthquakes and distant volcanic eruptions (3) are undoubtedly transient contributors to the low-frequency background of the deep sea.

Oceanic turbulence The role of turbulence in the sea as a source of ambient noise has been considered by Wenz (4). Conceivably, turbulence, in the form of irregular random water currents of large or small scale, is capable of creating a noise background in several ways. First, such currents may shake or rattle the hydrophone and its mounting and so produce a form of self-noise, rather than a part of the ambient noise of the sea. Second, the pressure changes associated with the turbulence may be radiated to a distance, and so appear as part of the background at places distant from the turbulence itself. However, the radiated noise of ocean turbulence is not likely to be of significance because of its quadrupole origin and its consequently rapid falloff with distance. The third, and most important, acoustic effect of turbulence is the turbulent pressure changes created inside the turbulent region. The turbulence gives rise to varying dynamic pressures that are picked up by a pressure-sensitive hydrophone located in the turbulent region. The magnitude of such pressures can be estimated, as was done by Wenz, from a guess as to the magnitude of the turbulent currents in the deep sea. If the turbulent component of the flow is written as u, then the associated dynamic pressure is ρu^2, where ρ is the fluid density. If u is taken to be 5 percent of the steady (dc) component of the current, then in a 1-knot steady current, the turbulent component would be 0.05 knot or 2.5 cm/s, and the dynamic pressure of the turbulent motion would then be, with $\rho = 1$, equal to 6.3 dyn/cm², or 116 dB above 1 μPa. Figure 7.3 shows estimates of the spectrum of turbulent pressures for three values of steady current speed $\bar{u}$, as derived by Wenz from theoretical and experimental relations of turbulent flows. The magnitudes and the slope of these estimated spectra of ambient oceanic turbulence for $\bar{u} = 2$ cm/s are in remarkable agreement with observed noise spectra in the decade 1 to 10 Hz. As a consequence, it may be inferred that, although no direct observational evidence has come to light, the turbulence of deep-ocean currents is a likely cause of low-frequency noise.

Nonlinear wave interactions An interesting nonlinear effect occurs when two progressive surface waves travel in opposite directions and interfere with one another so as to form a "standing" wave. Although a single progressive wave produces a pressure in the sea that dies out rapidly with depth (Fig. 7.2), it can be shown, by both a physical argument (5) and by theory that takes account of nonlinear terms in the hydrodynamic equations, that this is not true for two waves progressing in opposite directions and thereby forming a standing wave. Such a standing wave, formed by interacting progressive

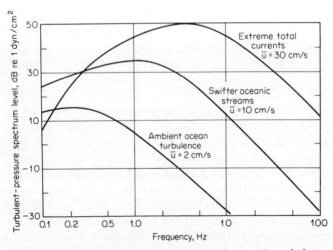

fig. 7.3 *Spectra of the pressures produced by oceanic turbulence, derived theoretically by Wenz. (Ref. 4.)*

waves, produces a pressure that is the same at all depths and has a frequency twice that of the generating surface waves. This process has been examined several times (6–9) as a possible source of the noise background of the sea. Marsh (6) finds that this theory yields noise levels agreeing remarkably well with observed data in both deep and shallow water. Because of this, and the directional effects which, according to Kuo (8), the theory successfully predicts, it would seem that this is a promising mechanism for the generation of noise by the surface of the sea at low frequencies.

Ship traffic Various pieces of evidence suggest that distant ship traffic is a dominant source of noise at frequencies around 100 Hz. First, ambient noise is known to arrive at a deep hydrophone in a preferentially horizontal direction at frequencies near 100 Hz and to be independent of wind and weather. Second, there is a "plateau," or a flattening, in observed ambient-noise spectra in this frequency region that coincides remarkably closely with the maximum in the radiated-noise spectra of ships. Finally, and most important, ambient-noise measurements in areas of high ship activity are higher and less wind-dependent at frequencies from 50 to 500 Hz than they are in areas where shipping is sparse.

The importance of ship traffic as a source of noise may be appreciated from the estimate that there are, on the average, 1,100 ships under way in the North Atlantic Ocean at any one moment of time. Quantitatively, Wenz (4) has made calculations using reasonable estimates of ship traffic and propagation conditions that yield noise spectra similar in shape and level to those measured in the sea.

All this evidence suggests that distant ship traffic is a principal source of noise in the decade 50 to 500 Hz. Such traffic may occur at distances of 1,000 miles or more from the measurement hydrophone. The competitor of shipping as a noise source is *distant storms,* in which a part of the kinetic energy of the wind is coupled to the sea as underwater sound.

Surface waves At still higher frequencies, ambient noise is governed by the roughness of the sea surface. During World War II many observations (10) of deep-water ambient noise between 500 Hz and 25 kHz indicated a direct connection between sea state or wind force and the level of ambient noise. Based on these observations, the well-known *Knudsen spectra* were obtained, having sea state or wind force as a parameter. Later data indicated a better correlation of noise with wind speed than with sea state, probably because of the difficulty of estimating the latter. The noise level has been shown to be correlated with the *local* wind speed over the measurement hydrophone.

Although it is clear that the sea surface must generate the major portion of the ambient noise in this frequency range, the process by which it does so is still uncertain.

Several processes come to mind, and a number of others have been advanced in the literature. Perhaps the most obvious of these are *breaking whitecaps,* which must produce crash noise when breaking occurs. Yet whitecaps cannot be the sole source of noise, since measured noise levels increase with sea state or wind speed well below the sea state in which whitecaps begin to appear.

Another possible source of noise is the *flow noise* produced by the wind in blowing over the rough sea surface. In so doing, it produces turbulent pressures that are transmitted to the water below and appear as noise at a sensor below the surface.

Another possibility is *cavitation,* or the collapse of air bubbles formed by turbulent wave action in the air-saturated near-surface waters. Such bubbles grow by rectified diffusion and, after having reached a critical size, collapse suddenly and generate a short pulse of sound. This process has been examined theoretically (11) and has been alleged to produce the observed broad peak in the spectrum of wind noise occurring in the frequency range 0.1 to 1 kHz (Fig. 7.5).

Still another possible noise producing process is the *wave-generating action* of the wind on the surface of the sea (12). Since the wind is always turbulent near the surface of the sea, it generates waves of different wavelengths that travel along the surface with a speed that depends on wavelength. Very long waves can travel with a speed greater than that of sound in the sea and will, in theory, radiate propagating pressure waves into the depths of the sea. This mechanism is held to account for the principal characteristics of wind noise, such as its level and its variation with wind speed and frequency. But whatever the physical process may be, it is clear that the rough sea surface, not too far from the location of the measurement hydrophone, is the dominant noise source at frequencies between 1 and 30 kHz.

Thermal noise In 1952, Mellen (13) showed theoretically that thermal noise of the molecules of the sea places a limit on hydrophone sensitivity at high frequencies. Reasoning that since the average energy per degree of freedom is kT (k = Boltzmann's constant, T = absolute temperature) and since the number of degrees of freedom in a large volume of sea is the same as the number of compressional modes possible in that volume, Mellen was able to compute the equivalent plane-wave pressure for thermal noise in water. For a nondirectional hydrophone which is perfectly efficient in converting acoustic to electric energy, the equivalent noise spectrum level is, for ordinary temperatures,

$$NL = -15 + 20 \log f \quad \text{dB re 1 } \mu\text{Pa}$$

where f is the frequency in kilohertz. For example, at 100 kHz the limiting spectrum level is $-15 + 40$, or $+25$ dB. For a hydrophone of directivity index DI and efficiency E, expressed in decibels, the equivalent thermal noise spectrum level becomes

$$NL = -15 + 20 \log f - DI - 10 \log E \quad \text{dB re 1 } \mu\text{Pa}$$

This noise is the same as the Nyquist noise developed in the radiation resistance of the hydrophone in water, and has been measured experimentally by Ezrow (14). The spectrum level of this kind of acoustic background in underwater hydrophones rises with frequency at the rate of 6 dB/octave ($+20 \log f$) and places a threshold on the minimum observable pressure levels in the sea.

7.2 Deep-Water Spectra

Figure 7.4 is an example of the spectrum of ambient noise as it might, conceivably, be observed at one deep-sea location. The spectrum is composed of segments of different slope having a different behavior under different conditions. As stated above, this complexity is interpreted as being due to a multiplicity of noise sources over the entire frequency range. A number of regions, or frequency bands, in the spectrum may be identified and associated with the principal sources described in the previous section. Five frequency bands are shown in Fig. 7.4. Band I, lying below 1 Hz, is a largely unknown portion of the spectrum at the present time, in which the noise is likely to be of hydrostatic origin (tides and waves), or originates in the earth as seismic unrest. Valid measurements in this Band and in Band II are extremely difficult to make because of the "self-noise" of the hydrophone and its supporting structure caused by currents, even near the seabed in deep water.

Band II is characterized by a spectral slope of -8 to -10 dB/octave (about

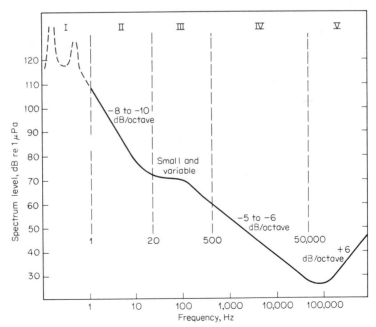

fig. 7.4 *Sample spectrum of deep-sea noise showing five frequency bands of differing spectral slopes. The slopes are given in decibels per octave of frequency.*

−30 dB/decade), with only a slight wind-speed dependence in deep water, and in which the most probable source of noise in deep water appears to be oceanic turbulence. In Band III, the ambient-noise spectrum flattens out into a "plateau," shown by nearly every measurement, and the noise appears to be dominated by distant ship traffic. Band IV contains the *Knudsen spectra* having a slope of −5 to −6 dB/octave (about −17 dB/decade) in which the noise originates at the sea surface at points not very far from the point of measurement. It should be noted that the process of generation of the noise in this spectral region must be different from that in Band II, where there is also a wind-speed dependence, since the spectra in the two frequency bands have different slopes with a flat, or reverse slope between, as is evident from Fig. 7.5. Band V is overwhelmed by thermal noise originating in the molecular motion of the sea and is uniquely characterized by a positive (rising) spectrum having a slope of +6 dB/octave.

For prediction purposes, average representative ambient-noise spectra for different conditions are required. Such average working curves are shown in Fig. 7.5 for different conditions of shipping and wind speed. In the infrasonic region below 20 Hz, only a single line is drawn, since the wind speed dependence is slight and uncertain. The ambient-noise spectrum at any location at any time is approximated by selecting the appropriate shipping and wind curves and fairing them together at intermediate frequencies where more than one source is important. The "heavy-shipping" curve is used for locations near the shipping lanes of the North Atlantic; the "light-shipping" curve is appropriate for locations remote from ship traffic.

Nearly all existing measurements of the ambient noise have been made in the North Atlantic Ocean and its seas, coasts, and harbors. Recently, however,

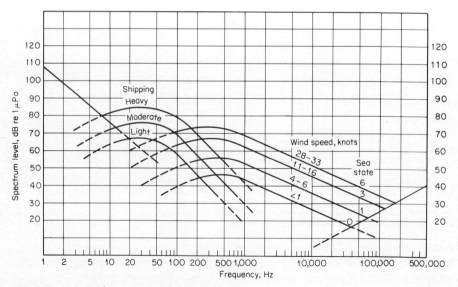

fig. 7.5 *Average deep-water ambient-noise spectra.*

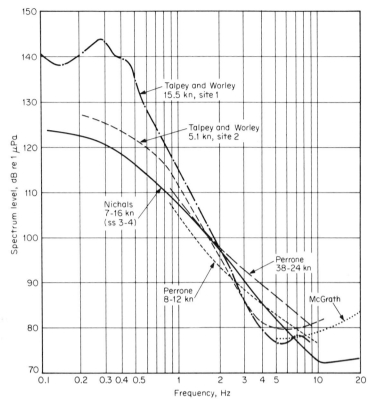

fig. 7.6 *A compilation of infrasonic ambient noise levels. Solid curve: data of Nichols (Ref. 16), depths 300 and 1,300 m averaged, off Eleuthera, octave band analyses. Dotted curves, data of Talpey and Worley, according to Nichols (Ref. 16), depth 4,300 m at site 1, 3,500 m at site 2, 25 miles apart south of Bermuda, 0.0043-Hz band analyses. Dashed curve: data of Yen and Perrone (Ref. 17), depth 900 m, south of Bermuda. Dots: data of McGrath (Ref. 18), depth 2,400 m, Mid-Atlantic Ridge, third-octave bands.*

the South Pacific and Indian Oceans have begun to receive their fair share of attention in three papers by Cato (15) that describe the noise levels and interesting biological sounds occurring in the waters near Australia.

In recent years, attention has been given to the levels of ambient noise in the so-called infrasonic or subsonic region of the spectrum lying below 20 Hz. Fig. 7.6 is a compilation of measurements in this region. The measurements are in fair agreement above 1 Hz, and show a marked change in slope in the vicinity of 10 Hz where shipping noise begins to dominate the spectrum.

7.3 Shallow-Water Ambient Noise

In contrast to the relatively well-defined levels of deep-water ambient noise, the ambient levels in coastal water and in bays and harbors are subject to wide variations. In such locations, the sources of shallow-water noise are highly variable, both from time to time and from place to place.

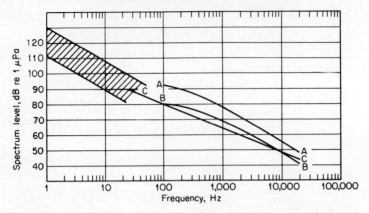

fig. 7.7 *A summary of noise levels in bays and harbors. World War II data. AA: A high noise location; entrance to New York Harbor in daytime. BB: An average noise location; upper Long Island Sound. CC: Average of many World War II measurements. Shaded area: Subsonic background measurements. [Largely from Knudsen, Alford, and Emling. (Ref. 10.)]*

At a given frequency in shallow water the noise background is a mixture of three different types of noise: (1) *shipping and industrial noise,* (2) *wind noise,* and (3) *biological noise.* At a particular time and place, the "mix" of these sources will determine the noise level, and, because the mix is variable with time, the existing noise levels will exhibit considerable variability from time to time and from place to place. As a consequence, only a rough indication can be given of the levels that might be found in bays and harbors and at offshore coastal locations.

Bays and harbors A great many measurements of noise inside different bays and harbors were made during World War II. These measurements, motivated by the needs of acoustic mines and harbor-protection sonars, were made in New York Harbor, lower Chesapeake Bay, San Diego Bay, and other bays and harbors. Some examples of the spectra resulting from these measurements are given in Fig. 7.7. Included in the figure is an average line *CC* showing the average of many determinations between 20 and 200 Hz and a shaded area giving the location of other measurements at subsonic frequencies. At the other end of the frequency scale, data at 30, 60, and 150 kHz for five harbor locations may be found in a paper by Anderson and Gruber (19).

In addition to the sources contributing to deep-water noise, other noise sources are important in bays and harbors. At such locations the noises of industrial activity of human origin, and the noises produced by marine life and by the turbulence of tidal currents, all conspire to create a noisy ambient environment.

Coastal waters In coastal waters, such as on the continental shelves, wind speed again appears to determine the noise level over a wide frequency range.

This has been demonstrated by measurements of Piggott (20) over a 1-year period on the Scotian shelf in water about 150 ft in depth. This work showed a dependence of noise level on wind speed at all frequencies between 10 and 3,000 Hz. The increase of level with wind speed was found to be 7.2 dB per wind speed doubled, or an increase of intensity slightly greater than the square of the wind speed. The various processes by which the wind conceivably generates noise—hydrostatic effects of wind-generated waves, whitecaps, and direct sound radiation from the rough sea surface—all must play a part in determining the level of the noise in offshore locations.

Figure 7.8 shows the spectra of Piggott for several wind speeds, together with two older average spectra measured in open water 700 ft deep 5 miles off Fort Lauderdale, Florida. It will be seen by comparison with the Knudsen spectra in deep water (Fig. 7.5) that the noise levels in coastal water are 5 to 10 dB higher than in deep water far from shore at frequencies greater than about 500 Hz. Shallow-water noise spectra at a depth of 200 m published by the Russians for an unstated location are given in Fig. 7.9. Evidently the noise is dependent on wind speed over the entire frequency range. The levels are generally higher than those of Fig. 7.8 at the same wind speed.

On the other hand, when shipping and biological noise sources are absent, and when wind noise is sole contributor to the background, the shallow-water noise levels reported in the literature are surprisingly concordant, not only among themselves, but with the ambient noise levels in deep water. Figure 7.10 is a compilation of measurements made at 1 kHz by a variety of investigators at different shallow-water locations. The noise levels refer to "quiet" conditions, that is, when neither ship noise nor biological noise could be identified by listening to the noise. The measured levels apparently depend only on wind speed, and are independent of hydrophone depth, water depth, and other site characteristics. The level increases at the rate of 6 dB per wind

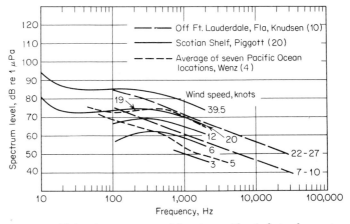

fig. 7.8 *Noise spectra at coastal locations with wind speed as a parameter.*

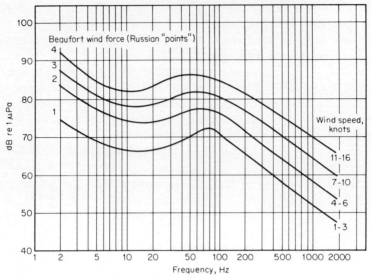

fig. 7.9 *Soviet spectra of ambient noise measured with a bottom-mounted hydrophone at a depth of 200 m. Analysis bandwidth 7%. Location not stated. The wind speed is stated in "points" which, from internal evidence, are believed to be Beaufort Scale numbers. (Ref. 21).*

speed doubled, or the intensity increases as the square of the wind speed, above a wind speed of about 5 knots. The "deep-water average" in Fig. 7.10 is taken from Fig. 7.5 at a frequency of 1 kHz.

At low frequencies and at low wind speeds, shallow water can be appreciably quieter than deep water. This is due to the shielding effect of poor sound propagation conditions in screening out noise originating at great distances. On the other hand, when shipping or other noises of human origin exist, or

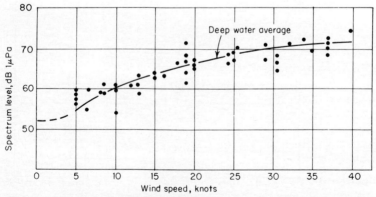

fig. 7.10 *Results of a compilation from the literature of ambient-noise measurements at 1 kHz in shallow water. The curve shows level versus wind speed at 1 kHz in deep water, plotted from Fig. 7.5.*

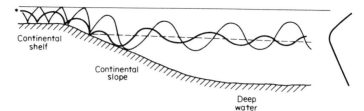

fig. 7.11 *The noise of coastal shipping can propagate to long ranges in deep water in the deep sound channel. The velocity profile is shown at right. Downward-refracted rays become upward-refracted once they cross the channel axis.*

when biological organisms contribute to the noise background, shallow water is a noisy and exceedingly variable environment for most sonar operations.

The noise of the numerous ships operating in the shallow waters of the continental shelf is in all likelihood a strong contributor to the noise in deep water far from shore. Thus, much of the noise at a deep hydrophone at Bermuda is believed to originate in the ship traffic in the coastal waters off the East Coast of United States. As seen in Fig. 7.11, this is a result of the negative velocity profile on the continental slope, causing a progression of sound down the slope until the axis of the deep sound channel is reached. After that, a reversal of refraction occurs, and from there on, the excellent propagation in the deep sound channel transmits the noise of coastal shipping to long distances (22).

7.4 Variability of Ambient Noise

Like most underwater sound parameters, ambient noise is eminently characterized by variability. Much of this variability arises from changes in the dominant sources of noise, such as changing wind speed and amount of shipping. To the extent possible with present knowledge, this variability has been accounted for in spectra such as those of Figs. 7.5 and 7.8. But a residual variability remains that creates an uncertainty of perhaps 5 to 10 dB between an estimate based on average values and an amount that might be measured at a given location over a short period of time. Transient sound sources of short or extended duration, such as those of biological origin, are prime offenders. Another source of variability is changing sound-transmission conditions, which affect the transmission of sound from distant sources. For example, Wilkinshaw (23) found from a 4-year period of observation with a bottomed hydrophone at Bermuda and at the Bahama Islands that the average ambient level was 7 dB higher in winter than in summer; this difference was attributed to better sound transmission during the winter season.

Other examples of the variability of ambient noise are the peculiar periodic variations observed by Wenz (24) in data collected at six locations spaced over 45° of longitude in the Pacific Ocean. The analysis of data in the frequency band 20 to 100 Hz revealed periodicities of 12 and 24 hours, with a maximum of noise at midnight and noon, local time, at all locations, independent of longitude. Although the change in level was small (normally 1.5 to 5 dB), it was undoubtedly real, but the origin of these changes in noise, which are apparently free from biological activity and synchronized to the mean solar day at diverse locations, remains a mystery.

Both tidal and human activity are sources of the variability of the ambient noise in bays and harbors. In Narragansett Bay, for example, the turbulence produced by tidal currents was believed (25) to be the source of observed changes in 25-Hz noise. Industrial activity has obvious periodicities that are likely to be reflected in the observed ambient levels in and near busy harbors. At a location 15 miles off the mouth of San Diego Bay, for example, some 12 dB more noise was observed (26) by day than by night, and 10 dB more during the day on weekdays than on weekends—changes directly associated with the tempo of the shipping and industrial life of the harbor.

In short, the variability of ambient noise levels is a consequence of the variability of the sources of noise. In a dynamic, active area where shipping creates an important and changeable form of environmental noise pollution, the standard deviation of a series of noise levels as measured over a period of time is relatively large. For example, at a location off St. Croix, Virgin Islands, a series of hourly levels in octave bands below 100 Hz had a standard deviation amounting to 5 to 7 dB (27). On the other hand, at higher frequencies and in the absence of biological noises, the variability of the wind speed during the observation period is an important factor. For example, at the site just mentioned, the hourly levels had a standard deviation of only 2 to 3 dB in the range 1 to 5 kHz, merely because the wind speed did not vary greatly during the time that the data were obtained (28).

In the South Fiji Basin northeast of New Zealand 1-minute samples of ⅓ octave band noise showed standard deviations less than 2.5 dB at all frequencies when no shipping was present (29). But when ship noise occurred, presumably caused by a single ship in this isolated area, the standard deviation was higher, being 5.5 to 7 dB at 10 Hz although it was only 2 dB above 200 Hz. At higher frequencies, where the noise is caused by the wind, the standard deviations of the data samples were very low (less than 0.1 dB).

7.5 Intermittent Sources of Ambient Noise

By intermittent noise sources are meant those that do not persist over periods of hours or days, but are of transient occurrence. Although some forms of biological noise cannot be called truly intermittent at some locations, but are, rather, steady and characteristic parts of the ambient background, such noises are included in this category of intermittent noises for convenience.

Biological sounds The sounds produced by biological organisms in the sea are many and varied, and have been extensively studied. Only a brief description can be given here. Additional information can be found in a book (30) devoted to the subject. Although the different sounds are extremely diverse, ranging from the calls of porpoises to the frying noise of a mass of soniferous shrimp, only three groups of marine animals are known to make sound: certain kinds of shellfish (Crustacea), certain kinds of true fish, and the marine mammals (Cetacea), such as whales, dolphins, and porpoises. Among the Crustacea, the most important are snapping shrimp, which are ubiquitous inhabitants of shallow tropical and semitropical waters having a bottom of rock, shell, or weed that offers the animals some concealment. These animals make noise by snapping their claws together, as one snaps thumb and forefinger, and thereby produce a broad spectrum of noise between 500 Hz and 20 kHz. Among the fish should be mentioned the croakers of Chesapeake Bay and other East Coast locations, which make intermittent series of tapping noises like that of a woodpecker by means of the contraction of drumming muscles attached to the air bladder; the same principle is involved in beating a drum. The Cetacea include whales and porpoises, which create noise by blowing air through the larynx. Porpoises, for example, produce frequency-modulated whistles that are associated with certain behavior patterns of these animals. Many peculiar chirps, grunts, boops, groans, yelps, and barks heard in the sea are of biological origin.

Although the sounds of marine organisms have been much studied by marine biologists, very little quantitative data are available to the engineer on the spectra and levels to expect in the natural environments. The bulk of available data dates from World War II, when measurements were made on a variety of biological sounds at different locations. Some of the spectra obtained are given in Figs. 7.12 and 7.13.

Although many species of fish make noise, unfortunately no species of any commercial importance appears to do so. As a result, passive sonars are useless for fish finding. For other sonar applications, the noise of fish and other forms of life in the sea is an annoying, though readily identified, hindrance to

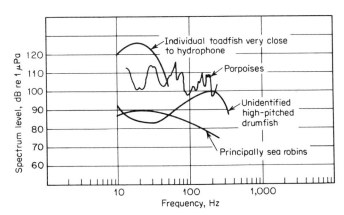

fig. 7.12 Sample spectra of the noise made by marine animals as observed at sea. (Ref. 10.)

218 / *principles of underwater sound*

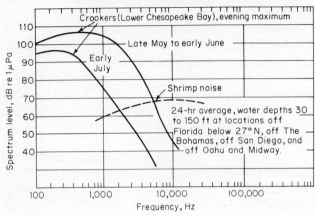

fig. 7.13 *Observed spectra of the noise of croakers and snapping shrimp measured in Chesapeake Bay and at various shallow-water locations in low latitudes. (Ref. 10.)*

sonar operations and forms an erratic and, in large part, unpredictable part of the ambient background in which the sonars must operate.

Dolphins make a wide variety of sounds for a wide variety of reasons of their own. Since dolphins have such a close affinity to man, they have been much studied, acoustically and otherwise, and a vast literature on their behavior exists (31). Also some quantitative information on the signals they make for echo location is available (32).

Twenty-cycle pulses Some peculiar transient sounds have been heard with low-frequency bottomed hydrophones over widespread areas of the Atlantic and Pacific Oceans. These have been called 20-*cycle pulses* (33, 34), and consist of trains of nearly sinusoidal 20-Hz pulses of about 1 second in duration, lasting for several minutes. The repetition rate is remarkably regular, with one pulse approximately every 10 seconds. The pulse trains, lasting anywhere from 6 to 25 min, are separated by silent periods 2 to 3 min long, and repeat for many hours. The apparent sources of these sounds have been tracked by triangulation using two or more hydrophones and have been found to move in a random manner with speeds typically 2 or 3 knots. The total radiated acoustic power of a single pulse has been estimated to lie between 1 and 25 watts. Because of this relatively high power, the pulses were found to be detectable to ranges of 35 miles in the shallow water of the continental shelf. The cause of these noises is believed by marine biologists to be a species of whale, although the mechanism of production of sound of such high acoustic intensity and low frequency is uncertain. Heartbeats and some sort of respiratory mechanism have been suggested (35).

Figure 7.14 is a sound spectrogram showing graphically some of the characteristics of the 20-cycle pulses.

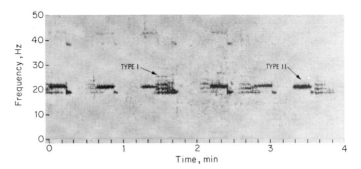

fig. 7.14 *Sound spectrogram of a number of 20-Hz pulses as recorded at a location in the Pacific Ocean. Two types of pulses, called type I and type II, were identified. (Ref. 36.)*

Rain Falling rain may be expected to increase ambient noise levels to an extent depending on the rate of rainfall and perhaps on the area over which the rain is falling. For example, an increase of almost 30 dB in the 5- to 10-kHz portion of the spectrum has been noted in a "heavy" rain (37); and at 19.5 kHz in a steady, though not torrential, rain in sea state 2, the levels increased to those corresponding to sea state 6—an increase of 10 dB (38). Spectra of the ambient background during rain have been published by Heindsman, Smith, and Arneson (39) from measurements made in 120 ft of water near the eastern end of Long Island Sound. Figure 7.15 shows three spectra of ambient noise in rain of different intensity taken from this paper, together with the no-rain spectrum estimated from Fig. 7.8 for the existing wind speed of 20 to 40 knots. It will be observed that the spectrum of the noise of heavy rain is nearly "white" between 1 and 10 kHz, with a noise increase of 18 dB in a "heavy" rain at 10 kHz over the no-rain spectrum level. The work of Franz (40), in which the noise produced by individual water droplets falling

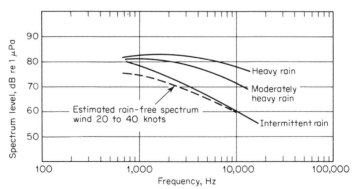

fig. 7.15 *Ambient-noise spectra in rain, observed in Long Island Sound. (Ref. 39.) The dashed curve is a spectrum in the absence of rain for the wind speed applying to the measured data.*

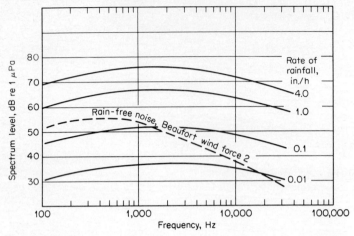

fig. 7.16 *Theoretical spectra of rain noise for different rates of rainfall. Dashed curve is for deep-sea noise at wind force 2 without rain. (Ref. 40.)*

in air upon a water surface was studied experimentally and theoretically, has led to an estimate of spectra of rain noise that should be expected at sea in terms of rate of rainfall (Fig. 7.16).

The agreement in shape and general magnitude between these basically derived spectra and those observed by Heindsman, Smith, and Arneson in the field is noteworthy. In connection with noise production by falling droplets, it may be noted in passing that the efficiency of a falling water droplet as a source of sound was found by Franz to be $2M^3$, where the efficiency is defined as the ratio of total radiated underwater acoustic energy to the kinetic energy of the droplet, and where M (a kind of Mach number) is the ratio of the velocity of the falling droplet in air to the velocity of sound in water.

Bom (41) subsequently has made measurements of the noise of falling rain at the center of a lake 250 meters in diameter with a maximum depth of 10 meters. He found the rain-noise spectrum to be mostly flat, in agreement with the theory of Franz, but to be 5 to 10 dB higher, perhaps because of the higher level to be expected in a shallow enclosed basin.

Explosions The increasing use of explosives for offshore seismic exploration and for research work at sea has caused distant explosive sounds to become common occurrences in the recorded noise background in deep water. Some 19,801 shots were identified in noise recordings taken during a one-year period with deep hydrophones off California (42). This is another example of how human activity affects the noise environment of the deep sea.

Seaquakes and volcanoes Earthquakes at sea ("seaquakes") produce intermittent sounds lasting for periods of seconds and occurring at infrasonic frequencies. Small seaquakes are of frequent and almost continuous occurrence in seismically active regions such as the Mid-Atlantic Ridge and the East Pacific Rise (43). Other natural sources of low-frequency noise are active underwater volcanoes. One of them, near the Mariana Islands in the Pacific, was studied acoustically using deep hydrophones many miles away (44).

7.6 Effect of Depth

The simplest possible noise model of the ocean is an infinite layer of uniform water with a plane surface along which the sources of noise are uniformly distributed. With this elemental model, one could predict the ambient-noise level to be independent of depth, except at frequencies above about 10 kHz, where absorption would cause the level to decrease with increasing depth (45). A more realistic model would allow for refraction and for multipath propagation over a lossy bottom. To compute the noise level from distant shipping three inputs are required: (1) the density of shipping, or number of ships per unit area of ocean out to several hundred miles from the receiver, (2) the source level for the radiated noise of each ship, and (3) the transmission loss as a function of range between a near-surface source and the depth of the receiver. By means of a suitable computer program, contributions from successive range rings about the receiver can be added up to obtain the level of shipping noise at the receiving location. The theory behind this kind of modeling, plus some observational data on the density of ship traffic in the North Atlantic, appears in a paper by Dyer (46).

Direct measurements of the depth dependence of noise can be made only with special instrumentation and through the use of a hydrophone whose calibration is known as a function of pressure and temperature. Both a slowly sinking hydrophone connected to a ship by a very light, flexible cable (47) and the manned bathyscaphe "Trieste" (48) have been used for this purpose.

In the latter work done in the Mediterranean at frequencies from 10 to 240 Hz, although no decrease of noise was found with depth in sea state zero, a decrease of about 15 dB was found at the highest frequency in sea state 2 between a shallow depth and the bottom at 3,000 meters. This depth effect was found to become less at lower frequencies. Although the magnitude of the effect appears extremely high, these findings are consistent with the view that high-frequency high-sea-state noise is of surface origin. In the work with a slowly sinking hydrophone, no depth dependence was found in the range 20 to 295 Hz down to 4,000 feet.

At one location north of St. Croix, Virgin Islands, a vertical string of six hydrophones placed 2,000 feet apart was used in an attempt to determine the ambient noise profile (27). The hydrophones were attached to a cable, like beads on a string, and were calibrated *in situ* by means of sound pulses sent down through the water to each hydrophone. A total of 114 noise samples taken hourly over two 2½-day periods was recorded. The noise profiles obtained are given in Fig. 7.17. They show that the noise level decreases with depth at all frequencies, with the greatest decrease in the first 2,000 feet. The gradient of noise level was found to be highest just below the surface; a hydrophone at 300 feet was 4 dB quieter than one at 50 feet within the 150-foot mixed layer. Subsequent work has been done by Morris (49) at two locations in the northeastern Pacific Ocean. In Fig. 7.17 the circles show average observed levels at three frequencies over a period of 51 hours at a site 400 miles west of San Diego. The agreement with the St. Croix data is surprisingly good considering the great differences in the two sites. Other data at the St. Croix location show little or no depth dependence at frequencies between 1,000 and 5,000 Hz, where the local wind is the source of noise, apart from the higher noise levels in the surface duct, as will be evident in Fig. 7.18.

Very close to the ocean surface, however, profound effects occur. If the hydrophone is less than one-quarter wavelength (acoustically) beneath the surface, pressure release takes place, and a pressure-sensitive hydrophone will observe a lower ambient level. On the other hand, the hydrostatic effect of the ocean waves becomes important at depths just below the surface. Another effect of the close proximity of the sea surface is the occurrence of a more "spiky" character of the noise, as will be mentioned in the following section.

Considerable data on the level of ambient noise in deep water have been obtained in recent years with deep-bottomed hydrophones (4, 67). These measurements indicate spectrum levels at frequencies above 500 Hz some 5 to 10 dB lower than those predicted by the "classic" Knudsen curves derived

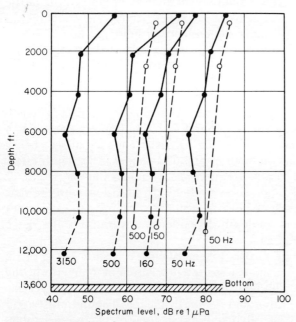

fig. 7.17 *Variation of ambient-noise level with depth. Dots: data obtained north of St. Croix, Virgin Islands, water depth 13,600 ft. (Ref. 27.) Circles: data from a site west of San Diego, California, water depth 14,700 ft. (Ref. 49.)*

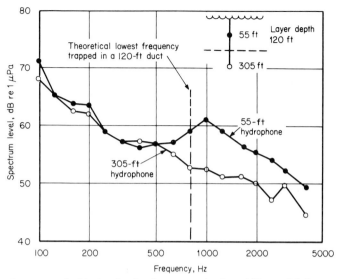

fig. 7.18 Ambient-noise levels at two depths within and below a mixed layer or surface duct. (Ref. 50.)

from measurements made at shallower depths. Much of this quieting at the bottom is the result of the bottom topography whose peaks screen out distant noise sources.

It might be surmised that, if the surface duct indeed is a trap for signals transmitted within it (Sec. 6.1), it would also be a trap for ambient noise (50). Because of this trapping, the surface duct should be more noisy than the waters below it. That this is indeed the case is illustrated by the observations plotted in Fig. 7.18, where evidently at frequencies above the cutoff frequency for duct transmission, the in-duct hydrophone is 5 to 10 dB more noisy than the below-duct hydrophone. This is of some operational importance, for it is often tacitly and erroneously assumed that there is no difference in the ambient background level in and below a surface duct. Time-delay correlograms (Fig. 7.19) show two sources of noise in the duct, one noise originating overhead and giving a correlogram that moves out in delay with increasing separation, and the other traveling within the duct and producing a correlogram remaining centered at zero delay.

7.7 Amplitude Distribution

Ambient noise has been found by probability density analyses (51) of data in one deep- and two shallow-water areas to have a gaussian amplitude distribution at moderate depths. This is consistent with the view that the noise originates through a great many sources of random amplitude and phase. Near the sea surface, however, as with a hydrophone at a depth of a few feet below the surface, ambient noise has been observed to be more spiky than gaussian,

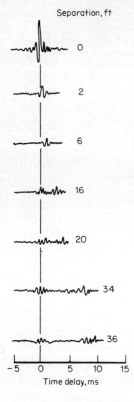

fig. 7.19 Ambient-noise correlograms for two hydrophones different distances apart vertically in a surface duct. Frequency band, 1–2 kHz. The autocorrelogram at zero separation (i.e., one hydrophone) serves to calibrate the vertical scale and to establish the zero point of the time delay scale. (Ref. 50.)

and to contain individual spikes or crashes of sound originating at close-by sources such as breaking wavelets. Similarly, under an ice cover, the noise is at times spiky because of the creation of tensile cracks in a continuous cover and because of the rubbing of ice masses together in a noncontinuous ice sheet.

Although ambient noise is gaussian over short time periods, it is clearly nonstationary over longer time periods because of the variability of the sources of noise, as we have already noted (Sec. 7.4).

7.8 Noise in Ice-covered Waters

Under an ice cover, ambient noise is notably different in character and level from the noise under ice-free conditions. When the ice is not continuous, as in a brashy ice pack, noise levels 5 to 10 dB higher than those measured at the same sea state in ice-free waters have been observed. On the other hand, with a continuous shore-fast ice cover under rising temperatures, very low noise levels occur. Such levels demand extreme care in electronic design to achieve the necessary low self-noise levels for their measurement.

In ice-covered waters, the level and character of the noise are highly variable and depend upon ice conditions, wind speed, snow cover, and air-temperature changes. With a solid shore-fast ice cover, Milne and Ganton (52) found that the noise is spiky or impulsive when the air temperature decreases and tensile cracks in the ice are formed. Under rising temperatures, the spiky character of the noise disappears and a gaussian amplitude distribution reappears. Another source of noise under an ice sheet is the wind, which appears to make noise through its own turbulence and the motion of drifting granular snow impinging on the rough sea ice. Wind noise is more prominent under a noncontinuous ice cover than it is under a continuous ice cover, when, according to the observations of Macpherson (53), the noise levels become independent of wind speed. A final source of noise that should be mentioned is the bumping and scraping together of ice flows in a noncontinuous ice cover.

At a frequency equal to the reciprocal of the round-trip travel time between surface and bottom and at its harmonics, line components have been identified (54) in the ambient-noise spectrum in shallow water under an ice cover. These tonal components in the noise were attributed to the existence of low-order propagation modes in the shallow-water duct. Similar line components in ice-free ambient noise in shallow water appear not to have been reported.

It is not possible at the present time, as Milne and Ganton state, to predict the spectra and amplitude distribution of the noise to be expected under an ice cover. Very low levels, well below the lowest Knudsen curve for sea state zero, have been found under a continuous ice sheet with calm winds and rising temperatures; on the other hand, levels some 40 dB higher have been observed when the ice had been actively cracking under falling air temperatures.

A sampling of various published noise spectra under different conditions in the Arctic is given in Fig. 7.20, along with curves of levels found to be ex-

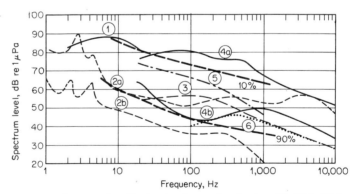

fig. 7.20 *Spectra of ambient noise observed under ice: (1) 80 to 90 in. ice, no cracking noises, Barrow Strait, April 1959. (Ref. 56.) (2) Old polar ice with frozen leads and pressure ridges, April 1961. (2a) Noisy periods, (2b) quiet periods. (Ref. 52.) (3) 70 percent 1-year ice 3 to 5 ft thick, 30 percent polar floes, September 1961. (Ref. 52.) (4) Shore-fast ice like that of (2), February 1963. Cracking noises prevalent (4a); absent (4b). (Ref. 52) (5) Average over a 2-week period in Beaufort Basin. (Ref. 57.) (6) Ice-free deep water, wind force 0 (Fig. 7.5). Curves labeled 10 and 90 percent show the estimated percentage of the time that a given level will be exceeded, based on all available Arctic data.*

226 / *principles of underwater sound*

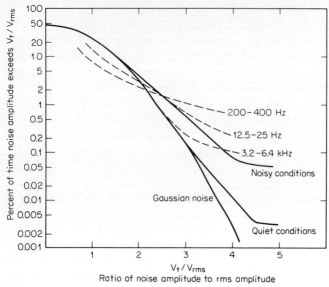

fig. 7.21 *Amplitude-distribution curves of ambient noise under an ice cover. Dashed curves, highly impulsive noise under decreasing temperature. (Ref. 52.) Solid curves, broadband noise under "quiet" and "noisy" conditions. (Ref. 57.)*

ceeded 10 and 90 percent of the time by these and other measurements. A summary of information on under-ice noise based almost entirely on U.S. references has appeared in the Russian literature (55). The non-gaussian spiky amplitude distribution of ice-cover noise, as measured in the Arctic by Milne (52, 56) and by Greene and Buck (57), is shown in Fig. 7.21. In the Antarctic the noise background has been observed by Kibblewhite and Jones (58) at a location in McMurdo Sound. Strangely enough, the noise of biological activity—by seals and humpback whales—was a prominent and almost continuous characteristic of the noise environment. When such activity was absent, under the quietest conditions, the noise levels below 1 kHz were 15 to 20 dB below those of sea state zero in the open sea, due to the absence of the turbulent interaction of wind and sea that occurs in open water.

An interesting kind of natural sea noise is made by icebergs as they melt in the sea. This form of noise appears to have a curious origin. Inside the ice forming an iceberg there exist numerous tiny bubbles of air under pressure that had been entrapped in the ice as it was formed. When the iceberg melts in the sea, the pressurized gas in a bubble is suddenly released as the melting ice wall reaches it, and the air rushes out with a sharp crack of sound. This kind of noise has been investigated using sonobuoys dropped near a number of bergs (59), and was found to have a flat (i.e., white) spectrum of such intensity that icebergs may be surmised to be the dominant source of noise in areas of the ocean where they are prevalent and are actively melting.

The edge of an ice-sheet, where waves slap and break against the edge of the ice, is a strong source of noise. Near a compact, sharp ice edge the level of the noise has been found to be 12 dB higher in the decade 100 to 1,000 Hz than in open water, and to be 20 dB higher than well inside the ice field (60). The ice boundary apparently acts as a quasi-infinite line source radiating cylindrically out to great distances in the open sea.

7.9 Directional Characteristics of Deep-Water Ambient Noise

We can approach the subject of the vertical directionality of deep-water ambient noise by means of a simple model. Consider a bottomless uniform ocean, without refraction or attenuation, having a surface covered with a dense, uniform distribution of noise sources. Let these noise sources be nondirectional (that is, let them be *monopoles*), and let each unit area of the surface radiate an intensity I_0 at one yard. Then at a point P in Fig. 7.22a, the incremental intensity dI produced by a small circular annulus of area dA at horizontal range r is

$$dI = \frac{I_0 dA}{l^2} = \frac{I_0 \cdot 2\pi r \cdot dr}{l^2}$$

But $r = h \tan \theta$, so that $dr = h \sec^2 \theta \, d\theta$; also, $l = h \sec \theta$. On substituting and canceling out, we obtain

$$dI = 2\pi I_0 \tan \theta \, d\theta$$

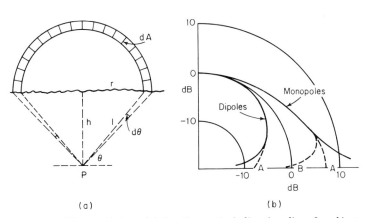

fig. 7.22 The simplest model for the vertical directionality of ambient noise: (a) Geometry with straight-line paths. (b) Directional patterns for surface distributions of monopoles and dipoles. The dashed segments near the horizontal show the effect of attenuation and refraction near ocean bottom (A) and near channel axis (B).

But if ψ is a solid angle, $d\psi = 2\pi \sin\theta\, d\theta$, so that the noise intensity per unit solid angle at angle θ becomes

$$N(\theta) = \frac{dI}{d\psi} = I_0 \sec\theta$$

This is the ambient-noise intensity per unit solid angle for a surface distribution of *monopole* soures. For a distribution of *dipole* sources, for which the intensity radiated at an angle θ is $I_0 \cos^2\theta$, the "beam pattern" of the noise received at a depth below the surface can be shown in the same way to be

$$N(\theta) = I_0 \cos\theta$$

These two functions $N(\theta)$ are plotted on polar coordinates in Fig. 7.22b. At angles near the horizontal, where $\theta = 0$, the effects of attenuation, refraction, and bottom-surface multipaths prevent the curves from going to either zero or infinity. In particular, a receiver located well within the deep-ocean sound channel cannot receive noise arriving along paths near the horizontal since such paths do not exist. For such a receiver, the beam pattern of the noise from monopole sources would therefore have a maximum at an oblique angle above or below the horizontal, with a minimum of $\theta = 0$. A strong noise minimum in near-horizontal directions would not be expected to occur with a receiver located on the bottom near the lower limit, or outside, of the deep-ocean sound channel. More sophisticated models than this simple one include the effects of refraction, bottom reflection, and multipath arrivals, and require a computer program for their evaluation.

The advent of modern processing techniques and the availability of deep bottomed hydrophone arrays have revealed a reasonably consistent picture of the directional nature of the ambient noise at a bottomed hydrophone in deep water.

A relatively large number of independent field determinations have been made. Anderson (61) and Becken (62) used a volume array with simultaneous preformed beams hung at depth from a surface ship; Forster (63) and Watson (64) used a bottom-mounted "fixed acoustic buoy" consisting of a vertical buoyed array in which narrow beams were formed electronically. VonWinkle (65), Fox (66), and Axelrod, Schoomer, and VonWinkle (67) all employed a 300-ft-long 40-element vertical array, located south of Bermuda, in which successive vertical beams were formed and steered electronically on shore; Cron, Hassel, and Keltonic (68) and Linnette and Thompson (69) correlated pairs of vertical hydrophones together and interpreted the correlograms in terms of a directional radiation function of the noise.

Figure 7.23a illustrates the findings of Axelrod, Schoomer, and VonWinkle (67) at frequencies of 112 and 1,414 Hz. These are polar plots of the ambient intensity per unit solid angle $N(\theta)$ arriving at a bottomed hydrophone as a function of vertical angle θ. At 112 Hz, more noise appears to arrive at the hydrophone from the horizontal than from the vertical, with a difference that *diminishes* with increasing wind force. At 1,414 Hz, the reverse is the case,

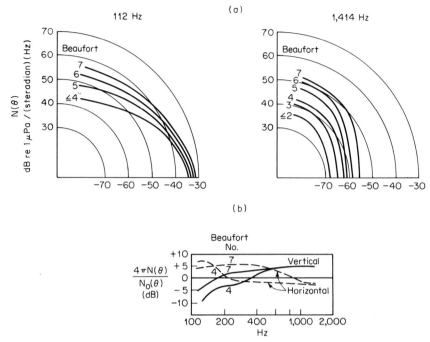

fig. 7.23 (a) Distribution of ambient noise in the vertical at a bottomed hydrophone at two frequencies. (Ref. 67.) (b) Curves of the ratio $4\pi N(\theta)/N_0(\theta)$ for noise arriving from the horizontal and the vertical at two wind speeds. (Ref. 67.)

with more noise arriving from overhead than horizontally, with a difference that *increases* with increasing wind force. This behavior is illustrated in Fig. 7.23b, where the ordinate is $N(\theta)$ normalized to an isotropic pattern by the factor 4π, the number of steradians in a unit sphere, and divided by the measured omnidirectional intensity N_0. When plotted in this way, the curves show the deviation from isotropicity of the noise.

This directional behavior is consistent with the view that low-frequency noise originates at great distances and arrives at the hydrophone via primarily horizontal paths, whereas high-frequency noise originates at the sea surface more nearly overhead.

The polar plots of Fig. 7.23a may be used to find the ambient-noise output of any directional hydrophone having a beam pattern described by the function $B(\theta,\varphi)$, where θ and φ are angles in polar coordinates. If the ambient-noise intensity per steradian is $N(\theta)$, the ambient-noise power appearing at the terminals of the hydrophone will be

$$k \int_0^{4\pi} N(\theta) B(\theta,\varphi)\, d\Omega = k \int_0^{2\pi} \int_{-\pi/2}^{\pi/2} N(\theta) B(\theta,\varphi) \cos\theta\, d\theta\, d\varphi$$

where $d\Omega$ = infinitesimal solid angle

k = proportionality factor between sound intensity and electric power at hydrophone terminals

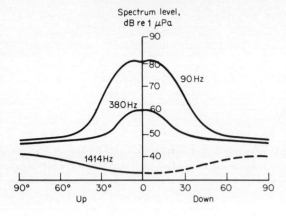

fig. 7.24 Directionality of ambient noise in the vertical. The curve for 1,414 Hz is copied from the previous figure (Beaufort 3) and is symmetrically extended into downward directions. Note that at a frequency between 380 and 1,414 Hz, ambient noise would be vertically isotropic.

The fact that at frequencies greater than 1 kHz the noise comes dominantly from the vertical suggests that the sea surface, by whatever process it produces noise, radiates preferentially downward and that each small area of the sea surface has a beam pattern of its own in the vertical. Most investigators (61, 62, 65) of sea-surface sound radiation have suggested a function of the form $I(\theta) = I_0 \cos^m \theta$, where I_0 is the intensity radiated by a small area of sea surface in the downward direction ($\theta = 0$), and m is an integer. Values of $m = 1, 2$, or 3 have been obtained, depending on conditions and method of measurement. Most measurements center roughly about $m = 2$, a value consistent with the hypothesis of a dipole source formed by the actual source and its image in the sea surface.

The rough agreement between the observations (Fig. 7.23) and the findings of our simple model (Fig. 7.22) supports this view, and suggests that a distribution of monopoles ($m = 0$) adequately models the low-frequency noise originating in distant shipping.

On Cartesian coordinates, Fig. 7.24 shows vertical directional patterns as observed with an array at a depth of several hundred feet in deep water at 90 and 380 Hz. Shown dashed is the curve for 1,414 Hz at Beaufort wind number 3 taken from Fig. 7.23. We notice the strong directionality about the horizontal at 90 Hz, where the noise comes from distant shipping. It arrives narrowly about the horizontal because long-range propagation paths are nearly horizontal. If the array were lowered to a greater depth, this single lobe would split into two, with maxima near ±15° and a minimum between, again because of the existing propagation paths. With increasing frequency the horizontal directionality diminishes because of the admixture of wind noise, until, at 1,414 Hz, the directional pattern is reversed. Thus, there is only a single frequency where ambient noise is sensibly isotropic, and this depends upon the existing shipping density and wind speed.

All of the above is in a vertical plane. Few or no observations have been made in the horizontal plane. At low frequencies in the region of dominance of distant shipping noise, the horizontal beam pattern can be obtained by convolving the beam pattern of the directional array with the levels and directions of distant ship sources. At higher frequencies in the regime of wind noise, the horizontal pattern in all likelihood would show more noise coming from the upwind and downwind directions than crosswise to the wind. However, as just mentioned, the facts must await some at-sea observations.

7.10 Spatial Coherence of Ambient Noise

The design of hydrophone arrays to minimize the pickup of ambient noise centers basically around the spatial coherence of the noise between separated receivers. By spatial coherence is meant the time-averaged crosscorrelation coefficient of the noise observed with hydrophones a distance apart in the sea. It is shown in Sec. 3.1 that the mean-square noise output of an array can be expressed in terms of the spatial correlation between all pairs of hydrophones of the array.

These functions have been worked out by Cron and Sherman (70) for pairs of hydrophones in an isotropic noise field and in the noise field produced by sources distributed along an infinite plane surface and having a radiation pattern of the $\cos^m \theta$ type mentioned above. For isotropic noise at a single frequency, the correlation coefficient, or normalized time-averaged product of the output of two hydrophones spaced a distance d apart, can be easily shown to be

$$\rho(d) = \frac{\sin kd}{kd}$$

where $k = 2\pi/\lambda$. For a surface-noise distribution of sources with $m = 2$, that is, for $I(\theta) = I_0 \cos^2 \theta$, Fig. 7.25, taken from the work of Cron and Sherman, shows the theoretical correlation between hydrophones separated horizontally and vertically, as well as for isotropic single-frequency noise. Somewhat different spatial correlation curves apply for band-limited noise and for values of m different from 2. Observations on the coherence of noise made with a bottom-mounted vertical array have shown (68) excellent agreement with theoretical predictions for $m = 2$ at frequencies between 400 and 1,000 Hz.

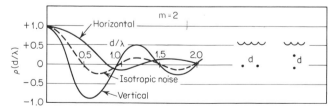

fig. 7.25 Spatial-correlation curves for isotropic noise at a single frequency and for a surface distribution of sources radiating like $\cos^2 \theta$, as received with hydrophones spaced a distance d apart horizontally and vertically. $\rho(d/\lambda)$ is the correlation coefficient as a function of the ratio of spacing to wavelength. (Ref. 70.)

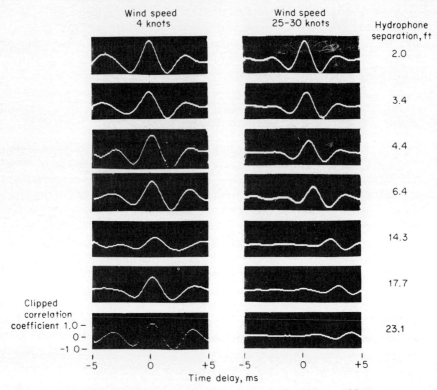

fig. 7.26 *Time-delay correlograms of ambient noise in the 200- to 400-Hz octave received by two vertically separated hydrophones spaced at different distances apart. Horizontal scale is relative time delay; a positive delay means that the upper hydrophone of the vertical pair is delayed electronically relative to the lower. (Ref. 72.)*

Time-delay correlograms show the corelation coefficient between two hydrophone outputs as a function of the time delay between them. For deep-sea ambient noise, time-delay correlograms have been obtained by Arase and Arase (71) and by Urick (72). Some examples are shown in Fig. 7.26 for different pairs of vertically separated hydrophones near the bottom in deep water off Bermuda at two different wind speeds. The correlograms are seen to be different in the two cases. At the lower wind speed, the correlograms have a peak that remains high and centered near zero delay as the hydrophone separation is increased; at the higher wind speed, the correlogram peak becomes weaker and shifts to the right with increasing hydrophone separation. The behavior is consistent with the view that the noise originates in distant shipping or storms at low wind speeds, and becomes dominated by sea-surface noise at high wind speeds. Values of noise correlation coefficient measured or read from such correlograms are useful for the design of noise-discriminating arrays steered in different vertical directions (Sec. 3.8).

7.11 Summary

The characteristics of the spectrum of ambient noise in deep water indicate the existence of a number of different causes of the noise in different portions of the spectrum (Fig. 7.4). These include tides, waves, and earth seismicity at the low-frequency end, distant shipping and rough sea surface at moderate frequencies, and molecular thermal motion of the sea at very high frequencies. By means of the component spectra of Fig. 7.5, the level of the deep-sea ambient noise at any frequency can be estimated with a moderate degree of assurance. By contrast, the magnitude of the noise in bays and harbors (Fig. 7.7) and in coastal locations (Fig. 7.8) is not nearly so predictable, as in indicated by the scatter of the reported data. Intermittent sources of the ambient background include biological sounds (Figs. 7.12 and 7.13), rain (Figs. 7.15 and 7.16), and the once-mysterious 20-cycle pulses. Under an ice cover, the ambient background (Figs. 7.20 and 7.21) has some peculiar properties of its own, such as the transient sounds of thermal cracks in the ice and the rubbing and bumping of ice blocks together. In recent years, the directional nature (Figs. 7.23 and 7.24) and spatial coherence (Fig. 7.25) of deep-sea ambient noise has been investigated, in large part through the use of bottom-mounted hydrophone arrays. These studies have shown that in deep water the noise is not isotropic, but has a directionality and a coherence in space consistent with presently accepted ideas about the origin of the noise in deep open water.

Of the various nonintermittent sources of noise in the sea, *shipping noise* and *wind noise* dominate much of the useful sonar spectrum. Their characteristics are summarized and compared in Table 7.1.

table 7.1 Summary of the Characteristics of the Two Principal Types of Ambient Noise

Shipping noise	Wind noise
Originates in distant shipping.	Originates locally on the sea surface.
Is dominant at low frequencies (100 Hz and below) in deep water.	Is dominant at high frequencies (1,000 Hz and above).
Travels over deep refracted and reflected paths.	Travels over direct paths.
Is not dependent on wind speed.	Is wind-speed dependent at the rate of 6 dB per wind speed doubled.
Arrives principally from low angles.	Arrives principally from high angles.
Time delay correlograms show a peak near zero time delay.	Time delay correlograms show a peak at time delays corresponding to upward angles between 45 and 80°.
In shallow water, is relatively unimportant except for nearby ship sources.	In shallow water, tends to dominate the spectrum when identifiable ship traffic and biological noise sources are absent.

REFERENCES

1. Revie, J., and D. E. Weston: Hydrophone Signals Due to Tidal and Wave Effects, *Deep Sea Res.*, **18**:545 (1971).

2. Urick, R. J.: Sea Bed Motion as a Source of the Ambient Noise Background of the Sea, *J. Acoust. Soc. Am.*, **56**:1010 (1974).
3. Snodgrass, J. M., and A. F. Richards: Observations of Underwater Volcanic Acoustics, *Trans. Am. Geophys. Union,* **37**:97 (1956).
4. Wenz, G. M.: Acoustic Ambient Noise in the Ocean: Spectra and Sources, *J. Acoust. Soc. Am.*, **34**:1936 (1962).
5. Longuet-Higgins, M. S.: Can Sea Waves Produce Microseisms?, National Academy of Sciences, *Nat. Res. Counc. Rep.* 306, 1956. See also, A Theory of Microseisms, *Trans. Roy. Soc. London,* **A243**:1 (1950).
6. Marsh, H. W.: Origin of the Knudsen Spectra, *J. Acoust. Soc. Am.*, **35**:409 (1963).
7. Brekhovskikh, L. M.: Underwater Sound Waves Generated by Surface Waves in the Ocean, *Akad. Nauk. SSR Izv., Atmos. Oceanic Phys.*, **2**:582 (1966).
8. Kuo, E.: Deep Sea Noise Due to Surface Motion, *J. Acoust. Soc. Am.*, **43**:1017 (1968).
9. Brekhovskikh, L. M., and V. V. Goncharov: Noise Generation by Surface Waves, *Akad. Nauk. SSR Izv., Atmos. Oceanic Phys.*, **5**:347 (1969).
10. Knudsen, V. O., R. S. Alford, and J. W. Emling: Underwater Ambient Noise, *J. Mar. Res.*, **7**:410 (1948).
11. Furduev, A. V.: Undersurface Cavitation as a Source of Noise in the Ocean, *Akad. Nauk. SSR Izv., Atmos. Oceanic Phys.*, **2**:523 (1966).
12. Isakovich, M. A., and B. F. Kuryanov: Theory of Low Frequency Noise in the Ocean, *Sov. Phys. Acoust.*, **16**:49 (1970).
13. Mellen, R. H.: Thermal-Noise Limit in the Detection of Underwater Acoustic Signals, *J. Acoust. Soc. Am.*, **24**:478 (1952).
14. Ezrow, D. H.: Measurement of the Thermal Noise Spectrum of Water, *J. Acoust. Soc. Am.*, **34**:550 (1962).
15. Cato, O.: Ambient Sea Noise in Waters near Australia, *J. Acoust. Soc. Am.*, **60**:320 (1976). Also, Marine Biological Choruses Observed in Tropical Waters near Australia, *J. Acoust. Soc. Am.*, **64**:736 (1978). Also, Some Unusual Sound of Apparent Biological Origin Responsible for Sustained Background Noise in the Timor Sea, *J. Acoust. Soc. Am.*, **68**:1056 (1980).
16. Nichols, R. H.: Infrasonic Ambient Ocean Noise Measurements, *J. Acoust. Soc. Am.*, **69**:974 (1979).
17. Yen, N., and A. J. Perrone: Mechanisms and Modelling of Wind-induced Low Frequency Ambient Sea Noise, *Nav. Underwater Systems Center Tech. Rep.* 5833 (1979).
18. McGrath, J. R.: Infrasonic Sea Noise at the Mid-Atlantic Ridge near 37° N., *J. Acoust. Soc. Am.*, **60**:1290 (1976).
19. Anderson, A. L., and G. J. Gruber: Ambient Noise Measurements at 30, 90 and 150 kHz in Five Ports, *J. Acoust. Soc. Am.*, **49**:928 (1971).
20. Piggott, C. L.: Ambient Sea Noise at Low Frequencies in Shallow Water of the Scotian Shelf, *J. Acoust. Soc. Am.*, **36**:2152 (1965).
21. Masterov, E. R., and S. P. Shorokhava: Results of an Experimental Study of the Energy Spectral Characteristics of Sea Noise, *Sov. Phys. Acoust.*, **19**:139 (1973).
22. Wagstaff, R. A.: Low Frequency Ambient Noise in the Deep Sound Channel—the Missing Component, *J. Acoust. Soc. Am.*, **69**:1009 (1981).
23. Wilkenshaw, H. M.: Low-Frequency Spectrum of Deep Ocean Ambient Noise (abstract), *J. Acoust. Soc. Am.*, **32**:1497 (1960).
24. Wenz, G. M.: Some Periodic Variations in Low-Frequency Acoustic Ambient Noise Levels in the Ocean, *J. Acoust. Soc. Am.*, **33**:64 (1961).
25. Willis, J., and F. T. Dietz: Some Characteristics of 25 cps Shallow-Water Ambient Noise, *J. Acoust. Soc. Am.*, **37**:125 (1965).
26. Wenz, G. M.: Informal Communication, *U.S. Navy Electron. Lab. Rep.* 338, 1962.
27. Urick, R. J., G. R. Lund, and T. J. Tulko: Depth Profile of Ambient Noise in the Deep Sea North of St. Croix, Virgin Islands, *U.S. Navy. Ord. Lab. Tech. Rep.* 72-176, 1972.

28. Urick, R. J., G. R. Lund, and T. J. Tulko: St. Croix Revisisted: Further Measurements of the Ambient Acoustic Background at Different Depths at a Deep Sea Location, *U.S. Nav. Ord. Lab. Tech. Rep.* 73-31, 1973.
29. Bannister, R. W., and others: Variability of Low-Frequency Ambient Sea Noise, *J. Acoust. Soc. Am.*, **65:**1156 (1979).
30. Tavolga, W. N. (ed.): "Marine Bioacoustics," Pergamon Press, New York, 1964.
31. Truitt, D. (ed.): "Dolphins and Porpoises: a Comprehensive Annotated Bibliography of the Smaller Cetacea," Gale Res. Co., Detroit, 1974.
32. Au, W., R. W. Floyd, and J. E. Haun: Propagation of Atlantic Bottlenose Dolphin Echo Location Signals, *J. Acoust. Soc. Am.*, **64:**411 (1978).
33. Walker, R. A.: Some Intense Low-Frequency Underwater Sounds of Wide Geographic Distribution, Apparently of Biological Origin, *J. Acoust. Soc. Am.*, **35:**1816 (1963).
34. Patterson, B., and G. R. Hamilton: Repetitive 20 cps Biological Hydroacoustic Signals at Bermuda, in "Marine Bioacoustics," p. 125, Pergamon Press, New York, 1964.
35. Schevill, W. E., W. A. Watkins, and R. H. Backus: The 20 Cycle Signals and Balaenoptera (Fin Whales), in "Marine Bioacoustics," p. 147, Pergamon Press, New York, 1964.
36. Thompson, P. O.: Marine Biological Sounds West of San Clemente Island, *U.S. Navy Electron. Lab. Rep.* 1290, 1965.
37. Teer, C. A.: Informal report, (British) Underwater Detection Establishment, 1949.
38. Richard, J. D.: Underwater Ambient Noise in the Straits of Florida and Approaches, *Univ. Miami Mar. Lab. Rep.* 56-12, June 1956.
39. Heindsman, T. E., R. H. Smith, and A. D. Arneson: Effect of Rain upon Underwater Noise Levels, *J. Acoust. Soc. Am.*, **27:**378 (1955).
40. Franz, G. J.: Splashes as Sources of Sound in Liquids, *J. Acoust. Soc. Am.*, **31:**1080 (1959).
41. Bom, N.: Effect of Rain on Underwater Noise Levels, *J. Acoust. Soc. Am.*, **45:**150 (1969).
42. Spiess, F. N., J. Northrop, and E. Werner: Location and Enumeration of Underwater Explosions in the North Pacific, *J. Acoust. Soc. Am.*, **43:**640 (1968).
43. Northrop, J., T. E. Stixrud, and J. R. Lovett: Sonobuoy Measurements of Seaquakes on the East Pacific Rise near San Benedicts Island, *Deep Sea Research,* **23:**519 (1976).
44. Northrop, J.: Detection of Low-Frequency Underwater Sounds from a Submarine Volcano in the Western Pacific, *J. Acoust. Soc. Am.*, **56:**837 (1974).
45. Urick, R. J.: Some Directional Properties of Deep-Water Ambient Noise, *U.S. Nav. Res. Lab. Rep.* 3796, 1951.
46. Dyer, I.: Statistics of Distant Shipping Noise, *J. Acoust. Soc. Am.*, **53:**564 (1973).
47. Berman, A., and A. J. Saur: Ambient Noise as a Function of Depth (abstract), *J. Acoust. Soc. Am.*, **32:**915 (1960). Also, Hudson Laboratories Contribution No. 77.
48. Lomask, M., and R. Frassetto: Acoustic Measurements in Deep Water Using the Bathyscaphe, *J. Acoust. Soc. Am.*, **32:**1028 (1960).
49. Morris, G. B.: Depth Dependence of Ambient Noise in the Northeastern Pacific Ocean, *J. Acoust. Soc. Am.*, **64:**581 (1978).
50. Urick, R. J.: Ambient Noise in the Surface Duct, *U.S. Navy J. Underwater Acoust.*, **26:**79 (1976).
51. Calderon, M. A.: Probability Density Analysis of Ocean Ambient and Ship Noise, *U.S. Navy Electron. Lab. Rep.* 1248, November 1964.
52. Milne, A. R., and J. H. Ganton: Ambient Noise under Arctic Sea Ice, *J. Acoust. Soc. Am.*, **36:**855 (1964). Also, *Pac. Nav. Lab. Can. Rep.* GS-1, January 1965.
53. Macpherson, J. D.: Some Under-Ice Acoustic Ambient Noise Measurements, *J. Acoust. Soc. Am.*, **34:**1149 (1962).
54. Milne, A. R., and S. R. Clark: Resonances in Seismic Noise under Arctic Sea Ice, *Bull. Seismol. Soc. Am.*, **54:**1797 (1964).
55. Bogorodskii, V. V., and A. V. Gusev: Under-Ice Noise in the Ocean, *Sov. Phys. Acoust.*, **14:**127 (1968).

56. Milne, A. R.: Shallow Water under Ice Acoustics in Barrow Strait, *J. Acoust. Soc. Am.*, **32:**1007 (1960).
57. Greene, C. R., and B. M. Buck: Arctic Ocean Ambient Noise, *J. Acoust. Soc. Am.*, **36:**1218 (1964).
58. Kibblewhite, A. C., and D. A. Jones: Ambient Noise Under Arctic Sea Ice, *J. Acoust. Soc. Am.*, **59:**790 (1976).
59. Urick, R. J.: The Noise of Melting Icebergs, *J. Acoust. Soc. Am.*, **50:**337 (1970).
60. Diachok, O., and R. S. Winokur: Spatial Variability of Underwater Ambient Noise at the Arctic Ice-Water Boundary, *J. Acoust. Soc. Am.*, **55:**750 (1974).
61. Anderson, V. C.: Arrays for the Investigation of Ambient Noise in the Ocean, *J. Acoust. Soc. Am.*, **30:**470 (1958).
62. Becken, B. A.: Directional Distribution of Ambient Noise in the Ocean, *Scripps Inst. Oceanogr. Rep.* 61-4, 1961.
63. Forster, C. A.: Ambient Sea Noise Directivity (abstract), *J. Acoust. Soc. Am.*, **34:**1986 (1962).
64. Watson, J. G.: Deep Ocean Noise Model Determined by a Deep Line Array (abstract), *J. Acoust. Soc. Am.*, **34:**1986 (1962).
65. VonWinkle, W. A.: Vertical Directionality of Deep Ocean Noise, *U.S. Navy Underwater Sound Lab. Rep.* 600, 1963.
66. Fox, G. R.: Ambient Noise Directivity Measurements, *J. Acoust. Soc. Am.*, **36:**1537 (1964).
67. Axelrod, E. H., B. A. Schoomer, and W. A. VonWinkle: Vertical Directionality of Ambient Noise in the Deep Ocean at a Site near Bermuda, *J. Acoust. Soc. Am.*, **37:**77 (1965).
68. Cron, B. F., B. C. Hassel, and F. J. Keltonic: Comparison of Theoretical and Experimental Values of Spatial Correlation, *J. Acoust. Soc. Am.*, **37:**523, 1965.
69. Linnette, H. M., and R. J. Thompson: Directivity Study of the Noise Field in the Ocean Employing a Correlative Dipole, *J. Acoust. Soc. Am.*, **36:**1788 (October 1964).
70. Cron, B. F., and C. H. Sherman: Spatial-Correlation Functions for Various Noise Models, *J. Acoust. Soc. Am.*, **34:**1732 (1962).
71. Arase, E. M., and T. Arase: Correlation of Ambient Sea Noise, *J. Acoust. Soc. Am.*, **40:**205 (1966).
72. Urick, R. J.: Correlative Properties of Ambient Noise at Bermuda, *J. Acoust. Soc. Am.*, **40:**1108 (1966).

eight

scattering in the sea: reverberation level

The sea contains, within itself and on its boundaries, inhomogeneities of many different kinds, ranging in size from the tiny particles of dust that cause the deep sea to be blue, to schools of fish within the volume of the sea, and to pinnacles and sea mounts on the sea-bed. These inhomogeneities form discontinuities in the physical properties of the medium and thereby intercept and reradiate a portion of the acoustic energy incident upon them. This reradiation of sound is called *scattering,* and the sum total of the scattering contributions from all the scatterers is called *reverberation.* It is heard as a long, slowly decaying, quivering tonal blast following the ping of an active sonar system, and is particularly obnoxious in systems of high power and/or low directivity. Since it often forms the primary limitation on system performance, a necessary part of the design process of new active sonars is to make an estimate of the reverberation level to be encountered under the conditions in which the system will be used. In this chapter some useful formulas will be given for this purpose, together with a description of the characteristics and sources of the reverberation observed in the sea.

8.1 Types of Reverberation

The reverberation-producing scatterers in the sea are of three basically different classes. One type of scatterer occurs in the volume, or body, of the sea and produces *volume reverberation.* Examples of scat-

238 / *principles of underwater sound*

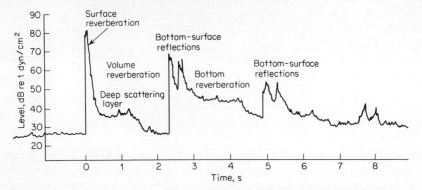

fig. 8.1 *Reverberation following a 2-lb explosive charge detonating at 800 ft. in water 6,500 ft deep, as observed with a nearby hydrophone at a depth of 135 ft. Filter band 1 to 2 kHz.*

terers producing volume reverberation are the marine life and inanimate matter distributed in the sea and the inhomogeneous structure of the sea itself. *Sea-surface reverberation* is produced by scatterers located on or near the sea surface, and *bottom reverberation* originates at scatterers on or near the sea bottom. The last two forms of reverberation may be analytically considered together as *surface reverberation,* since a two-dimensional distribution of scatterers is involved. Figure 8.1 is an example of the reverberation that is observed to follow an explosive charge in deep water and is received on a shallow, nearby nondirectional hydrophone. In this example, the reverberation occurring immediately after the direct blast from the charge is surface reverberation, which quickly dies away into volume reverberation originating in large part in the deep scattering layer. The remainder of the return is caused by the sea bottom, which, together with the sea surface, produces a long and irregular reverberation tail caused by multiple reflection and scattering at the two boundaries.

8.2 The Scattering-Strength Parameter

The fundamental ratio upon which reverberation depends is called *scattering strength*. It is the ratio, in decibel units, of the intensity of the sound scattered by a unit area or volume, referred to a distance of 1 yd, to the incident plane-wave intensity. In symbols, if I_{scat} is the intensity of the sound scattered by an area of 1 yd^2 or a volume of 1 yd^3, when the intensity is measured at a greater distance and reduced to 1 yd, and if I_{inc} is the intensity of the incident plane wave, then the scattering strength is defined to be

$$S_{s,v} = 10 \log \frac{I_{scat}}{I_{inc}}$$

The scattering strength parameter may be conceptually visualized as illustrated in Fig. 8.2*a* and *b* for volume reverberation and for surface reverberation, respectively. The direction of scattering is normally *back* toward the source, in which case the quantity S is more specifically termed *backscattering strength*. The point of reference for S is the point P distant 1 yd from the unit volume or area in the backward direction. In "bistatic" sonars, having a widely separated source and receiver, other directions will be of interest. From their manner of definition, it follows that scattering strength and target strength are analogous sonar parameters.

In radar (1) it is customary to use the *backscattering cross section* of a unit area or volume as the basic parameter involving scattering. This unit was adopted for sonar during World War II and was given the symbols m_v and m_s. It is the ratio of the scattered *power*, referred to 1 yd, to the *intensity* incident on a unit volume or area. Since

$$\text{Power} = \text{intensity} \times \text{area}$$

and, if it is assumed arbitrarily that the scattering is uniformly distributed over a sphere and a hemisphere, respectively, it follows that

$$S_v = 10 \log \frac{m_v}{4\pi}$$

$$S_s = 10 \log \frac{m_s}{2\pi}$$

where m_v and m_s are the backscattering cross sections of a unit volume and area. These cross sections may be viewed as the areas which, when multiplied by the intensity, equal the power removed from the incident wave by a unit volume or area and reradiated in all directions. However, the scattering strength parameter, first introduced in 1954 (2), has replaced m_v and m_s in most sonar usage.

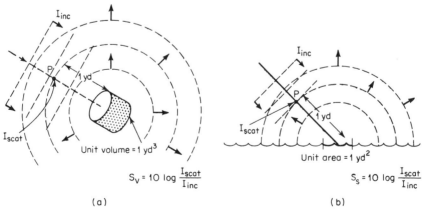

fig. 8.2 *Conceptual definitions of scattering strength for volume and surface scattering.*

8.3 Equivalent Plane-Wave Reverberation Level

In the sonar equations as previously written (Chap. 2), the term RL refers to the *equivalent plane-wave reverberation level*. This is the level of the axially incident plane wave which produces the same hydrophone voltage across the hydrophone terminals as that produced by the received reverberation. RL is therefore directly analogous to the echo level EL, with which it may be compared to obtain the echo-to-reverberation ratio across the hydrophone terminals.

In seeking an explicit, instead of purely formal, general expression for RL, we need certain simplifying assumptions. Although these may seem to restrict the results to purely idealized situations, the resulting expressions for reverberation have been found to be useful for design and prediction purposes under many actual conditions. The assumptions necessary are the following:

1. Straight-line propagation paths, with all sources of attenuation other than spherical spreading neglected. The effect of absorption of sound in the sea on the reverberation level can, in any case, be readily allowed for.
2. A random, homogeneous distribution of scatterers throughout the area or volume producing reverberation at any one instant of time.
3. A density of scatterers so large that a large number of scatterers occur in an elemental volume dV or area dA.
4. A pulse length short enough for propagation effects over the range extension of the elemental volume or area to be neglected.
5. An absence of multiple scattering: that is, the reverberation produced by reverberation is negligible.

8.4 Volume-Reverberation Theory

Consider, in Fig. 8.3a, a directional projector in the ideal medium implied by the above assumptions, containing in its volume a large number of uniformly distributed scatterers. Let the beam pattern of the projector be denoted by $b(\theta,\varphi)$, and let the axial intensity at unit distance be I_0 (the source level SL = 10 log I_0). By the definition of the beam-pattern function, the intensity at 1 yd in

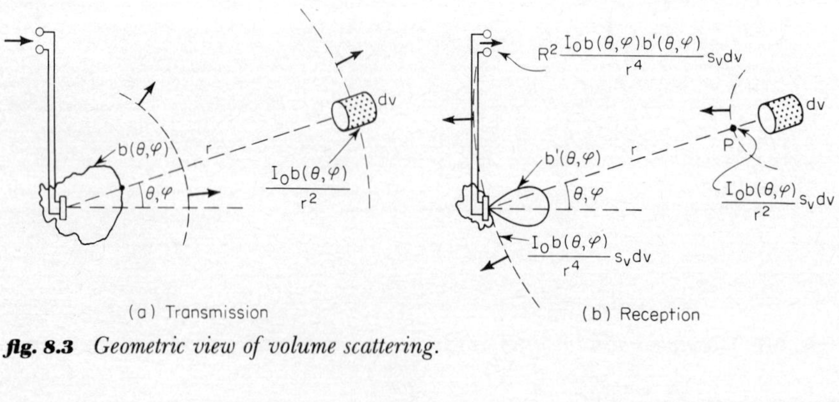

(a) Transmission (b) Reception

fig. 8.3 *Geometric view of volume scattering.*

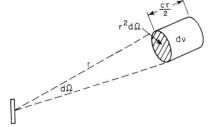

fig. 8.4 *Selection of the elemental volume for volume reverberation.*

the (θ,φ) direction is $I_0 b(\theta,\varphi)$. At a distance r in this direction, let there be a small volume dV of volume scatterers. The incident intensity at dV will be, by assumption 1 above, $I_0 b(\theta,\varphi)/r^2$. The intensity of the sound backscattered by dV, at point P distant 1 yd back toward the source (Fig. 8.3b), will be $[I_0 b(\theta,\varphi)/r^2]s_v\, dV$, where s_v is the ratio of the intensity of the backscattering produced by a unit volume, at a distance of 1 yd from the volume, to the intensity of the incident sound wave. As defined in the previous section, the quantity $10 \log s_v$ is the backscattering strength for volume reverberation and is denoted by the symbol S_v. S_v is dependent on the type and density of scatterers that give rise to the reverberation, as will be discussed thoroughly later on. Back in the neighborhood of the source, the reverberation contributed by dV will have the intensity $(I_0/r^4)b(\theta,\varphi)s_v\, dV$ and will produce a mean-squared voltage output of $R^2(I_0/r^4)b(\theta,\varphi)b'(\theta,\varphi)s_v\, dV$ at the terminals of a hydrophone having a receiving beam pattern $b'(\theta,\varphi)$ and voltage response R. (In simple sonars, the same transducer is used for projection and reception, and the patterns b and b' are identical; in many modern sonars, however, the projector and receiver beam patterns are different.) By assumption 3, the elemental volumes dV can be made so small that their total contribution can be summed up by integration, and by assumption 2, the coefficient s_v is a constant and can be removed from under the integral sign. The total reverberation hydrophone output will accordingly become $(R^2 I_0/r^4)s_v \int_v b(\theta,\varphi)b'(\theta,\varphi)\, dV$. Defining now the *equivalent plane-wave reverberation level RL* as the intensity, in decibel units, of an axially incident plane wave producing the same hydrophone output as the observed reverberation, we obtain

$$\mathrm{RL} = 10 \log \left(\frac{I_0}{r^4} s_v \int_v bb'\, dV \right)$$

where the θ and φ symbols have been dropped for economy in writing.

To proceed from here, we must now have a closer look at the elemental volumes dV. As shown in Fig. 8.4, let us take dV to be an infinitesimal cylinder of finite length with ends normal to the incident direction. The area of the end face of this infinitesimal cylinder dV may be written $r^2\, d\Omega$, where $d\Omega$ is the elemental solid angle subtended by dV at the source. The extension in range of dV is such that, with a pulsed sonar, the scattering produced by all portions of dV arrives back near the source at the same instant of time. This requirement means that if one imagines a sound pulse to be incident upon the

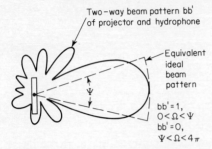

fig. 8.5 Actual and equivalent beam patterns.

scatterers within dV, the scattering of the "front" end of the pulse by the "rear" scatterers in dV will arrive back at the source at the same instant as the scattering of the "rear" of the pulse by the "front" scatterers in dV. The extension in range must accordingly be $c\tau/2$, where τ is the pulse length and c is the velocity of sound. Hence the elemental volume becomes

$$dV = r^2 \frac{c\tau}{2} d\Omega$$

and

$$RL = 10 \log \left(\frac{I_0}{r^4} r^2 \frac{c\tau}{2} s_v \int bb' \, d\Omega \right)$$

where, by assumption 4, the range extension $c\tau/2$ is small compared to the range r.

The integral $\int bb' \, d\Omega$ can be interpreted as the equivalent beam width of the projector-hydrophone combination for volume reverberation. Let the actual two-way beam pattern bb' be replaced by an ideal pattern of unit relative response within a solid angle Ψ, and zero relative response beyond Ψ, as shown in Fig. 8.5. Then the width Ψ of this equivalent ideal beam is given by

$$\int_0^{4\pi} bb' \, d\Omega = \int_0^{\Psi} 1 \times 1 \, d\Omega = \Psi$$

The solid angle Ψ may therefore be regarded as the opening angle of the ideal beam pattern, having a flat response within Ψ and none beyond, which, for reverberation, is equivalent to the actual two-way beam pattern.

In terms of Ψ, the expression for the equivalent plane-wave reverberation level becomes

$$RL_v = 10 \log \left(\frac{I_0}{r^4} s_v \frac{c\tau}{2} \Psi r^2 \right)$$

which may be written as

$$\boxed{\begin{aligned} RL_v &= SL - 40 \log r + S_v + 10 \log V \\ V &= \frac{c\tau}{2} \Psi r^2 \end{aligned}}$$

The quantity V may be termed the *reverberating volume*, and is the volume of scatterers returning reverberation at any one instant of time.

In this geometric view, the reverberation is caused by those scatterers lying within the equivalent beam at the range of interest and contained in the volume V, determined by the equivalent beam width Ψ of the projector-hydrophone combination, the pulse length τ, and the range r. Expressions for Ψ for some simple transducers are given in Table 8.1 in terms of the transducer dimensions and the width, in degrees, of the two-way beam pattern. For other projector-hydrophone combinations, Ψ can be found by integration of the pattern function bb', as indicated by the integral expressions of Table 8.1.

table 8.1 Equivalent Two-Way Beam Widths in Logarithmic Units, Based on Ref. 92

Array	$10 \log \Psi$ dB re 1 steradian	$10 \log \Phi$ dB re 1 radian
Integral expression	$10 \log \int_0^{2\pi} \int_{-\pi/2}^{\pi/2} b(\theta,\varphi) b'(\theta,\varphi)$ $\times \cos\theta \, d\theta \, d\varphi$	$10 \log \int_0^{2\pi} b(0,\varphi) b'(0,\varphi) \cdot d\varphi$
Circular plane array, in an infinite baffle of radius $a > 2\lambda$	$20 \log \left(\dfrac{\lambda}{2\pi a}\right) + 7.7$ or $20 \log y - 31.6$	$10 \log \dfrac{\lambda}{2\pi a} + 6.9$ or $10 \log y - 12.8$
Rectangular array in an infinite baffle, side a horizontal, b vertical, with $a, b \gg \lambda$	$10 \log \dfrac{\lambda^2}{4\pi ab} + 7.4$ or $10 \log y_a y_b - 31.6$	$10 \log \dfrac{\lambda}{2\pi a} + 9.2$ or $10 \log y_a - 12.6$
Horizontal line of length $l > \lambda$	$10 \log \dfrac{\lambda}{2\pi l} + 9.2$ or $10 \log y - 12.8$	$10 \log \dfrac{\lambda}{2\pi l} + 9.2$ or $10 \log y - 12.8$
Nondirectional (point) transducer	$10 \log 4\pi = 11.0$	$10 \log 2\pi = 8.0$

y is the half angle, in degrees, between the two directions of the two-way, or product, beam pattern in which the response is 6 dB down from the axial response. That is, y is the angle from the axis of the two-way beam pattern such that $b(y)b'(y) = 0.25$. For the rectangular array, y_a and y_b are the corresponding angles in planes parallel to the sides a and b.

For a sonar having a nondirectional projector and hydrophone, $b = b' = 1$, and

$$\Psi = \int bb' \, d\Omega = 4\pi$$

The ratio of the reverberation received by such a sonar to that received by a directional sonar of the same source level is $4\pi/\int bb' \, d\Omega$. During World War II the quantity $10 \log (4\pi/\int bb' \, d\Omega)$ was termed (3) the *directivity index for volume reverberation*, in analogy with its counterpart for isotropic noise, and given the symbol J_v.* In this alternative formulation the additional 4π is absorbed into the term s_v, and the expression for reverberation level becomes

---

* Originally both J_v and DI were defined so as to be *negative*. Current practice is to define DI and J_v as *positive* quantities.

$$RL = SL - 40 \log r - J_v + 10 \log m_v + 10 \log \left(\frac{c\tau}{2} r^2\right)$$

where $m_v = 4\pi s_v$ is the backscattering cross section of a unit volume.

J_v is the analog for reverberation of the parameter DI that pertains to isotropic noise (Chap. 3). The definitions of these two quantities are

$$J_v = 10 \log \frac{4\pi}{\int bb' \, d\Omega}$$

$$DI = 10 \log \frac{4\pi}{\int b' \, d\Omega}$$

We observe from these expressions that J_v includes both the transmitting and receiving beam patterns b and b' of the source and receiver while DI involves only the receiving pattern b'. J_v applies for an *isotropic*, or homogeneous, distribution of scatterers while DI applies for *isotropic* noise. Both can be imagined to be defined in terms of an *equivalent ideal solid-angle beam width* Ψ defined by $\int bb' \, d\Omega$ in one case and $\int b' \, d\Omega$ in the other. Both these solid angles reduce to 4π for a nondirectional source and receiver, and both J_v and DI become equal in that case to 0 dB.

A geometric, rather than a purely mathematical, view of reverberation is valuable in practical sonar problems. The magnitude and dimensions of the reverberating volume, its location in the sea, and the kind of scatterers it might contain are all useful aids for intelligent performance predictions of active sonar equipment.

It might be wondered, in the light of the severe restrictive assumptions made at the outset in Sec. 8.3, whether the resulting expressions based on such a simple model have any validity. This question has been examined for both single-frequency and FM pings over a range of pulse lengths and sweep widths (4). It was found that the simple model had negligible error for pulse lengths of 10 ms or less, but the reverberation level computed by it was in error by about 5 dB for 1,000-ms pulses, probably because of a nonhomogeneous distribution of scatterers over the great interval of range represented by the long pulse. We conclude that our simple model is indeed valid, at least for short pulses and for relatively short ranges where the multipath effects that complicate a reverberation calculation (Sec. 8.17) are absent.

8.5 Surface-Reverberation Theory

By surface reverberation is meant reverberation produced by scatterers distributed over a nearly plane surface, rather than throughout a volume. Most notably, such scattering surfaces are the surface and the bottom of the sea.

The derivation of an expression for the equivalent plane-wave level of surface reverberation proceeds as that for volume reverberation. The result is

$$RL_s = 10 \log \left(\frac{I_0}{r^4} s_s \int b(\theta,\varphi) b'(\theta,\varphi) \, dA\right)$$

where dA is an elemental area of the scattering surface and $10 \log s_s = S_s$ is the *scattering strength for surface reverberation*. If dA is taken as a portion of a circular annulus in the plane of the scatterers, with center directly above or below the transducer, we may write

$$dA = \frac{c\tau}{2} r \, d\varphi$$

where $d\varphi$ is the subtended plane angle of dA at the center of the annulus. The reverberation level becomes

$$RL_s = 10 \log \frac{I_0}{r^4} s_s \frac{c\tau}{2} r \int_0^{2\pi} b(\theta,\varphi) b'(\theta,\varphi) \, d\varphi$$

This integral is difficult to evaluate analytically at high angles between the transducer beam and the scattering surface. Some useful graphs for this have been given in a paper by Urick and Hoover (5). In the more usual sonar case, where the axis of the beam is only slightly inclined toward the scattering surface, the scattering surface nearly corresponds to the $\theta = 0$ plane of the transducer beam pattern, so that

$$RL_s = 10 \log \left(\frac{I_0}{r^4} s_s \frac{c\tau}{2} r \int_0^{2\pi} b(0,\varphi) b'(0,\varphi) \, d\varphi \right)$$

The plane angle of the equivalent ideal beam is such that

$$\int_0^{2\pi} b(0,\varphi) b'(0,\varphi) \, d\varphi = \int_0^{\Phi} 1 \times 1 \, d\varphi = \Phi$$

and, therefore,

$$RL_s = 10 \log \left(\frac{I_0}{r^4} s_s \frac{c\tau}{2} r \Phi \right)$$

or

$$\boxed{\begin{aligned} RL_s &= SL - 40 \log r + S_s + 10 \log A \\ A &= \frac{c\tau}{2} \Phi r \end{aligned}}$$

Expressions for Φ are included in Table 8.1. The area A is the area of the surface of scattering strength S_s lying within the ideal beam width Φ which produces the same reverberation as that actually observed. In Fig. 8.6, the reverberating area A is illustrated for a downward-pointing transducer receiving bottom reverberation. The point P, directly beneath, is the center of a circular annulus of which only the portion A, of width Φr and range extension $c\tau/2$, returns reverberation at any one instant of time. S_s, in general, varies with the angle θ of the incident sound beam upon the scattering surface; for a transducer at a constant distance above or below the scattering surface, S_s will

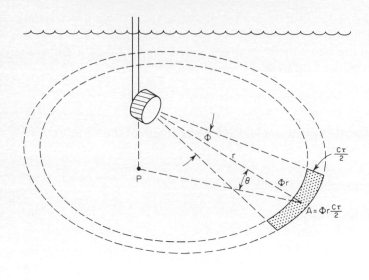

fig. 8.6 *The reverberating area for surface reverberation. The transducer is visualized as a downward-pointing circular plane array receiving bottom reverberation.*

therefore vary with range or time after emission of the pulse. As in volume reverberation, it is important that the design engineer be able to visualize where in the sea the sound emitted by the active sonar is going and what scatterers it is likely to encounter.

8.6 Target Strength and Scattering Strength

As mentioned above, the scattering strength for volume or surface reverberation is defined as 10 times the logarithm to the base 10 of the ratio of the scattered intensity produced by 1 yd^3 of the ocean or 1 yd^2 of scattering surface, respectively, to the incident intensity. The scattered intensity is referred to a distance of 1 yd from the unit volume or surface in the direction back toward the source of sound, unless some other direction is specified. When so defined, the scattering strength of the sea and its boundaries and the target strength of sonar targets are essentially identical parameters. Scattering strength is the target strength, for scattering, of a unit volume or area.

Accordingly, in reverberation backgrounds, it becomes easy, in simple situations, to find the echo-to-reverberation ratio for a target of known target strength. The level of the echo from the target is (Chap. 2)

$$EL = SL - 2TL_E + TS$$

where SL = source level
$\quad$ TL_E = transmission loss for echo
$\quad$ TS = target strength

The reverberation level, or the level of the axial plane wave producing the same mean-squared hydrophone voltage as the reverberation, is

$$RL = SL - 2TL_R + S_{s,v} + 10 \log A, V$$

where TL_R is the transmission loss for reverberation, and $S_{s,v}$ and A,V are the scattering strength and the area or volume of the reverberating area or volume, respectively. If, as is usually the case, the scattering volume or area surrounds the target so that the target lies in the midst of the reverberation-producing area or volume, TL_R and TL_E are the same, and the echo-to-reverberation ratio becomes

$$EL - RL = TS - (S_{s,v} + 10 \log A, V)$$

The echo-to-reverberation ratio is thus simply the difference of two "strengths," one for the echo and the other for the scattering, without reference to source or medium characteristics. The target must, however, lie in the vicinity of the reverberation-producing scatterers in order for the two transmission losses to be the same and in order to avoid beam-pattern complications. As an example, let a target of target strength $+10$ dB lie on a sea bottom of scattering strength -30 dB. The echo-to-reverberation ratio is desired for conditions of range, pulse duration, and equivalent beam width such that $10 \log A$ equals 15 dB. For these conditions, the echo-to-reverberation ratio becomes simply $10 + 30 - 15 = +25$ dB. At other ranges and times, the echo-to-reverberation ratio will be different because both the angle of incidence of the sound beam upon the bottom and the size of the reverberating area will vary with range or time after emission of the sonar pulse.

8.7 Surface Scattering by a Layer of Volume Scatterers

Some kinds of scatterers in the sea, such as the deep scattering layer or a layer of air bubbles just below the sea surface, lie in layers of finite thickness instead of being diffusely distributed throughout an irregular volume. Such *layered reverberation* is best considered as a form of surface reverberation. If the average volume scattering strength of the layer is S_v, the corresponding surface scattering strength is

$$S_s = S_v + 10 \log H$$

where H is the layer thickness in yards. In general, if the volume scattering strength is not constant within the layer,

$$S_s = 10 \log \int_0^H s_v(h)\, dh$$

where $s_v(h)$ is the volume scattering strength at a depth h within the layer, expressed in linear units ($S_v = 10 \log s_v$).

When H is made indefinitely great, the quantity S_s becomes the scattering

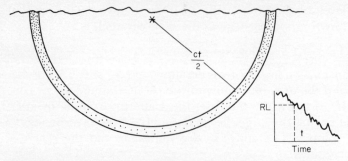

fig. 8.7 *The quantity column or integrated scattering strength refers to a spherical shell of radius ct/2. The reverberation level at time τ is RL. Column scattering strength refers to all the scatterers within the shell; most of these occur at shallow depths. Strong refraction in the sea can distort the shell and produce erroneous results.*

strength of the entire water column, and is called *column* or *integrated* scattering strength. It can be readily measured in the field by means of explosive sound signals and sonobuoys, and it is a conveient single number in survey work for the total amount of volume reverberation existing at the time and place the data were obtained (6). When measured in this way, column scattering strength pertains to all the scatterers lying in a large hemispherical shell surrounding the source and receiver (Fig. 8.7). If the shell is large, the scatterers will lie at a relatively shallow depth—hence the term *column* scattering strength for this quantity.

8.8 Reverberation Level for Short Transients

Hitherto we have considered the reverberation from a sonar pulse having a constant average intensity I_0 during a fixed definite duration τ. Although these conditions apply for many sonars, they do not apply for short transients of indefinite waveform and duration.

For such waveforms, it can be shown that the product $I\tau_0$ occurring in the reverberation equations can be replaced by the integral $\int_0^\infty i(t)\,dt$, in which $i(t)$ is the instantaneous intensity of the source at 1 yd, and the integration is taken for the entire duration of the transient. A restriction to short transients is necessary in order for the range r and scattering strength S in the reverberation equations to be taken as constants. The integral, which reduces to $I_0\tau$ for flat-topped pulses of duration τ, is the energy flux density of the source at a distance of 1 yd, in units of the energy flux density of a plane wave having an rms pressure of 1 μPa taken for a period of 1 second. This unit amounts to -152 dB below 1 erg/cm². A source of energy flux density E in these units at 1 yd can be said to have an energy-flux-density source level SL_E defined by

$$SL_E = 10 \log E$$

The reverberation equations become

$$RL_v = SL_E - 20 \log r + S_v + 10 \log \Psi + 10 \log \frac{c}{2}$$

$$RL_s = SL_E - 30 \log r + S_s + 10 \log \Phi + 10 \log \frac{c}{2}$$

The quantity $10 \log (c/2)$, where c is the velocity of sound, is a kind of conversion factor necessary to equate intensity level, on the left, to an energy-density level SL_E, on the right; with 1 yd as the unit of distance, c must be taken in yards per second so that $10 \log (c/2) = 29$ dB. For nondirectional sonars, $\Psi = 4\pi$, and $\Phi = 2\pi$.

These expressions for the reverberation level in terms of the energy-density of a short transient were first obtained by Urick (7) and have been independently derived and used for the measurement of sea-surface scattering strengths with explosive sources by Chapman and Harris (8). For explosives, the quantity SL_E has been established for certain conditions by the work of Weston (9), Stockhausen (10), and Christian and Blaik (11) (Sec. 4.4).

8.9 Air Bubbles in Water

Occurrence Air bubbles, or more generally gas bubbles, appear in various forms in the sea. They occur immediately below the surface of the sea, where they are produced by the breaking of waves and are carried by turbulence beneath the surface. They appear within certain biological organisms, as noted above, as, for example, the swim bladders of fish. They are generated in the wakes of ships, where they are known to persist for long periods of time. Free air bubbles in the sea are quite small, since the larger bubbles tend to rise quickly to the surface. They form only a very small percentage, by volume, of the sea in which they occur. Nevertheless, because air has a markedly different density and compressibility than seawater, and because of the resonant characteristics of bubbles, the suspended air content of seawater has a profound effect upon underwater sound. Some of these effects are considered in the following sections.

Physical processes When a sound wave strikes an air bubble, the bubble partakes, to an extent, in the compression and rarefactions of the incident sound wave. This response of the bubble to the excitation of the incident sound wave depends on the frequency of the sound wave and the size of the bubble. At a certain frequency, a resonant response occurs when the inertial properties of the system—the mass of the bubble plus a portion of the surrounding water—become matched to the compressibility of the system. At resonance, a maximum oscillation of the bubble size develops, and a maximum amount of energy is extracted from the incident sound wave. A portion of this energy is scattered in all directions by the pulsating bubble, and the remainder is converted to heat—a process brought about by heat conduction inside the bubble, the viscosity of the surrounding water, and the surface tension at the bubble surface. The oscillating bubble may therefore be viewed as intercepting a portion of the incident sound wave characterized by the *extinction cross section* of the bubble and reradiating it as scattered sound in all directions, as well as converting it to heat. These processes depend upon the frequency of oscillation, the size of the bubble, and the physical and thermodynamic properties of the two mediums involved.

The theory of the resonant behavior of air bubbles in water was worked out in World War II, and may be found in one of the Summary Technical Reports of the NDRC (12), as well as in the book by Albers (13). This theory will not be repeated here, but only the results will be given, along with an occasional reference to subsequent literature.

Sound velocity in a bubbly medium When air is *dissolved* in water, measurements show (14) that the effect on the velocity of sound is completely negligible, even when the water is completely saturated with air. When, however, air is *suspended* as tiny bubbles, its effect is profound; a minute amount of air substantially reduces the velocity of sound in the bubbly fluid. When the bubbles are all much smaller than the resonant size, the velocity of sound has been found experimentally (15) to be given by the simple mixture theory. By this theory (16), the compressibility and the density of air and water, which determine the sound velocity, are weighted by the proportionate amount of the two mediums present in a given volume of mixture. If the compressibilities of mixture, air, and water are denoted by κ, κ_a, and κ_w, respectively, and if the corresponding densities are ρ, ρ_a, and ρ_w, then the sound velocity of a mixture of air and water containing a fraction β of air by volume is, by the mixture theory, for small β,

$$v = \left(\frac{1}{\rho\kappa}\right)^{1/2} = \left\{\frac{1}{[\beta\rho_a + (1-\beta)\rho_w][\beta\kappa_a + (1-\beta)\kappa_w]}\right\}^{1/2}$$

$$= \left[\frac{1}{\rho_w\kappa_w(1 + \beta\kappa_a/\kappa_w)}\right]^{1/2} = v_w\left[\frac{1}{1 + \beta\kappa_a/\kappa_w}\right]^{1/2}$$

$$= v_w\left(\frac{1}{1 + 2.5 \times 10^4\beta}\right)^{1/2}$$

since $\kappa_a \gg \kappa_w$, $\rho_a \ll \rho_w$, and $\beta \ll 1$. The formula indicates that a volume fraction of air of only 0.01 percent ($\beta = 10^{-4}$) reduces the sound velocity to 53 percent of the velocity v_w of air-free water.

By contrast, at high frequencies well beyond the resonant frequency of the smallest bubble present in the mixture, the effect of suspended air content is negligible. In this region of bubble size, large compared to the resonant size, the two velocities are related by (17)

$$c^{-2} = c_0^{-2} - \frac{4\pi n a}{(2\pi f)^2}$$

where a = bubble radius
n = number of bubbles in a unit volume

The second term on the right is small compared to the first for small bubble concentrations.

In the vicinity of resonance, large changes in sound velocity take place. The velocity in a bubbly mixture over a wide frequency range surrounding the resonant frequency is illustrated in Fig. 8.8, which shows the smoothed velocity curve obtained experimentally by Fox, Curley, and Larson (18) in a cloud of uniform air bubbles having an average diameter of 0.011 cm and a volume concentrations β of 2×10^{-4}. The velocity passes through its value for bubble-free water at the resonant frequency.

Resonant frequency At a certain frequency, a bubble of a given size is in resonance with the exciting sound wave and a maximum extinction cross section occurs. This frequency is (Ref. 12, eq. 18)

$$f_r = \frac{1}{2\pi a} \sqrt{\frac{3\gamma P_0}{\rho}}$$

where γ = ratio of specific heats of the gas
P_0 = hydrostatic pressure
ρ = density of the water

For air ($\gamma = 1.4$) at zero depth in the sea ($P_0 = 1$ atm) this becomes

$$f_r = \frac{326}{a}$$

where f_r = resonant frequency, Hz
a = bubble radius, cm

At a depth of d ft in water, the corresponding expression is

$$f_r = \frac{326}{a}\sqrt{1 + 0.03d}$$

The value of $k_r a = 2\pi a/\lambda_r$ for a resonant bubble at zero depth is 0.0136.

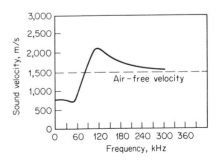

fig. 8.8 *Curve showing the measured sound velocity in a cloud of bubbles of uniform size (diameter 0.011 cm) over a wide range of frequency. (Ref. 18.)*

Damping constant A measure of the above-mentioned dissipation processes is the damping constant δ associated with bubble pulsation. δ is the reciprocal of the Q of the bubble oscillation and is dimensionless. In terms of free oscillations, δ is the reciprocal of the number of cycles required for the amplitude of motion to decrease to $e^{-\pi}$ of its original value; for forced oscillations, δ is defined as

$$\delta = \frac{f_2 = f_1}{f_r}$$

where f_2 and f_1 are the frequencies above and below resonance at which the square of the amplitude diminishes to one-half its value at the resonant frequency f_r (half-power points).

The damping constant δ of air bubbles in water has been determined by various investigators using different methods, with none-too-concordant results. A survey made by Devin (19) of the theory and observations of oscillatory damping of air bubbles in water shows that the damping constant predicted by hydrodynamic and thermodynamic theory is valid. Figure 8.9 shows the theoretical damping constant δ for resonant air bubbles, together with the contributions to the total damping produced by (1) heat conduction during bubble pulsations, (2) radiation of sound by the pulsating bubble, and (3) viscosity of the surrounding water.

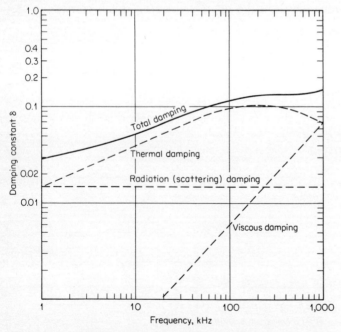

fig. 8.9 *Theoretical damping of resonant air bubbles in water.* [After Devin (Ref. 19).]

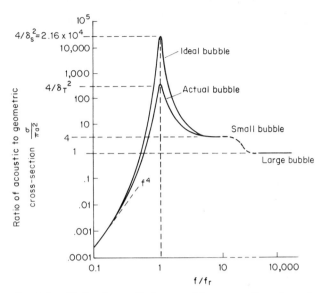

fig. 8.10 Ratio of acoustic to geometric cross section of an air bubble in water as a function of normalized frequency f/f_r.

Extinction, scattering, and absorption cross sections In terms of the damping constant δ, the extinction cross section at frequency f of a bubble of radius a resonant at frequency f_r is (Ref. 12, eq. 43)

$$\sigma_e = \frac{4\pi a^2 \delta/k_r a}{(f_r^2/f^2 - 1)^2 + \delta^2}$$

where $k_r a = 2\pi a/\lambda_r = 2\pi f_r a/c_0$. The absorption cross section is (Ref. 12, eq. 45)

$$\sigma_a = \frac{4\pi a^2 (\delta/k_r a - 1)}{(f_r^2/f^2 - 1)^2 + \delta^2}$$

and the *scattering cross section* is (Ref. 12, eq. 34)

$$\sigma_s = \frac{4\pi a^2}{(f_r^2/f^2 - 1)^2 + \delta^2}$$

so that

$$\sigma_e = \sigma_a + \sigma_s$$

These cross sections have a maximum at the resonant frequency and fall off with frequency away from resonance in the same manner as the response curve of a tuned electric circuit containing a resistive loss.

The behavior of the acoustic cross section of a small air bubble in water is illustrated in Fig. 8.10. A "small" bubble is defined as one for which $a \ll \lambda$. An "ideal" small bubble is one that extracts energy from a sound wave only by scattering, while an "actual" bubble creates additional loss through heat conduction (thermal damping) and the viscous drag of the surrounding fluid (viscous damping). From Fig. 8.10 we observe that at frequencies well below resonance, the acoustic cross section of a small bubble increases as the fourth power of the frequency and reaches a peak at the resonance frequency f_r. At resonance, the ideal bubble has a ratio of acoustic cross section to geometric cross section equal to $4/\delta_s^2 = 2.16 \times 10^4$, where $\delta_s = 1.36 \times 10^{-2}$ is the damping constant due to scattering (see Fig. 8.9).

An actual bubble, however, has a lower acoustic cross section than the ideal bubble because of its greater damping; its cross-sectional ratio amounts to $4/\delta_T^2$, where δ_T is the total damping constant whose value depends on frequency. Beyond resonance the ratio of cross sections approaches the value 4. This value applies only for "small" bubbles. Acoustically "large" bubbles ($a \gg \lambda$), for which small bubble theory does not apply, have a ratio of cross sections equal to unity; that is, their acoustic cross section equals their geometric cross section πa^2.

Target strength and scattering strength The target strength of a single bubble becomes

$$\text{TS} = 10 \log \frac{\sigma_s}{4\pi} = 10 \log \frac{a^2}{(f_r^2/f^2 - 1)^2 + \delta^2}$$

and the scattering strength of a unit volume (1 yd^3) of water containing n bubbles/yd^3, each of scattering cross section σ_s, is (multiple scattering neglected)

$$S_v = 10 \log \frac{n\sigma_s}{4\pi} = 10 \log \frac{na^2}{(f_r^2/f^2 - 1)^2 + \delta^2}$$

where the units of the radius a must be taken in yards.

Attenuation coefficient The attenuation of a plane wave propagating through water containing bubbles is expressed by an attenuation coefficient defined by

$$\alpha = 10 \log \frac{I_1}{I_2}$$

where I_1 and I_2 are the intensities of the wave at two points located a unit distance apart in the direction of propagation. If there are n resonant bubbles/yd^3, each with extinction cross section σ_e yd^2, the attenuation coefficient is (Ref. 12, eq. 53)

$$\alpha = 4.34\, n\sigma_e \quad \text{dB/yd}$$

Distributions of bubbles When, as is usual, the bubbles are not of uniform size, the effective cross section must be obtained by integration over the range of sizes of the bubbles. If the number of bubbles per cubic yard with radii between a and $a + da$ is denoted by $n(a) \, da$, the effective scattering cross section is

$$\sigma = \int_0^\infty n(a)\sigma(a) \, da$$

where $\sigma(a)$ = extinction, absorption, or scattering cross section
$n(a)$ = size-distribution function of bubble population

Because only bubbles of near-resonant size will make a large contribution to σ, and because $n(a)$ does not vary rapidly with a, the integral can be easily evaluated by approximate methods for a given bubble-distribution function.

8.10 Volume Reverberation: The Deep Scattering Layer

At the time when the first analytic studies in reverberation were made, it was realized that under certain conditions the reverberation observed in sonars must come from the body of the sea. The conditions involved downward refraction in deep water when the sound beam of a horizontally pointing sonar was bent downward deep into the body of the sea. It was also observed that the volume scatterers, whatever they may be, were not uniformly distributed in depth, but tended to be concentrated in a diffuse layer called, at the time, the ECR layer, in recognition of its three discoverers, C. F. Eyring, R. J. Christensen, and R. W. Raitt (20). The present name for this layer is the *deep scattering layer,* abbreviated DSL, which is now known to be more complex than previously realized.

The scatterers responsible for volume scattering appear undoubtedly to be biological in nature; that is, they are a part of the marine life existing in the sea. Nonbiological sources such as dust and sand particles, thermal microstructure (21), and turbulence of natural or human origin, such as ships' wakes, can be shown by analysis to be usually insignificant contributors to the scattering strength observed at sea.

Although the source of volume scattering in the sea has been definitely established as biological, the organisms responsible for it are of many different kinds. Many studies, such as those of Johnson, Backus, Hersey, and Owen (22) using short pulses of sound, deep-towed net hauls, and underwater photography, have been made of the organisms most likely responsible for the volume scattering of the deep scattering layer. Shrimplike euphausids, squid, and copepods have been postulated; and fish—more particularly their gas-filled swim bladders (23)—are the most likely source of the frequency-selective acoustic behavior of the layer at lower frequencies. Undoubtedly, the layer is a complex aggregate of different biological organisms, and therefore may be expected to possess a scattering strength varying with frequency, location, season, and even time of day.

At frequencies in excess of 20 kHz, the scatterers responsible for the DSL are likely to be *zooplankton,* or the smaller marine animals that feed upon the *phytoplankton* and are in turn fed upon by small pelagic fish. Examples are siphonophores and cephalopods. At lower frequencies from about 2 to 10 kHz, the dominant scatterers are the various types of *fish* that possess a *swim bladder*—an air-filled sac used by a fish to maintain and adjust its buoyancy. It amounts to an internal air bubble that becomes resonant at a frequency which depends on the size and depth of the fish. The resonance effect of the swim bladder is far less, however, than that of a free air bubble (Sec. 8.9) because it is not spherical in shape and, moreover, it is surrounded by lossy fish tissue. When they are sufficiently numerous, small fish without swim bladders can contribute to volume scattering, but the types that possess this organ appear responsible for the scattering at the lower end of the spectral range.

Variation with depth: diurnal migration The variation of volume scattering strength with depth has been studied by using depth-sounding sonars pointing vertically downward to measure the return as a function of time (and therefore depth) before the onset of the bottom return. The results of such measurements show an overall decrease of scattering strength with depth, consistent with general distribution of biological organisms with depth in the sea, but with an often well-marked increase of 5 to 15 dB within the DSL. The DSL itself generally exhibits a diurnal migration in depth, being at a greater depth by day than by night, and with a rapid depth change near sunrise and sunset. This diurnal depth migration often extends over several hundreds of feet and tends to be just sufficient to keep the intensity of light illumination constant at the depth of the layer (24).

Figure 8.11 shows scattering strength profiles for two Pacific Ocean areas at a frequency of 24 kHz. The average rate of decrease of S_v with depth is about

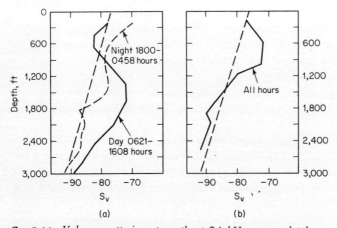

fig. 8.11 *Volume scattering strength at 24 kHz versus depth as measured in two Pacific Ocean areas: (a) Guadalupe Island area, latitude 29°N; (b) Queen Charlotte Island area, latitude 51°N. Dashed lines are estimated minimum values. (United States Navy Electronics Laboratory data quoted in Ref. 91.)*

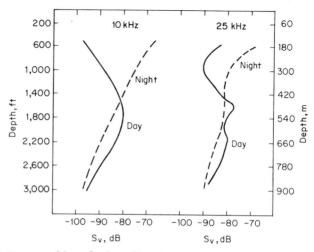

fig. 8.12 Mean depth profiles of S_v at two frequencies observed at six locations between Hawaii and California. (Ref. 25.)

5 dB/1,000 ft, as shown by the dashed lines. Additional profiles, obtained at several locations between Hawaii and California (25), are shown in Fig. 8.12. Here the difference between the nighttime and daytime profiles is striking, and represents the diurnal migration of the DSL. At a still lower frequency, and in another experiment (26), the same behavior is evident, again for an area off the California coast, as may be seen in Fig. 8.13.

The result of migration is to cause S_v to become markedly greater at shallow depths by night than by day—so much so that the shape of the profile is entirely different. This is illustrated vividly by the oscilloscope photographs using a 41-kHz depth sounder shown in Fig. 8.14. During the day there are almost no scatterers at shallow depths; at night they produce intense scattering within the first 200 ft or so below the surface. With an echo-sounder using short pulses at a high frequency, the DSL can be resolved into a fine structure of many sublayers, not all of which take part in the diurnal migration (27).

Variation with frequency At frequencies in excess of 20 kHz, the variation of S_v with frequency is slight or absent. Although no intensive study of S_v with frequency has been made in this frequency range, measurements at various times by different investigators in different areas show no clear frequency dependence of S_v at high kilohertz frequencies. This nondescript behavior is consistent with the view that the scattering originates from a variety of organisms of a variety of sizes and densities, many of the same dimensions as the wavelength, and all deformable and free to move in the sound field. Rayleigh fourth-power law scattering (Sec. 9.3), which applies for small fixed, rigid particles, can hardly be expected to apply under these conditions.

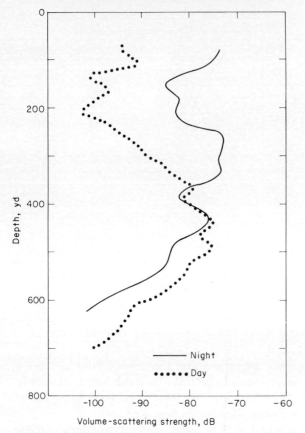

fig. 8.13 *Profiles of S_v against depth showing vertical migration. Frequency 5 kHz. The column scattering strength was -54.5 dB during the day and -50.5 dB at night. (Ref. 26.)*

On the other hand, at lower frequencies, observations with broadband short explosive pulses have revealed the existence of numerous and highly variable peaks in the volume reverberation spectra, each corresponding to a different depth in the sea. Different frequencies are found to peak at different depths—thereby indicating a multilayered structure within the DSL. This was most clearly indicated by the observations of Hersey, Backus, and Hellwig (28) using explosive sound sources at three locations in the western North Atlantic. At all locations, scattering layers with different resonant frequencies between 5 and 20 kHz were found. Most, but not all, layers showed a diurnal vertical migration, accompanied by a change in frequency. The depth, intensity, and resonant frequencies varied greatly from locality to locality.

Subsequent studies by Marshall and Chapman (29) using explosives confirmed this complex frequency-depth structure of the layer. For example, at one location northwest of Bermuda, a frequency analysis of the explosive reverberation showed a peak frequency of 5.4 kHz by day which shifted suddenly at sunset to a frequency of 4.7 kHz. The depth of the resonant layer (corresponding to the time of arrival of the onset of the peak frequency) changed from 2,200 ft by day to 1,600 ft by night, with a layer thickness of about 500 ft. The frequency change was interpreted as consistent with the behavior with pressure of a resonant air bubble of constant volume—suggesting the swim bladders of bathypelagic fish.

The result is that, at frequencies below about 15 kHz, S_v has a complex spectrum with a shape that is altered by the diurnal migration of the scatterers. Figure 8.15 shows examples of daytime and nighttime spectra of the column scattering strength (Sec. 8.7) as obtained by Chapman and Marshall (30) through analyses of explosive reverberation at a number of North Atlantic locations. In these spectra the shift toward lower frequencies at night is noteworthy, and is believed to be the result of a lowering of the resonance frequency of the air-filled swim bladders of fish in the water column as they migrate by night upward toward lower hydrostatic pressures.

More recent, and generally confirming, measurements of volume reverberation have been obtained at six locations off the coast of South America (31). Figure 8.16 shows the spread of the data obtained. Notice the fall-off below

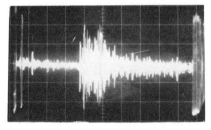

Day
Time: 1200 hours local time

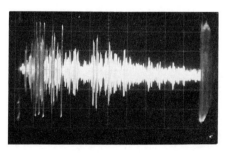

Night
Time: 2400 hours local time

Time or depth

fig. 8.14 *Volume reverberation received on a depth sounder. The bottom echo is seen at the right. The water depth was 600 ft in Dabob Bay, Puget Sound. The depth sounder had a pulse length of 2 ms at a frequency of 41 kHz. The horizontal scale is 23.4 ms per division.*

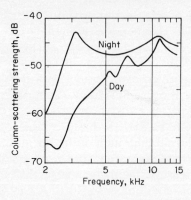

fig. 8.15 Spectra of column scattering strength to a depth of 850 m at a location between Nova Scotia and Bermuda, latitude 36°41'N. (Ref. 30.)

about 3 kHz and the large difference in scattering strength between night and day in this frequency range.

In summary, at frequencies of several kilohertz, the spectrum of volume reverberation originating at depths above about 3,000 ft is likely to contain one or more broad peaks of different frequency at different depths, locations, and time of day. Such frequency peaks apparently do not occur below about 1 kHz, since the lowest frequency thus far observed (32) is 1.6 kHz.

Deep-scattering-layer characteristics   Since the DSL will often be the dominant source of volume reverberation in broad-beamed or downward-tilted sonars, its acoustic characteristics will be of interest to the sonar engineer. These may be summarized as follows:

1. The DSL is a regular acoustic and biological feature of the world's oceans. As measured by the quantity column scattering strength, it is variable

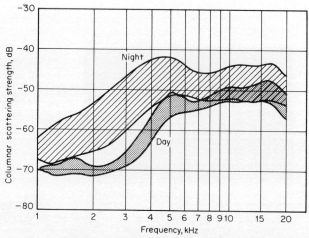

fig. 8.16 Columnar scattering strength versus frequency as measured at six locations in the tropical Western Atlantic by Gold and Renshaw (31). Nighttime data fell in the crosshatched region; daytime data in the stippled region.

over broad regions of the oceans, reflecting changes in biologic content in different areas. Canadian observations (33), made on a cruise between the two coasts of Canada via Cape Horn at the southern tip of South America, indicated that the column scattering strength had a maximum near latitude 45° North and 45° South, and a minimum near 20° North and South, with a secondary maximum near the Equator. The measurements between Nova Scotia and Puerto Rico are seen in Fig. 8.17. However, in another series of observations (34), the opposite latitude effect was found in the Pacific northward from Hawaii, where a decreasing column scattering strength occurred between latitudes 20 and 45°N.

2. The DSL tends to rise at sunset and descend at sunrise, and remains at constant depth during daytime and nighttime hours.

3. The depth of the DSL can be expected to lie between 600 and 3,000 ft (180 and 900 m) by day in midlatitudes, and to be shallower by night. In the Arctic, the deep scattering layer is no longer "deep," but lies just underneath the ice cover (35).

4. At frequencies near 24 kHz, the volume scattering strength within the layer is -70 to -80 dB, but is variable from place to place within the layer. At lower frequencies, S_v within the layer is variable and unpredictable.

5. S_v is frequency-selective at frequencies between 1.6 and 12 kHz, with different frequencies occurring at different depths. The DSL is therefore multilayered and contains layers of different resonant frequency at different depths, not all of which participate in the diurnal depth migration.

6. Layers of scatterers occur in shallow water as well. Most notable are schools of fish feeding at a particular depth. In the North Sea, a layer of scatterers, presumably biological, has been observed (36) at a depth of 15 fathoms, just at the top of a strong thermocline.

Wakes A target frequently encountered at sea, especially under conditions when ship maneuvers are involved, is the wake left behind by a ship or

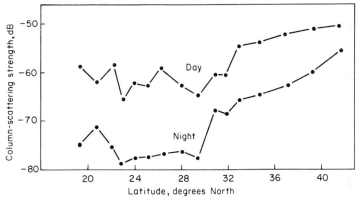

fig. 8.17 *Variation of column scattering strength $\int S_v\, dz$ with latitude by day and by night between Halifax, Nova Scotia, Canada, and Puerto Rico. Frequency band 1.6–3.2 kHz. (Ref. 33.)*

submarine. Because a ship's wake is an extended, rather than a discrete, target, its echo has some of the characteristics of reverberation, even though it may appear as a clear-cut, though extended, underwater target.

The acoustic characteristics of wakes were much studied during World War II, and a thorough discussion of the subject may be found in one of the Summary Technical Reports of the NDRC (12).

The wake of a ship is a length of bubbly turbulent water created by the ship's propeller. It extends an indefinite distance astern and gradually dissipates as the air bubbles rise to the surface or dissolve. Its width is initially that of the wake-laying ship, but gradually increases with distance astern. The thickness of the wake of a surface ship is initially about twice the draft and varies with distance astern in an uncertain manner.

The acoustic backscattering of wakes is described by a parameter termed *wake strength,* defined as the scattering strength *per yard of wake length.** Thus, wake strength is defined per unit length of wake in a manner analogous to volume scattering strength, per unit volume and area, respectively. The level EL of the echo from a wake of strength W is accordingly given by a pair of equations similar to those for surface reverberation:

$$EL = SL - 40 \log r + W + 10 \log L$$
$$L = \Phi r$$

where the term $40 \log r$ implies spherical spreading in the absence of absorption, and the echoing length L of the wake is the product of the equivalent plane-angle beam width Φ, and r is the range in yards. This formulation applies only for long pulse lengths such that the dimensions of the wake returning sound at any one instant of time are determined by the beam width and range alone.

Some typical values of W, measured at distances a few hundred yards astern of the wake-producing vessel, are given in Table 8.2. W was found to decrease with the "age" of the wake (in effect, the distance astern) at the rate of about 1 dB/min and to increase with the size of the vessel (for surface ships). No clear variation with frequency over the range 15 to 60 kHz was evident.

8.11 Sea-Surface Reverberation

Because of its roughness and the possibility of the occurrence of entrapped air bubbles just beneath it, the sea surface is a profound scatterer of sound. This scattering has long been noted to be responsible for the reverberation received in horizontally pointing sonars under conditions such that the sound beam is not carried down into the depths of the sea by refraction. It is analogous in many ways to the "sea clutter" of radar.

---

* W was originally defined (12) per *foot* of wake. A reference length of 1 yd is used here for dimensional conformity with scattering strength. Values of W quoted in Ref. 12 have accordingly been increased by $10 \log 3 = 5$ dB.

**table 8.2 Wake Strengths W for Various Ships and Submarines
(Ref. 12, Tables 4 and 6)**

Ship	W, dB	Frequency, kHz	Speed, knots	Depth, ft
Surface ships				
CVE's and AP's	−9	24		
DD's and DE's	−11	24		
Laboratory yachts ("Scripps" and "Jasper")	−15	24		
Small boats	−19	24		
Submarines				
USS-S23 (SS-128)	−13	60	9.5	Surfaced
	−21	60	6	90
USS-S34 (SS-139)	−8	45	9.5	Surfaced
	−18	45	6	90
USS "Tilefish" (SS-307)	−8	45	9.5	Surfaced
	−15	45	6	90
USS-S18 (SS-123)	−28	45	6	45

Sea-surface scattering strengths have been measured by nondirectional (mostly explosive) sources and receivers and by directional sonars in which a sound beam is pointed upward to intercept the sea surface at a desired angle. The scattering strength of the sea surface has been found to vary with angle, frequency, and the roughness of the surface, measured by the speed of the wind above it.

Variation with angle and wind speed The first systematic postwar investigation of sea-surface backscattering appears to have been published by Urick and Hoover (5), through measurements of the sea return at 60 kHz made with a tiltable directional transducer hung beneath the surface from a float. The results of this study are shown in Fig. 8.18. S_s is seen to increase with wind speed at low grazing angles, but to decrease with wind speed at high angles near normal incidence, with a "crossover" near 80°. This peculiar behavior was attributed (37) to different processes responsible for the backscattering over different ranges of angle. At low angles (<30°), the scattering was attributed to a layer of air bubbles just beneath the sea surface, for which observational evidence was presented; at intermediate angles (30 to 70°), scattering was invoked as the dominant process, and at high angles (70 to 90°), the return was postulated to originate as reflection from normally inclined wave facets acting as tiny acoustic mirrors. Scattering curves at 60 kHz similar to those of Fig. 8.18, but without the flattening at low angles attributed to a bubbly layer, have been derived by Garrison, Murphy, and Potter (38) from measurements in Dabob Bay, Puget Sound, Washington. These curves are shown in Fig. 8.19.

At lower frequencies, explosive sound sources together with a nearby nondirectional hydrophone have been used to obtain the scattering strength of

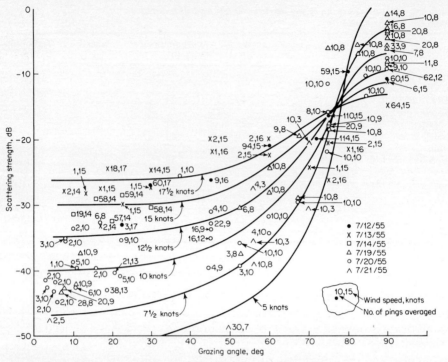

fig. 8.18 *Variation of sea-surface scattering strength at 60 kHz with angle at different wind speeds off Key West, Florida. Plotted points show the number of pulses averaged and the wind speed for individual determinations. (Ref. 5.)*

the sea surface. In this experimental arrangement, illustrated in Fig. 8.20, the time after detonation is related to the angle between the sea surface and the incident direction, an angle that becomes progressively smaller with increasing time.

A series of measurements of this kind was carried out by Chapman and Harris (8) over a 52-hour period, during which the wind speed ranged from zero to 30 knots. The data were analyzed in octave bands between 0.4 and 6.4 kHz, and the results were fitted by the empirical expressions

$$S_s = 3.3\beta \log_{10} \frac{\theta}{30} - 42.4 \log_{10} \beta + 2.6$$

$$\beta = 158 \, (vf^{1/3})^{-0.58}$$

where S_s is the surface scattering strength at grazing angle θ in degrees, wind speed v is in knots, and frequency f is in hertz. Figure 8.21 shows curves computed from this formula for several frequencies with wind speed as a parameter. The average difference between the observed and computed values of S_s was said to amount to only 2 dB. A subsequent series of measurements over a wider frequency range was made by Chapman and Scott (39).

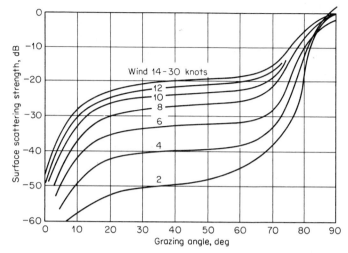

fig. 8.19 *Variation of sea-surface scattering strength at 60 kHz with angle at different wind speeds in Dabob Bay, Puget Sound, Washington. (Ref. 38.)*

Chapman and Marshall (30) also made additional measurements over an extended frequency range by the same method, and obtained similar results.

As at 60 kHz, the low-frequency scattering strength of the sea surface increases with wind speed and angle over most of the angular range. Most measurements, such as those of Richter (40), made with an explosive source and a nondirectional hydrophone, show S_s to be independent of angle at low angles and low wind speeds—somewhat like the measurements of Fig. 8.18, taken at 60 kHz with a directional transducer. More recently, it has become

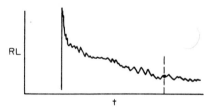

fig. 8.20 *Measurement of sea-surface scattering strength at different angles with a non-directional hydrophone and an explosive charge.*

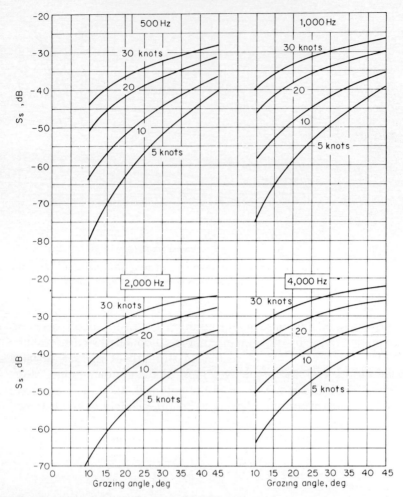

fig. 8.21 Low-frequency backscattering strength of the sea surface computed from empirical expressions given by Chapman and Harris. (Ref. 8.)

clear that under these conditions the deep scattering layer dominates the reverberation return and the reverberation received is not properly interpreted as arising at the sea surface.

Variation with frequency   A composite plot of a number of determinations of S_s over a wide frequency range is given in Fig. 8.22, showing measurements by Chapman and Scott (39), Urick and Hoover (5), and Garrison, Murphy, and Potter (38). It is clear that at low angles a strong frequency dependence exists, amounting to about 3 dB/octave. This dependence vanishes at angles approaching normal incidence. This behavior is consistent with present understanding of the origin of sea-surface scattering, as will be explained in the section to follow.

8.12 Theories and Causes of Sea-Surface Scattering

The theory of the scattering of sound by rough surfaces has received considerable attention, and a relatively abundant literature is available. An introduction to this literature may be found in the references quoted in a paper by Miles (41); an extensive theoretical background also exists for the rough-surface scattering of electromagnetic waves (42). Fortuin (43) has prepared an excellent summary, containing some 87 cited references, of the theoretical literature dealing with reflection and scattering of sound by the sea surface. Unfortunately, much of this literature is of interest only to the theoretician, and only two of the many theoretical discussions of the subject have been carried to the point where a comparison with observed data of sound scattering by the sea surface is possible.

One of these is the theory of Eckart (44) based on the Helmholtz solution of the wave equation in the form of a surface integral over the scattering surface. The result has been reduced to practical form by Chapman and Scott (39):

$$S_s = -10 \log 8\pi\alpha^2 - 2.17\alpha^{-2} \cot^2 \theta$$

where θ = grazing angle, deg
α^2 = mean-square slope of surface waves

From measurements of the sun's glitter, Cox and Munk (45) obtained the following empirical relationship between mean-square wave slope and wind speed:

$$\alpha^2 = 0.003 + 5.12 \times 10^{-3} W$$

where W is the speed of the wind in meters per second. Good agreement was

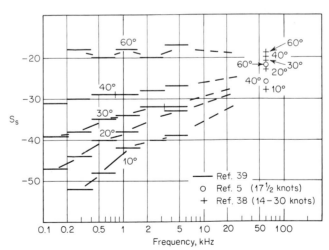

fig. 8.22 Variation with frequency of the surface scattering strength at a wind speed of 20 knots for various grazing angles.

found by Chapman and Scott between this theory and their field data, shown previously in Fig. 8.21, at angles greater than 60°. No frequency dependence is predicted by the Eckart relationship, and none is evident in the field data at angles of 60° and beyond.

Another theoretical result has been derived by Marsh (46, 48) and by Marsh, Schulkin, and Kneale (47). Here, the result of scattering back toward the source of sound is

$$S_s = 10 \log \left[\frac{\tan^4 \theta}{32} \frac{\omega^5 A^2(\omega)}{g^2} \right]$$

where θ = grazing angle
g = acceleration of gravity
$A^2(\omega)$ = "power" spectrum of sea-surface elevation at angular frequency ω

The sea surface is postulated to act as a diffraction grating; the scattering is produced by only those waves or wavelets of the sea-surface roughness that "match" the incident sound, that is, those that produce reinforcement in the sound scattered back toward the source. This process will be considered somewhat more fully in the section on bottom scattering. If, following Marsh, the sea-surface spectrum is taken to be that postulated by Burling (49), namely,

$$A^2(\omega) = 7.4 \times 10^{-3} g^2 \omega^{-5}$$

a particularly simple result is obtained. It is

$$S_s = -36 + 40 \log \tan \theta$$

This result was compared by Marsh (48) with measured explosive-source sea-surface scattering strengths in sea states ranging from 0 to 4. The comparison is shown in Fig. 8.23. The measured data points fell in the area shown shaded in the figure, with no apparent dependence on frequency or sea roughness. Although some degree of agreement is evident, it should be noted that the Marsh theory predicts no dependence of scattering on either frequency or wind speed, in contradiction to the bulk of measured data on sea-surface scattering, and fails to give meaningful results in the vicinity of normal incidence ($\theta = 90°$).

A more recent extension of this theory published by Bachmann (50) includes the effects of the high-amplitude long-wavelength portion of the wave spectrum, or the "swell." At low grazing angles, the swell acts to produce shadowing of wave crests as well as to increase the grazing angle at the exposed portions of the sea surface. The theoretical result is

$$S_s = 10 \log \pi^3 \mu A \tan^4 (\theta + \Delta\theta)$$

where $\mu = 2.2 \times 10^{-7} U^{7/2}$ with U the wind speed in meters per second; A is a shadowing factor less than unity that is negligible below 5° at all wind speeds; and $\Delta\theta$ is the effective increase of grazing angle due to the swell. At a wind speed of 16 knots and for $\theta = 5°$, $\Delta\theta$ amounts to 2°. In agreement with this

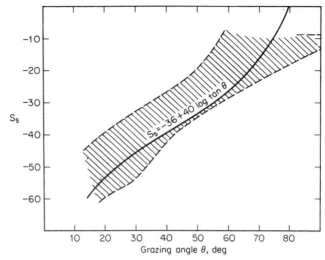

fig. 8.23 *Backscattering strengths (shaded area) measured at five locations in the Norwegian Sea, compared with theoretical curve based on the sea roughness spectrum. [After Marsh (Ref. 48).]*

theory are Russian measurements (51) at 1, 3, and 5 kHz using a directional transducer to exclude the contribution of the deep scattering layer to the backscattered return. It is noteworthy that the above theoretical result shows no frequency dependence, in contrast to the empirical findings of Chapman and Harris (Fig. 8.21).

It is almost intuitively apparent that a rough surface must become effectively smooth when the wavelength with which the surface is "viewed" becomes large enough or when the grazing angle becomes small enough. That is, the backscattering must vanish at long wavelengths and small grazing angles. In Fig. 8.24, the backscattering must depend on the ratio of distance $2AA'$ (the average path difference for backscattering between the peaks and troughs of the rough surface) to wavelength. This led Schulkin and Shaffer (52) to interpret essentially all the available published data on the sea-surface backscattering coefficient in terms of the ratio $2h(\sin \theta)/\lambda$, where h is the mean peak-trough roughness, λ the wavelength, and θ the grazing angle. In terms of the frequency f, this ratio becomes proportional to $fh \sin \theta$. The summary expression obtained was

$$S_s = 10 \log (fh \sin \theta)^{0.99} - 45.3$$

It was found that the data could be fitted by this expression with a standard deviation of 5 dB over a range of $fh \sin \theta$ between 4 and 200. However, this empirical expression must not be used near normal incidence, nor at angles so small that some source of scattering other than the sea surface may dominate the observed reverberation. As a practical aid to using this expression, a

270 / principles of underwater sound

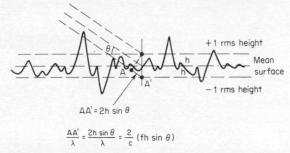

fig. 8.24 *Relationship of roughness, wavelength, and angle as a backscattering parameter.*

relationship between crest-to-trough wave height h, in feet, and wind speed V, in knots, was given by Schulkin and Shaffer as

$$h = 0.0026 V^{5/2}$$

From a physical point of view, the peculiar behavior of the scattering coefficient with angle suggests that different scattering processes operate in different regions of grazing angle in somewhat the same way that the peculiar spectra of ambient noise suggest different origins of the noise in different spectral regions. As mentioned above, three processes have been postulated (37) to account for the S-shape characteristic of curves of S_s versus angle. In Region I of Fig. 8.25, the backscattering is conjectured to be due to a layer of scatterers (probably air bubbles whipped into the sea at moderate and high sea states) just below the sea surface itself or, when a nondirectional system is involved, to scatterers in the deep scattering layer. In this angular region, the characteristics of the bubbly near-surface layer required to produce the observed backscattering strengths have been considered by Clay and Medwin (53). An alternative explanation of the flattening of the curves in Region I is the shadowing

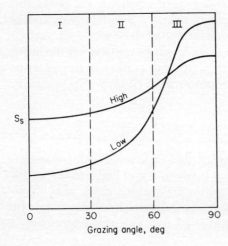

fig. 8.25 *Three regions of different sea-surface backscattering processes for two wind speeds.*

of portions of the sea surface by the troughs of intervening waves. This explanation is particularly appealing at low frequencies where the bubble sizes required for resonance in the bubbly layer are doubtless scarce or absent. The effect of shadowing on backscattering has been placed on a good theoretical and observational foundation by the work of Gardner (54). In Region II, the dominant process is one of scattering by roughnesses comparable in size to the acoustic wavelength, whereas in Region III, the process is essentially one of reflection (that is, coherent scattering) by wave facets inclined at right angles (for backscattering) to the direction of the incident sound. It may be noted that, from a scattering point of view, some interesting analogs exist, both as to processes and magnitude of the scattering, in the behavior of the sea surface in scattering electromagnetic waves incident from above and in scattering sound waves incident from below (55).

8.13 Sea-Bottom Reverberation

Like the sea surface, the bottom is an effective reflector and scatterer of sound and acts to redistribute in the ocean above it a portion of the sound incident upon it. This redistribution is illustrated in Fig. 8.26, which shows the "beam pattern" of the sea bottom for sound incident on it. In this diagram the radius of the pattern in any direction is proportional to the intensity of the sound scattered or reflected in that direction. Pattern A might be characteristic of smooth reflecting bottoms, while pattern B might be typical of rough scattering bottoms, for which the intensity is less in the specular direction OR and correspondingly greater in other directions. Patterns such as these have been

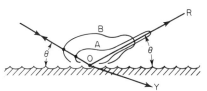

fig. 8.26 *"Beam pattern" for scattering of sound incident at angle θ. The radius of the pattern is proportional to the intensity, per unit solid angle, of sound scattered or reflected in any direction. OR is the specular direction. The arrow Y indicates that some sound enters the bottom as compressional and shear waves.*

much studied in connection with the glossiness of surfaces for light (56), but have not yet been obtained for sound scattering from the sea bottom. In the specular direction OR, the *reflection-coefficient* has been considered as a part of the subject of sound propagation in the sea (Sec. 5.8).

Scattering occurs three-dimensionally, as well as in the plane containing the source and the normal to the surface. In one study (57) this "side scattering"

272 / principles of underwater sound

was found to be uniform; a sand bottom at very low grazing angles was discovered to be an isotropic scatterer of 24-kHz sound in horizontal directions.

Variation with angle and bottom type It was observed at an early date that the reverberation from the seabed was greater over rocky bottoms than over mud bottoms. It has since become customary to relate bottom scattering strength to the type of bottom, such as mud, silt, sand, boulders, rock, even though it is realized that the size of the particles comprising a sedimentary bottom is only an indirect indicator of acoustic scattering.

A relatively large number of mostly discordant measurements have been reported in the literature. As for the sea surface, the scattering strength of a particular bottom is found, in practice, by measuring the backscattering it produces and converting to scattering strength by the appropriate form of the basic equation for reverberation. Measurements have been made at high frequencies with tiltable directional transducers, in which the angle at which the sound beam strikes the bottom is the angle of tilt, corrected if necessary for refractive bending, and at low frequencies with nondirectional sources and receivers, where the angle is determined by the time after emission of the ping or the explosive pulse.

Figure 8.27 is a compilation of high-frequency measurements in the range 24 to 100 kHz for a variety of different bottom types, as reported by different observers. Most of these observations were made in shallow coastal locations. Included are average values obtained during World War II (Ref. 58, p. 317), examples of individual determinations in Narrangansett Bay, Rhode Island (2), and average curves obtained from measurements in 16 different coastal locations (59).

Figure 8.28 is a similar compilation, but for low-frequency deep-water measurements, all made in water depths in excess of 350 fathoms. In most cases,

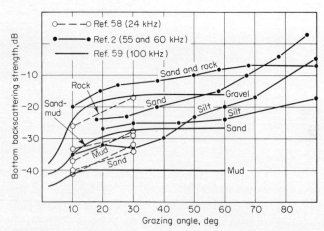

fig. 8.27 *Measured backscattering strength of the seabed at various coastal locations.*

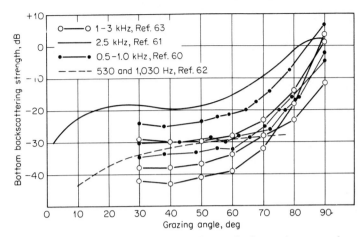

fig. 8.28 *Low-frequency deep-water bottom backscattering strength as reported in four sets of measurements.*

these data were obtained with nondirectional transducers, with explosive sources being used for two sets of measurements. The bottom at such deep-water locations is the fine mud of the deep abyssal sea. Included here are data reported by Urick and Saling (60) at three locations at the outer edge of the continental slope in water 2,100 to 2,800 fathoms deep, by Patterson (61) in water 12,000 ft deep, by Mackenzie (62) in 2,000-fathom water near San Diego, and by Burstein and Keane (63) at three locations in depths of 350, 1,500, and 2,050 fathoms.

A Lambert's law relationship (see below) between backscattering strength and angle appears to provide a good approximation to the data for many deep-water bottoms at grazing angles below about 45°. This was first pointed out by Mackenzie (62) on the basis of his own and older data. Figure 8.29 shows Mackenzie's plot, to which more recent data points by Schmidt (64) and

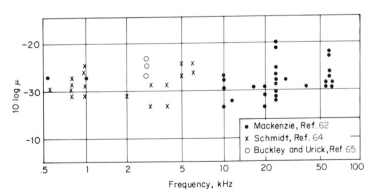

fig. 8.29 *The quantity $10 \log \mu$ in the Lambert's law relationship $S_B = 10 \log \mu + 10 \log \sin^2 \theta$, as found from various field measurements at different frequencies.*

Buckley and Urick (65) have been added. The ordinate in this figure is the quantity $10 \log \mu$ in the expression $S_B = 10 \log \mu + 10 \log \sin^2 \theta$, where θ is the grazing angle. According to Fig. 8.29, $10 \log \mu$ averages to be about -29 dB; hence, at an angle of $10°$ according to this expression, S_B would be equal to $-46\frac{1}{2}$ dB over a wide range of frequencies. Although this regular behavior is indeed satisfying, we must realize that in all likelihood it applies only for one acoustic bottom type (type III, see below), and then only for small to moderate grazing angles. For example, the results obtained by Merklinger (66) at five locations over three physiographic regions between Nova Scotia and Bermuda clearly fail to obey such a simple relationship.

It is evident from these figures that the engineer will have considerable difficulty in selecting a value of S_B for use in a practical problem. Bottom composition, as expressed conveniently by particle-size designators like "mud" and "sand," gives only an approximate indication of the backscattering strength at a given grazing angle. Some property of the seabed other than the size of its sedimentary particles apparently must govern the backscattered return.

Variation with frequency A comparison of Fig. 8.28 for low frequencies with the mud and sand curves of Fig. 8.27 at higher frequencies would indicate an absence of a strong frequency dependence of the bottom backscattering coefficient. This lack of frequency dependence was pointed out by Mackenzie (62), who, by comparison with older data, found coefficients at 530 and 1,030 Hz essentially identical with previously reported values at 10 kHz and above.

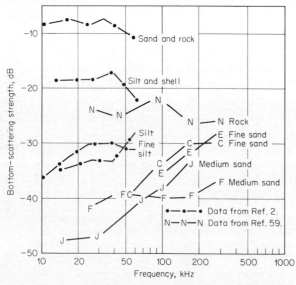

fig. 8.30 *Frequency variation of bottom backscattering. Dots, 30° grazing angle (Ref. 2); letters, 10° grazing angle. (Ref. 59).*

However, a closer look at the data would suggest a definite frequency variation at frequencies beyond 10 kHz for the smoother (mud and sand) bottoms. Figure 8.30 shows the data of Urick (2) at a grazing angle of 30° and of McKinney and Anderson (59) at an angle of 10° extending over a wide frequency range. The sand and silt bottoms show a rise in scattering strength with frequency at the rate of about 3 dB/octave (or a variation as the 1.0 power of the frequency), whereas little or no frequency dependence appears to exist for the "rock," "sand and rock," and "silt and shell" bottoms. This behavior may be attributed to a difference in the scale of bottom roughness. Bottoms with roughness large compared to a wavelength have a backscattering coefficient independent of frequency; bottoms having an appreciable portion of their roughness spectrum at roughnesses small compared to a wavelength have a scattering strength which increases with frequency.

This behavior has been confirmed by Russian observations (67) of deep-sea backscattering at 100 areas of the Atlantic Ocean, using small explosive charges and filtering in the band 1 to 30 kHz. The results, shown in Fig. 8.31, were broken down into three broad areas, corresponding to different degrees of bottom roughness with the results plotted against angle (at a number of fixed frequencies) and against frequency (at a number of fixed angles). Three basic bottom types were recognized. One, here called type I, is the deep-sea abyssal plain having little roughness, in which the angular variation and frequency variation (as $f^{3.8}$) are high. At the other extreme are areas which we may call type III, having heavily dissected bottoms with underwater ridges, in which a Lambert's law variation with angle is observed with no frequency dependence over the 1- to 30-kHz range. An intermediate type of bottom may be called type II, termed "hill regions" (in probable reference to the continental slope and the continental rise), in which an intermediate behavior with angle and frequency was observed. These observations were said to be in agreement with theoretical results in which the scattering coefficient is expressed as the sum of two terms, one frequency independent and contributed to by large irregularities, the second strongly frequency dependent (as f^3) and resulting from irregularities small compared to a wavelength.

A compilation of bottom scattering strength measurements from a large variety of sources, including new data at 48 kHz and at low grazing angles, shows no apparent dependence upon frequency over the range 0.53 to 100 kHz (68). From this compilation, the smooth curves of Fig. 8.32 have been drawn to assist the selection of a value for this troublesome sonar parameter.

Causes and theories of sea-bottom backscattering Although particle size, for sedimentary bottoms, serves as a first-cut means of classifying bottoms in terms of acoustic backscattering, the roughness of the sea bottom appears to be the dominant determining characteristic for backscattering. Just as wind speed is an indicator of surface roughness, so is bottom type an indicator of bottom roughness. However, since the sea bottom is partly transparent to

276 / **principles of underwater sound**

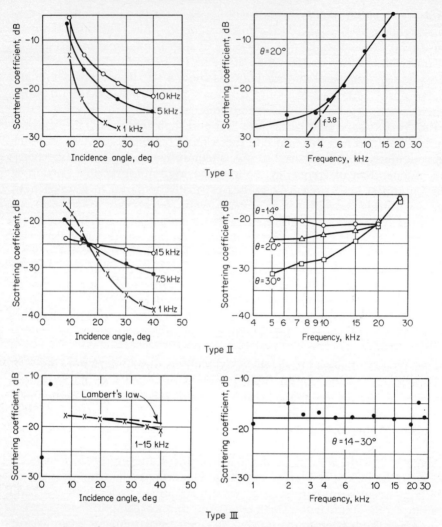

fig. 8.31 Bottom backscattering strength in deep-water areas of three different bottom types plotted against angle and frequency. In the left-hand figures the angle is the angle of incidence, with 0° normal incidence and 90° grazing incidence. (Ref. 67.)

sound, in the sense that a portion of the incident acoustic energy is transmitted to the earth below, bottom type must in part determine the partition of energy between the water above and the bottom beneath.

In addition to the roughness, other processes may, in some circumstances, play a significant role in the scattering process. One is the particulate nature of sedimentary bottoms, where each sedimentary particle may itself be imagined to be a scatterer of sound, and where the return of sound from the bottom is produced by a form of volume reverberation within the bottom itself. However, studies in model tanks having perfectly smooth bottoms (69)

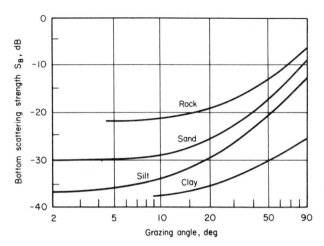

fig. 8.32 Smooth curves of bottom backscattering strength S_B as a function of grazing angle, based on the compiled data of Ref. 68.

show far too little backscattering in comparison with natural bottoms; moreover, only a little roughening of the smooth sand suffices to greatly increase the backscattered return. Penetration of sound into the bottom and its subsequent reradiation as backscattering may well be an important process at low frequencies and high grazing angles, where reflection and scattering from subsurface layers and other inhomogeneities may occur. At high frequencies, however, there appears little justification for invoking processes other than the irregular discontinuity of acoustic properties occurring at the bottom itself to explain the principal characteristics of the bottom return.

The function called Lambert's law is a type of angular variation which many rough surfaces appear to satisfy for the scattering of both sound and light. In Fig. 8.33, let sound of intensity I_i be incident at angle θ on the small surface area dA. The power intercepted by dA will be $I_i \sin \theta \, dA$. This power is *assumed*, by Lambert's law, to be scattered proportionately to the sine of the angle of scattering, so that the intensity at unit distance in the direction φ will be

$$I_s = \mu I_i \sin \theta \sin \varphi \, dA$$

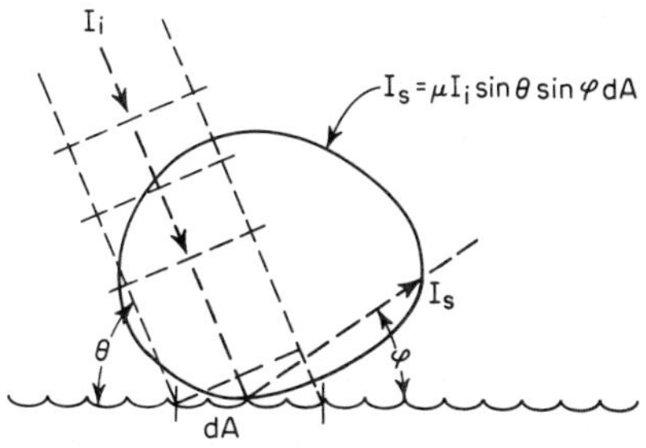

fig. 8.33 Lambert's law for a scattering surface.

where μ is a proportionality constant. For a unit area we have, on taking 10 times the logarithm of each side,

$$10 \log \frac{I_s}{I_i} = 10 \log \mu + 10 \log (\sin \theta \sin \varphi)$$

and in the backward direction, for which $\varphi = \pi - \theta$,

$$S_B = 10 \log \mu + 10 \log \sin^2 \theta$$

Thus, by Lambert's law, the backscattering strength must vary as the square of the sine of the grazing angle. If all the incident acoustic energy is redistributed into the upper medium, with none lost by transmission into the medium below, then it can be shown by integration that $\mu = 1/\pi$; the normal-incidence backscattering strength would therefore be $10 \log (1/\pi) = -5$ dB. It should be emphasized that the law rests on a particular assumption concerning the redistribution of the scattered energy in space. Although many materials follow Lambert's law closely in scattering light, none does so exactly. Lambert's law applies specifically to the radiation of light by radiant, absorptive materials (70); the "law" should properly be called Lambert's "rule" for scattering. Nevertheless, it is, as we have seen, a good description of the backscattering of sound by very rough bottoms.

Marsh's theory of scattering, previously referred to in connection with the sea surface, is applicable to the sea bottom as well. In terms of the wave numbers k and K, referring to the wavelengths of the incident acoustic wave and of a component of the bottom roughness, respectively, the Marsh formula for a bottom transmitting no energy to the solid medium below can be written

$$S_B = 10 \log \left[\frac{\sin^4 \theta}{\pi \cos \theta} k^3 A^2(K) \right]$$

where $A^2(K)$ is the amplitude squared of the irregularity in the bottom contour having wave number K. $A^2(K)$ is the bottom-roughness power spectral density. If λ_a is the wavelength of the incident sound wave, and λ_b the wavelength of a component of bottom roughness, then $k = 2\pi/\lambda_a$, and $K = 2\pi/\lambda_b$. The essential concept of the theory is that the rough bottom acts as a diffraction grating for scattering in the manner first described by Rayleigh (71), wherein the scattering is dominantly produced by that wavelength component of the bottom roughness for which phase reinforcement occurs for scattering back in the direction of the incident sound. In Fig. 8.34, the wavelength component λ_b of the roughness spectrum responsible for backscattering is that for which the scattering from points A and A' are in phase at a large

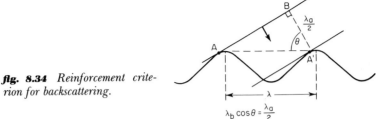

fig. 8.34 Reinforcement criterion for backscattering.

distance. The distance BA' must therefore be one-half wavelength, so that λ_a and λ_b become related by

$$\lambda_b \cos \theta = \frac{\lambda_a}{2}$$

or

$$K = 2k \cos \theta$$

Unfortunately, few data exist on the roughness power spectrum of the sea bottom. Marsh (72) has compiled the roughness spectra for various land, sea, and bottom surfaces, as shown in Fig. 8.35. It is interesting to observe that bottoms having a "white" spectrum, such as area J of Fig. 8.35, would, by this theory, have a scattering strength varying as the cube of the frequency; those

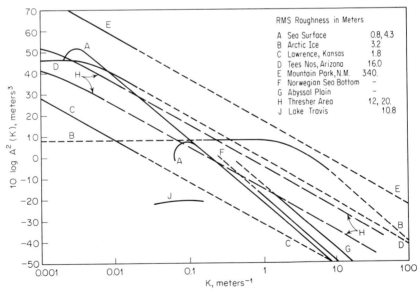

fig. 8.35 Geophysical elevation spectra. [Compiled by Marsh (72).] Included are spectra of the sea and ice surfaces (A, B), land areas (C, D, E), and the sea bottom (F, G, H, J).

for which $A^2(K)$ varies as K^{-3}, or as the inverse cube of the frequency, would show a scattering strength independent of frequency.

8.14 Under-Ice Reverberation

The undersurface of an ice cover may be expected to produce strong reverberation when it is rough and has a jagged, irregular under-ice topography. Figure 8.36 shows the results of measurements from the backscattering strength of the ice-covered sea for two Arctic locations at different times of year. The spring pack ice measured by Milne (73) during late April and early May was described as consisting "largely of broken and pressure-packed one-year ice intermixed with broken frozen leads." The summer polar ice, for which much smaller backscattering strengths were obtained by Brown (74) during the month of September, was not described, but was probably less irregular in its bottom contour. Included in Fig. 8.36 are curves of the backscattering of the ice-free sea at a wind speed of 25 knots. Both sets of data for under-ice scattering, although discordant, show an increase of scattering strength with frequency and grazing angle. Certain other data (75), however, indicate an absence of a frequency variation, and so may be characteristic of a third kind of under-ice topography. The engineer, accordingly, has only fragmentary data available on which to base a prediction of the reverberation to be expected under an ice cover, although it is clear that backscattering strengths higher than those for the ice-free sea at low and moderate wind speeds must prevail.

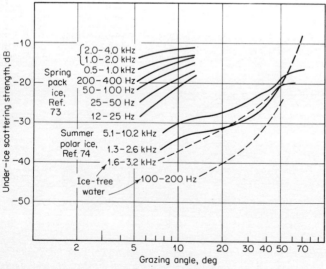

fig. 8.36 *Scattering strengths of an ice cover.*

8.15 Reverberation in Shallow Water

We have already seen, in Fig. 8.1, that the reverberation envelope in deep water contains a series of sharp peaks representing successive repeated bounces of sound between the sea surface and the sea bottom. These peaks can be eliminated, if desired, by incorporating sufficient vertical directivity in the system. In shallow water, on the other hand, they are crowded together in time and occur shortly after emission of the sonar pulse; at later times, the reverberation has a smoothly decaying envelope. It represents backscattering from the sea surface and from the sea bottom of sound that is propagated outward toward, and backward from, the scatterers.

When downward refraction occurs, the scattering from the bottom dominates the reverberant return and the reverberation level can be crudely predicted from a simple model once the bottom scattering strength is estimated (76). When upward refraction occurs, as in isothermal water, surface reverberation is more important than bottom reverberation, and the reverberation level is found to depend on wind speed and sea state. Under these circumstances it is often found, strangely enough, that when a target echo occurs, the echo-to-reverberation ratio tends to be sensibly independent of range; that is, the echo level falls off with range at the same rate as the reverberation.

Shallow-water reverberation tends to be strongly coherent in the vertical (77), with a correlation coefficient between vertically separated hydrophones that increases with time or range. This occurs because the high angle paths, along which the backscattered return arrives when the range is short, are eliminated in propagation to and from long ranges in shallow water, and the reverberation becomes more nearly coherent as time goes on.

It follows that *vertical* directivity provides little array gain over reverberation in shallow water. On the other hand, in the *horizontal*, reverberation is only weakly coherent (78) and a horizontal array is as effective a suppressor of reverberation in shallow water as it is in deep water.

In shallow waters where fish are prevalent, volume reverberation caused by fish and other organisms is important. Biological scattering results in an increased attenuation of transmitted sound and, when fish schooling takes place, produces discrete echolike blobs in the shallow-water reverberation envelope.

8.16 Characteristics of Reverberation

With a pulsed sinusoidal sonar, the reverberation that follows the emitted pulse is heard as an irregular, quivering, slowly decaying tone. Within the smooth decay there appear onsets of increased reverberation whenever the emitted sound intercepts the sea surface and bottom, as well as blobs of reverberation of roughly the same duration as that of the pulse. In this section, some of the characteristics of reverberation, other than its intensity, are briefly described.

Amplitude distribution Because the backscattered return originates at a large number of scattering obstacles in the sea and on its boundaries, such as the roughness irregularities of the seabed, the instantaneous reverberation amplitude fluctuates from instant to instant as the number and phase relationship of the scatterers change with range or time. Field observations long ago (79) showed that the instantaneous reverberation amplitude is Rayleigh-distributed; that is to say, the probability distribution of the amplitude is that found originally by Rayleigh (80) for a large number of sine waves of equal amplitude but random phase. In such a distribution, the probability that the instantaneous intensity, or amplitude squared, is greater than the value I is exp $(-I/\bar{I})$, where $\bar{I}$ is the average intensity, or mean-square amplitude. This means that the reverberation amplitude is greater than its rms value about 37 percent of the time and less than its rms value 63 percent of the time, with a marked tendency for the reverberation to be below the rms value at any given time. The Rayleigh distributed property of reverberation has been repeatedly confirmed in investigations since World War II (81, 82).

Coherence By *coherence* is meant the autocorrelation and crosscorrelation properties of reverberation. The reverberation observed with a pulsed sonar is well known to have an envelope containing blobs of approximately the same duration as that of the emitted pulse. This blobby character of reverberation is illustrated in Fig. 8.37. It is evidenced statistically by the finding (79) that the autocorrelation coefficient of the reverberation envelope decreases to nearly zero in a time interval equal to a pulse duration.

The coherence of the reverberation received at two vertically separated hydrophones has also been studied (32). It was found that bottom reverberation at the times near the onset of the bottom return is highly coherent and that that from the deep scattering layer is much less so, with a crosscorrelation coefficient diminishing for both sources of reverberation with increasing hydrophone separation and increasing frequency. As an example, the peak correlation coefficient, read from correlograms obtained with a clipper correlator, was found to be about 0.8 for the bottom return near normal incidence, but only 0.3 or less for surface or deep-scattering-layer reverberation at a vertical separation of hydrophones of 10 ft and in the frequency band 1 to 2 kHz. Basic information of this sort is of interest for the design of hydrophone arrays to suppress reverberation (Sec. 3.1).

Frequency distribution Observations show that the reverberation that follows a sinusoidal sonar pulse does not in general lie entirely at the frequency of the pulse, but is both shifted in frequency and spread out into a frequency band. The shift in center frequency is simply the doppler shift caused by the velocity of the platform on which the sonar is mounted, together with any uniform velocity of the reverberation-producing scatterers themselves.

The other frequency characteristic of reverberation, frequency spread, is due to several causes. One is the finite duration of the sonar pulse itself,

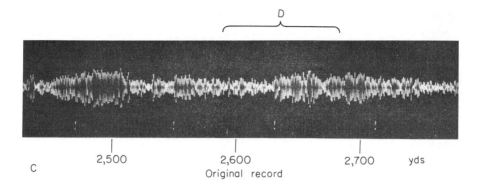

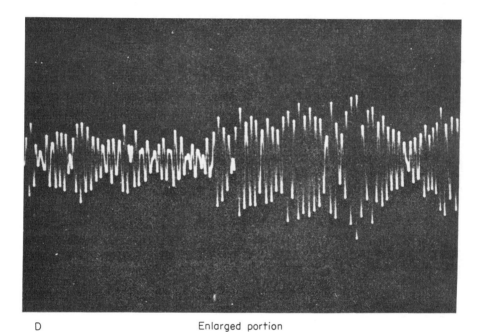

fig. 8.37 *A portion of an original recording of reverberation and an enlarged portion, illustrating the blobby, irregular envelope of pulsed sinusoidal reverberation. (Ref. 79, fig. 12.)*

which, if it is t seconds long, has a frequency spread of $1/t$ hertz about the center frequency. Another is the fact that on a moving ship the reverberation arriving from different directions experiences a different doppler shift because of the differing components of motion in different directions. For example, with a transducer of an equivalent plane beam width Φ facing in the forward direction on a vessel under way with velocity v, the reverberation emitted and arriving on the axis of the beam would have a doppler shift of $(2v/c)f$ hertz, whereas that on the edges would be shifted by $(2v/c)f \cos(\varphi/2)$. The reverberation band would therefore be $(2vf/c)[1 - \cos(\varphi/2)]$ hertz wide.

A third factor causing a band spread of reverberation is the doppler spread caused by the motion of the scatterers themselves. Although the first two of these causes are readily calculated and can be allowed for in sonar design, the motion of the reverberation-producing scatterers is a largely unknown oceanographic characteristic of the sea.

Early work during World War II was done with a frequency analyzer called a period meter (79), but the interpretation of the results obtained in terms of the spectrum of reverberation has remained obscure. Later work has been done in a laboratory tank (83) and at sea (84). The latter investigation concerned the frequency shift and the frequency spread of sea-surface reverberation. At 85 kHz and in a 20-knot wind, the mean reverberation frequency was shifted downward by 25 Hz when pointing downwind, suggesting a mean downwind scatterer velocity of 35 cm/s or a ratio of 0.035 of the wind speed. At the same time, the reverberation was spread over a band 25 Hz wide between the -3 dB points above and below the mean frequency, so as to suggest a range of scatterer velocities of 0 to 70 cm/s downwind. Such effects of scatterer motion would be absent for bottom reverberation, but would be expected to be relatively large for the mobile biological scatterers causing volume reverberation.

More recent work on the spectrum of surface reverberation as observed with a fixed transducer has been done by Igarashi and Stern (85) and by Swarts (86). Figure 8.38 shows some spectra of surface reverberation as observed by Igarashi and Stern with 110-ms pulses at 60 kHz and an upward

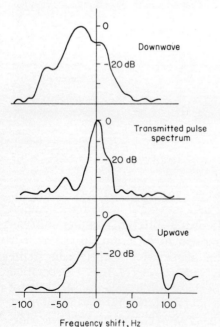

fig. 8.38 *Observed spectra of surface reverberation relative to the 110-ms transmitted pulse. Frequency 60 kHz, grazing angle 30°. (Ref. 85.)*

angle of 30°. The doppler shift and the frequency broadening of the backscattered return, relative to the spectrum of the transmitted pulse, are evident. Doppler shifts amounting to 10 and 20 Hz at 60 kHz, and 3 to 6 Hz at 15 kHz, were found in these experiments; Swarts observed shifts of 10 to 15 Hz at 28.2 kHz. The interpretation given to these results was that the backscattering is produced principally by those waves that "resonate," or form a diffraction grating, with the acoustic wave that impinges upon them, in the same way as we have already described for bottom reverberation (Fig. 8.34). If this is indeed the scattering process, the doppler shift should be independent of wind speed and should depend only on the velocity of forward motion of those waves that happen to have just the right wavelength to resonate with the impinging sound wave. An additional cause of doppler spreading of reverberation is tidal and other water currents that create movement of surface and volume scatterers relative to a fixed transducer.

Concerning the frequency spread of volume reverberation, as contrasted to surface reverberation, one would expect that at low frequencies, where the reverberation is principally due to resonant bladder-fish, the spectrum is relatively wide as a result of fish motion. That this is so is shown by measurements of volume reverberation at 15 kHz, where a frequency spread corresponding to a scatterer motion of 1 knot (about 10 Hz) was observed (87). On the other hand, at high frequencies, above 100 kHz, the reverberation spectra are observed to be very narrow (± 0.5 Hz) apparently because the scatterers are small plankton at rest relative to the water (88). This makes possible the detection and study of turbulence in the body of the ocean medium by measurements of the spectra of volume reverberation (89).

Both the mean frequency and the spread of the frequency of reverberation are of direct concern for the design of filters to suppress reverberation.

8.17 Reverberation Prediction

In predicting the reverberation to be expected in a particular echo-ranging sonar, it is necessary, as mentioned above, to visualize what kind of reverberation is likely to be encountered at different times after emission of the sonar pulse and where it is coming from. For this purpose, a rough ray diagram is a necessity, for it indicates not only the possible sources of reverberation but also the grazing angles involved in the encounters of the sonar beam with the surface and bottom.

In simple situations, such as at short ranges where only one form of reverberation is involved, a reverberation computation proceeds in a straightforward manner. After the appropriate form of the reverberation-level equation is selected, depending on the type of reverberation to be expected, the equivalent beam width for the transducer used can be found from Table 8.1 and a value of scattering strength can be selected for the kind of reverberation to be expected from the various curves given above. At different times and ranges, different sources of reverberation and different values of scattering strength, depending on grazing angle, will be involved.

Concerning a prediction of sonar reverberation in such a manner, particular attention should be given to the assumptions underlying the reverberation-level equations. In particular, the transmission loss to and from the scatterers was taken to be that of spherical spreading, and the effects of shadow zones and convergence zones in affecting the propagation of sound to the scattering region were neglected. Hence, the equations must be modified where necessary to take account of conditions peculiar to a prediction problem.

At long ranges, a reverberation computation becomes increasingly difficult because of the increasing number of paths and sources of reverberation. Referring to Fig. 8.39, for example, at some one instant of time, reverberation may be received from the surface at A, the deep-scattering layer at B, and the bottom at C. The contribution of each source must be evaluated separately and combined to give the level of reverberation at the instant of time being considered. At long ranges in shallow water, a reverberation prediction requires special techniques and computation on a digital computer such as those used by Mackenzie (90).

Under conditions that approximate those for which the expressions for RL were derived, a valid estimate of the expected reverberation can readily be made. As an example, we compute RL for an echo-ranging sonar operating at 50 kHz, with a source level of 220 dB and a pulse duration of 1 ms. The transducer is assumed to be a line transducer 1 ft long located 100 ft above a mud bottom. The bottom-reverberation level is desired for the instant of time corresponding to a diagonal range to the bottom of 200 yd. At 50 kHz the wavelength is 0.1 ft. By Table 8.1, $10 \log \Phi = 10 \log (0.1/2\pi) + 9.2 = -8.8$. The equivalent beam width is therefore 0.13 rad, or 7.5°. The reverberating area is therefore

$$A = \frac{c\tau}{2} \Phi r = (0.83)(0.13)(200) = 22 \text{ yd}^2$$

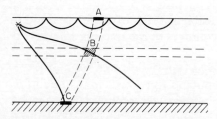

fig. 8.39 *Multiple sources of reverberation at long ranges.*

The grazing angle at the bottom is 9.5°. From Fig. 8.27 a value of S_B of -39 dB is selected. The reverberation level is therefore

$$RL = SL - 40 \log r + S_B + \log A$$
$$= 220 - 92 - 40 + 13$$
$$= +101 \text{ dB re } 1 \text{ } \mu\text{Pa}$$

REFERENCES

1. Skolnik, M. I.: "Introduction to Radar Systems," chap. 12, McGraw-Hill Book Company, New York, 1962.
2. Urick, R. J.: Backscattering of Sound from a Harbor Bottom, *J. Acoust. Soc. Am.*, **26:**231 (1954).
3. Physics of Sound in the Sea: Part II, Reverberation, *Nat. Def. Res. Comm. Div. 6 Sum. Tech. Rep.* 8, 1947.
4. Keane, J. J.: Volume Reverberation as a Function of Single Frequency Pulse Lengths and FM Sweep Rates, *J. Acoust. Soc. Am.*, **43:**566 (1968).
5. Urick, R. J., and R. M. Hoover: Backscattering of Sound from the Sea Surface: Its Measurement, Causes, and Application to the Prediction of Reverberation Levels, *J. Acoust. Soc. Am.*, **28:**1038 (1956).
6. Davis, E. E., J. B. Parham, and C. E. Kelly: Modification and Utilization of AN/SSQ-41 Sonobuoys for the Collection of Volume Reverberation Data from Aircraft, *U.S. Nav. Oceanogr. Office Tech. Rep.* 211, 1968.
7. Urick, R. J.: Generalized Form of the Sonar Equations, *J. Acoust. Soc. Am.*, **34:**547 (1962).
8. Chapman, R. P., and J. H. Harris: Surface Backscattering Strengths Measured with Explosive Sound Sources, *J. Acoust. Soc. Am.*, **34:**1592 (1962).
9. Weston, D. E.: Underwater Explosions as Acoustic Sources, *Proc. Phys. Soc. London*, **76:**233 (1960).
10. Stockhausen, J. H.: Spectrum of the Shock Wave from 1-lb TNT Charges Exploded Underwater, *J. Acoust. Soc. Am.*, **36:**1220 (1964).
11. Christian, E. A., and M. Blaik: Near Surface Measurements of Deep Explosions, *J. Acoust. Soc. Am.*, **38:**57 (1965).
12. Physics of Sound in the Sea: Part IV, Acoustic Properties of Wakes, *Nat. Def. Res. Comm. Div. 6 Sum. Tech. Rep.* 8, pp. 460–474, 1947.
13. Albers, V. M.: "Underwater Acoustics Handbook," 2d ed., chap. 7, The Pennsylvania State University Press, University Park, Pa., 1965.
14. Weissler, A., and V. A. Del Grosso: The Velocity of Sound in Sea Water, *J. Acoust. Soc. Am.*, **23:**219 (1951).
15. Urick, R. J.: A Sound Velocity Method for Determining the Compressibility of Finely Divided Substances, *J. Appl. Phys.*, **18:**983 (1947).
16. Wood, A. B.: "A Textbook of Sound," 2d ed., p. 360, G. Bell & Sons, Ltd., London, 1941.
17. Meyer, R. F., and B. W. Romberg: Acoustic Scattering in the Ocean, *Proj. Trident Tech. Rep.* 1360863, Arthur D. Little, Inc., 1963.
18. Fox, F. E., S. R. Curley, and G. S. Larson: Phase Velocity and Absorption Measurements in Water Containing Bubbles, *J. Acoust. Soc. Am.*, **27:**534 (1955).
19. Devin, C.: Damping of Pulsating Air Bubbles in Water, *J. Acoust. Soc. Am.*, **31:**1654 (1959).
20. Eyring, C. F., R. J. Christensen, and R. W. Raitt: Reverberation in the Sea, *J. Acoust. Soc. Am.*, **20:**462 (1948).
21. Pekeris, C. L.: Note on the Scattering of Radiation in an Inhomogeneous Medium, *Phys. Rev.*, **71:**268 (1947).

22. Johnson, H. R., R. H. Backus, J. B. Hersey, and D. M. Owen: Suspended Echo Sounder and Camera Studies of Mid-Water Sound Scatterers, *Deep-Sea Res.*, **3**:266 (1956).
23. Hersey, J. B., and R. H. Backus: New Evidence That Migrating Gas Bubbles, Probably the Swim Bladders of Fish, Are Largely Responsible for Scattering Layers on the Continental Rise North of New England, *Deep-Sea Res.*, **7**:190 (1954).
24. Moore, H. B.: Scattering Layer Observations, *Atlantic Cruise 151, Sci. Rep.* 3, Woods Hole Oceanographic Institution, Reference 49-2, 1949. Also, The Relation between the Scattering Layer and the Euphausiacea, *Biol. Bull.*, **99**:181 (1950).
25. Anderson, V. C.: Frequency Dependence of Reverberation in the Ocean, *J. Acoust. Soc. Am.*, **41**:1467 (1967).
26. Vent, R. J.: Acoustic Volume Scattering Measurements at 3.5, 5.0 and 12.0 kHz in the Eastern Pacific Ocean, *J. Acoust. Soc. Am.*, **52**:373 (1972).
27. Fisher, F. H., and E. D. Squier: Observation of Acoustic Layering and Internal Waves with a Narrow-Beam 87.5 kHz Echo Sounder, *J. Acoust. Soc. Am.*, **58**:1315 (1975).
28. Hersey, J. B., R. H. Backus, and J. Hellwig: Sound Scattering Spectra of Deep Scattering Layers in the Western North Atlantic Ocean, *Deep-Sea Res.*, **8**:196 (1962).
29. Marshall, J. R., and R. P. Chapman: Reverberation from a Deep Scattering Layer Measured with Explosive Sound Sources, *J. Acoust. Soc. Am.*, **36**:164 (1964).
30. Chapman, R. P., and J. R. Marshall: Reverberation from Deep Scattering Layers in the Western North Atlantic, *J. Acoust. Soc. Am.*, **40**:405 (1966).
31. Gold, B. A., and W. E. Renshaw: Joint Volume Reverberation and Biological Measurements in the Tropical Western Atlantic, *J. Acoust. Soc. Am.*, **63**:1809 (1978).
32. Urick, R. J., and G. R. Lund: Vertical Coherence of Explosive Reverberation, *J. Acoust. Soc. Am.*, **36**:2164 (1964).
33. Chapman, R. P., O. Z. Bluy, and R. H. Adlington: Geographic Variations in the Acoustic Characteristics of Deep Scattering Layers, *Proc. Int. Symp. Biol. Sound Scattering in the Ocean, Maury Center for Ocean Sci. Rep.* MC 005, p. 306, 1970.
34. Scrimger, J. A., and R. C. Turner: Backscattering of Sound from the Ocean Volume between Vancouver Island and Hawaii, *J. Acoust. Soc. Am.*, **54**:483 (1973).
35. Hansen, W. J.: Biological Causes of Scattering Layers in the Arctic Ocean, *Proc. Int. Symp. Biol. Sound Scattering in the Ocean, Maury Center for Ocean Sci. Rep.* MC 005, p. 508, 1970.
36. Weston, D. E.: Unpublished report, 1957.
37. Urick, R. J.: The Processes of Sound Scattering at the Ocean Surface and Bottom, *J. Mar. Res.*, **15**:134 (1956).
38. Garrison, G. R., S. R. Murphy, and D. S. Potter: Measurements of the Backscattering of Underwater Sound from the Sea Surface, *J. Acoust. Soc. Am.*, **32**:104 (1960).
39. Chapman, R. P., and H. D. Scott: Surface Backscattering Strengths Measured over an Extended Range of Frequencies and Grazing Angles, *J. Acoust. Soc. Am.*, **36**:1735 (1964).
40. Richter, R. M.: Measurements of Backscattering from the Sea Surface, *J. Acoust. Soc. Am.*, **36**:864 (1964).
41. Miles, J. W.: On Non-Specular Reflections at a Rough Surface, *J. Acoust. Soc. Am.*, **26**:191 (1954).
42. Beckman, P., and A. Spizzichino: "The Scattering of Electromagnetic Waves from Rough Surfaces," The Macmillan Company, New York, 1963.
43. Fortuin, L.: Survey of Literature on Reflection and Scattering of Sound Waves at the Sea Surface, *J. Acoust. Soc. Am.*, **47**:1209 (1970).
44. Eckart, C.: Scattering of Sound from the Sea Surface, *J. Acoust. Soc. Am.*, **25**:566 (1953).
45. Cox, C., and W. Munk: Measurement of the Roughness of the Sea Surface from Photographs of the Sun's Glitter, *J. Opt. Soc. Am.*, **44**:838 (1954).
46. Marsh, H. W.: Exact Solution of Wave Scattering by Irregular Surfaces, *J. Acoust. Soc. Am.*, **33**:330 (1961).
47. Marsh, H. W., M. Schulkin, and S. G. Kneale: Scattering of Underwater Sound by the Sea Surface, *J. Acoust. Soc. Am.*, **33**:334 (1961).

48. Marsh, H. W.: Sound Reflection and Scattering from the Sea Surface, *J. Acoust. Soc. Am.*, **35:**240 (1963).
49. Burling, R. W.: "Wind Generation of Waves on Water," Ph.D. dissertation, Imperial College, University of London, 1955.
50. Bachmann, W.: A Theoretical Model for the Backscattering Strength of a Composite-Roughness Sea Surface, *J. Acoust. Soc. Am.*, **54:**712 (1973).
51. Andreeva, I. B., A. V. Volkova, and N. N. Galybin: Backscattering of Sound by the Sea Surface at Small Grazing Angles, *Sov. Phys. Acoust.* **26:**265 (1980).
52. Schulkin, M., and R. Shaffer: Backscattering of Sound from the Sea Surface, *J. Acoust. Soc. Am.*, **36:**1699 (1964).
53. Clay, C. S., and H. Medwin: High Frequency Acoustical Reverberation from a Rough Sea Surface, *J. Acoust. Soc. Am.*, **36:**2131 (1964).
54. Gardner, R. R.: Acoustic Backscattering from a Rough Surface at Extremely Low Grazing Angles, *J. Acoust. Soc. Am.*, **53:**848 (1973).
55. Hoover, R. M., and R. J. Urick: Sea Clutter in Radar and Sonar, *Inst. Radio Eng. Conv. Rec.*, pt. 9, p. 17, 1957.
56. Harrison, V. G.: "Gloss: Its Definition and Measurement," Tudor Publishing Company, New York, 1945.
57. Urick, R. J.: Side Scattering of Sound in Shallow Water, *J. Acoust. Soc. Am.*, **32:**351 (1960).
58. Physics of Sound in the Sea: Part II, Reverberation, *Nat. Def. Res. Comm. Div. 6 Sum. Tech. Rep.* 8, p. 316, 1947.
59. McKinney, C. M., and C. D. Anderson: Measurements of Backscattering of Sound from the Ocean Bottom, *J. Acoust. Soc. Am.*, **36:**158 (1964).
60. Urick, R. J., and D. S. Saling: Backscattering of Explosive Sound from the Deep-Sea Bed, *J. Acoust. Soc. Am.*, **34:**1721 (1962).
61. Patterson, R. B.: Back-Scatter of Sound from a Rough Boundary, *J. Acoust. Soc. Am.*, **35:**2010 (1963).
62. Mackenzie, K. V.: Bottom Reverberation for 530 and 1030 cps Sound in Deep Water, *J. Acoust. Soc. Am.*, **33:**1498 (1961).
63. Burstein, A. W., and J. J. Keane: Backscattering of Explosive Sound from Ocean Bottoms, *J. Acoust. Soc. Am.*, **36:**1596 (1964).
64. Schmidt, P. B.: Monostatic and Bistatic Backscattering Measurements from the Deep Ocean Bottom, *J. Acoust. Soc. Am.*, **50:**327 (1971).
65. Buckley, J. P., and R. J. Urick: Backscattering from the Deep Sea Bed at Small Grazing Angles, *J. Acoust. Soc. Am.*, **44:**648 (1968).
66. Merklinger, H. M.: Bottom Reverberation Measured with Explosive Charges Fired Deep in the Ocean, *J. Acoust. Soc. Am.*, **44:**508 (1968).
67. Jitkovskiy, Yu. Yu., and L. A. Volovova: Sound Scattering from the Ocean Bottom, paper E67, *Proc. Fifth Int. Acoust. Congr. Liége Belg.*, 1965.
68. Wong, H.-K., and W. D. Chesterman: Bottom Backscattering Near Grazing Incidence in Shallow Water, *J. Acoust. Soc. Am.*, **44:**1713 (1968).
69. Nolle, A. W., et al.: Acoustical Properties of Water Filled Sands, *J. Acoust. Soc. Am.*, **35:**1394 (1963).
70. Houston, R. S.: "A Treatise on Light," p. 368, Longmans, Green & Co., Inc., New York, 1938.
71. Rayleigh, Lord: "The Theory of Sound," vol. II, pp. 89–96, Dover Publications, Inc., New York, 1945.
72. Marsh, H. W.: "Reflection and Scattering of Sound by the Sea Bottom," paper presented at the 68th meeting of the Acoustical Society of America, October 1964, Avco Marine Electronics Office, New London, Connecticut.
73. Milne, A. R.: Underwater Backscattering Strengths of Arctic Park Ice, *J. Acoust. Soc. Am.*, **36:**1551 (1964).
74. Brown, J. R.: Reverberation under Arctic Ice, *J. Acoust. Soc. Am.*, **36:**601 (1964).

75. Mellen, R. H., and H. W. Marsh: Underwater Sound Reverberation in the Arctic Ocean, *J. Acoust. Soc. Am.*, **35**:1645 (1963).
76. Urick, R. J.: Reverberation-derived Scattering Strength of the Shallow Sea Bed, *J. Acoust. Soc. Am.*, **48**:392 (1970).
77. Urick, R. J., and G. R. Lund: Vertical Coherence of Shallow Water Reverberation, *J. Acoust. Soc. Am.*, **47**:342 (1970).
78. Urick, R. J., and G. R. Lund: Horizontal Coherence of Explosive Reverberation, *J. Acoust. Soc. Am.*, **47**:909 (1970).
79. Principles of Underwater Sound, *Nat. Def. Res. Comm. Div. 6. Sum. Tech. Rep.*, **7**:90–98 (1947).
80. Rayleigh, Lord: "The Theory of Sound," vol. I, pp. 35–42, Dover Publications, Inc., New York, 1945.
81. Cron, B. F., and W. R. Schumacher: Theoretical and Experimental Study of Underwater Sound Reverberation, *J. Acoust. Soc. Am.*, **33**:881 (1961).
82. Olshevskii, V. V.: Probability Distribution of Sea Reverberation Levels, *Sov. Phys. Acoust.*, **9**:378 (1964).
83. Liebermann, L. W.: Analysis of Rough Surfaces by Scattering, *J. Acoust. Soc. Am.*, **35**:932 (1963).
84. Mellen, R. H.: Doppler Shift of Sonar Backscatter from the Surface, *J. Acoust. Soc. Am.*, **36**:1395 (1964).
85. Igarashi, Y., and R. Stern: Observation of Wind-Wave Generated Doppler Shifts in Surface Reverberation, *J. Acoust. Soc. Am.*, **49**:802 (1971).
86. Swarts, R. L.: Doppler Shift of Surface Backscatter, *J. Acoust. Soc. Am.*, **52**:457 (1972).
87. Garabed, E. P.: Measurements of the Frequency Spectrum of Pulsed Sine-Wave Reverberation, *U.S. Nav. Air Development Center Rep.* AE-6713 (1967).
88. Rasmussen, R. A., and N. E. Head: Characteristics of High-Frequency Sea Reverberation and Their Application to Turbulence Measurement, *J. Acoust. Soc. Am.*, **59**:55 (1976).
89. Rasmussen, R. A.: Remote Detection of Turbulence from Observations of Reverberation Spectra, *J. Acoust. Soc. Am.*, **63**:101 (1978).
90. Mackenzie, K. V.: Long-Range Shallow-Water Reverberation, *J. Acoust. Soc. Am.*, **34**:62 (1962).
91. Barakos, P. A.: Underwater Reverberation as a Factor in ASW Acoustics, *U.S. Navy Underwater Sound Lab. Rep.* 620, September 1964.
92. The Discrimination of Transducers against Reverberation, *Univ. Calif. War Res. Mem.* UT5, file 01.40, NDRC Div. 6.1-5530-968, May 1943.

nine

reflection and scattering by sonar targets: target strength

This chapter is concerned with the subject of *catacoustics*. According to Webster's "New International Dictionary of the English Language," 2d edition, the word "catacoustics" is defined as "that part of acoustics which treats of reflected sounds or echoes."

In active sonar the parameter *target strength* refers to the echo returned by an underwater target. Such targets may be objects of military interest, such as submarines and mines, or they may be schools of fish sought by fish-finding sonars. Excluded from the category of "targets" are inhomogeneities in the sea of indefinite extent, such as scattering layers and the ocean surface and bottom, which, because of their indefinite size, return sound in the form of *reverberation* instead of as *echoes*.

In the context of the sonar equations, target strength is defined as 10 times the logarithm to the base 10 of the ratio of the intensity of the sound returned by the target, at a distance of 1 yd from its "acoustic center" in some direction, to the incident intensity from a distant source. In symbols,

$$\text{TS} \equiv 10 \log \frac{I_r}{I_i}\bigg|_{r=1}$$

where I_r = intensity of return at 1 yd
I_i = incident intensity

Pictorially this is shown in Fig. 9.1, where P is the point at which I_r is imagined to be measured and C is the acoustic center of the target. This fictitious point,

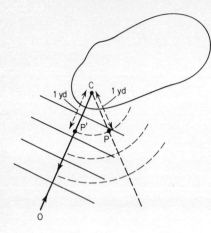

fig. 9.1 *Geometry of target strength. A plane wave is incident on the target in the direction OC. Target strength refers to the points P and P' 1 yd from the acoustic center C.*

inside or outside of the target itself, is the point from which the returned sound appears to originate on the basis of measurements made at a distance. In "monostatic" sonars, having the same, or closely adjacent source and receiver, the point P lies in the direction back toward the source of sound. In "bistatic" sonars, P can lie in any direction relative to the target, and target strength then becomes a function of both the incident direction and the direction of the receiver, both relative to some axis of symmetry of the target. Because most sonars are "monostatic," we will restrict most of our discussion to "back reflection" and "backscattering," in which the reference point lies at P' back in the direction of the incident sound.

Target strength measurements are always at a long range—that is, in the "far field" (Fig. 4.2)—where the target reradiates as a point source of sound. This virtual point souce is the acoustic center C.

Special mention should be made of the use of 1 yd as the reference distance for target strength. This arbitrary reference often causes many underwater objects to have *positive* values of target strength. Such positive values should not be interpreted as meaning that more sound is coming back from the target than is incident upon it; rather, they should be regarded as a consequence of the arbitrary reference distance. If, instead of 1 yd, 1 kyd was used, all customary targets would have a negative target strength. In metric units of length, where 1 meter is the reference distance instead of 1 yd, all target-strength values are *lower* by the amount 20 log 1.0936 or 0.78 dB; in other words, when a 1-meter reference distance is used, all target-strength values referred to 1 yd must be *reduced* by 0.78 dB.

The meaning of target strength can be shown by computing the target strength of a sphere, large compared to a wavelength, on the assumption that the sphere is an isotropic reflector; that is, it distributes its echo equally in all directions. Let a large, perfect, rigid sphere (Fig. 9.2) be insonified by a plane wave of sound of intensity I_i. If the sphere is of radius a, the power intercepted by it from the incident wave will be $\pi a^2 I_i$. On the assumption that the sphere reflects this power uniformly in all directions, the intensity of the reflected wave at a distance r yards from the sphere will be the ratio of this power to the area of a sphere of radius r, or

$$I_r = \frac{\pi a^2 I_i}{4\pi r^2} = I_i \frac{a^2}{4r^2}$$

where I_r is the intensity of the reflection at range r. At the reference distance of 1 yd, the ratio of the reflected intensity I_r to the incident intensity is

$$\left.\frac{I_r}{I_i}\right|_{r=1} = \frac{a^2}{4}$$

and the target strength of the sphere becomes

$$\text{TS} \equiv 10 \log \left.\frac{I_r}{I_i}\right|_{r=1} = 10 \log \frac{a^2}{4}$$

It is therefore evident that an ideal sphere of radius 2 yd ($a = 2$) has a target strength of 0 dB. In practical work, spheres make good reference targets for sonar when they can be used, because their target strengths are relatively independent of orientation.

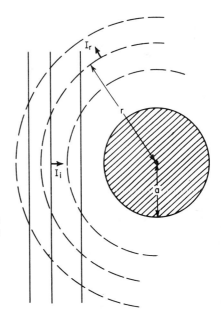

fig. 9.2 *A sphere as an isotropic reflector of an incident plane wave.*

9.1 The Echo as the Sum of Backscattered Contributions

For radar it was shown by Kerr (1) that the backscattering cross section of a radar target in integral form is

$$\sigma = \frac{4\pi}{\lambda^2} \left| \int_\alpha^\beta \frac{dA}{dz} e^{2ikz} dz \right|^2$$

where σ = ratio of scattered power to incident intensity
dA/dz = rate of change of cross-sectional area of body in direction of propagation z
k = wave number $2\pi/\lambda$
λ = incident wavelength
$z = \alpha$, $z = \beta$ are the range limits of the target

The return of an incident plane wave from the target can therefore be regarded as the sum of many wavelets, each originating at the changes in cross-sectional area of the target and added with respect to phase. The application of this expression to sonar is restricted to targets that are large, rigid, and immovable in the sound field, that is, to large targets that do not deform or move under the impact of the incident sound wave.

A thorough application of this approach to the backscattering of underwater sound targets has been made by Freedman (2). In his notation

$$TS = 10 \log |J|^2$$

where

$$J = \frac{1}{\lambda} \sum_{g=1}^\infty e^{-2ik(r_g - r_1)} \sum_{n=0}^\infty \frac{D_g^n(A)}{(2ik)^n}$$

and where the symbol $D_g^n(A)$ denotes the magnitude of the discontinuity of the nth derivative of the cross-sectional area of the object at range r_g, and r_1 is the range of the closest point of the target. The target-strength factor J is thus proportional to the sum of all the discontinuities of the derivatives of the cross-sectional area A of the object, measured at some range $r_g - r_1$ from the point nearest the source, weighted by the factor $1/(2ik)^n$, and then summed over all points of the object after allowance for phase by means of the factor $e^{-2ik(r_g - r_1)}$. Freedman also showed experimentally that echo envelopes can be predicted from the various contributions of the derivatives of the cross-sectional-area function occurring at times corresponding to the location of the cross sections along the object in the direction of propagation. A summation method for finding the reflection from irregular bodies has also been the subject of a paper by Neubauer (3).

Figure 9.3, taken from Freedman's work, shows the wave theory echo from a sphere, where each discontinuity in the derivatives give rise to a component of the echo. Wave theory yields a diffracted echo from the edge of the geometrical shadow in addition to the echo from the front of the sphere.

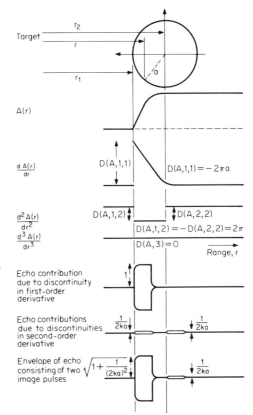

fig. 9.3 *Wave theory origin of the echo from a sphere. The echo is made up of contributions from the discontinuities in the derivatives of the cross-sectional area of the sphere. The theory applies only to objects that are smooth, large compared to λ, immovable, nondeformable, and impenetrable to sound. (Ref. 2.)*

9.2 Geometry of Specular Reflection

For objects of radii of curvature large compared to a wavelength, the echo originates principally by *specular reflection,* in which those portions of the target in the neighborhood of the point at which sound is normally incident give rise to a coherent reflected echo. One way to find the magnitude of the specular reflection is to construct Fresnel, or quarter-wave zones, on the surface of the body and to add their contributions, as has been done for underwater targets by Steinberger (4). A heuristic intuitive approach is to consider target strength as a measure of the spreading of an incident plane wave induced by specular reflection from a curved surface. If the power or energy-density contained within a small area A_i of the incident sound beam is spread, on reflection, over the area A_r at unit distance, then the target strength is 10 log (A_i/A_r). These areas can be determined by drawing rays. This geometric view of specular reflection will be illustrated by computing the target strength of a sphere and a general convex surface, both satisfying the requirement of large radii of curvature compared to a wavelength.

Figure 9.4 shows a perfect, large rigid sphere with a plane sound wave incident from the left. Adjacent incident rays intersect along the curve OQO',

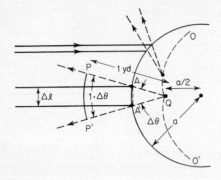

fig. 9.4 Target strength of a sphere. The acoustic energy contained in the pencil of diameter Δl is spread on reflection over a portion PP' of a 1-yd sphere.

having a cusp at Q halfway from the surface to the center of the sphere. This *caustic* is the locus of the acoustic centers of the adjacent intersecting rays. Consider a small cylindrical bundle of parallel rays normally incident upon the sphere at AA' and of cross section Δl. If the intensity of the incident wave is I_i, the power contained in the bundle is $I_i\pi(\Delta l)^2/4$. This power will, in effect, be reradiated within an angle $\Delta\theta$ from the acoustic center Q for these rays. At unit distance from Q, this power is distributed over a portion of a sphere PP' of area $\pi(\Delta\theta \times 1)^2/4$ if $\Delta\theta$ is small, and the intensity there becomes

$$I_r = \frac{I_i\pi(\Delta l)^2/4}{\pi(\Delta\theta \times 1)^2/4} = I_i\left(\frac{\Delta l}{\Delta\theta}\right)^2$$

But, referring to the triangle AQA', it will be seen that

$$AA' = \Delta l = \frac{a}{2}\Delta\theta$$

for small $\Delta\theta$. Hence $\Delta l/\Delta\theta = a/2$ and

$$\text{TS} \equiv 10\log\frac{I_r}{I_i}\bigg|_{r=1} = 10\log\frac{a^2}{4}$$

This is the same as before, and indicates that a large sphere reflects the incident plane wave in the backward direction *as though* it were a uniform, or isotropic, reflector of sound. For this simple expression to hold strictly, distances must be reckoned from the acoustic center of the sphere located halfway from the surface to the center. For practical purposes in sonar, where ranges much greater than the radius of the sphere are involved, the exact location of the acoustic center is not usually significant. It should be observed that we have considered a sphere that is (1) *perfect* in shape, without irregularities, depressions, or protuberances, (2) *rigid*, or nondeformable by the impinging sound beam, (3) *immovable*, or does not partake of the acoustic motion of the field in which it is embedded, and (4) *large* compared to a wavelength $(2\pi a/\lambda \gg 1)$.

This same method can be extended to the reflection at normal incidence from any convex surface having all radii of curvature large compared to a wavelength. This requirement carries with it the absence of protuberances, corners, and angles, all of which involve small radii of curvature and which serve as scatterers instead of reflectors of sound. Such a large, smooth, convex object is shown in Fig. 9.5a, with sound normally incident at point P along the line OP. Imagine a series of planes through OP intersecting the object. Two of them lie at right angles to each other and intersect the object in the *principal normal sections* having a maximum and a minimum radius of curvature. Figure 9.5b is a plan view at P in which the principal normal sections are AA' and BB'; Fig. 9.5c is a cross section through AA' in which R_1 is the radius of curvature, O the center of curvature, and Q the acoustic center located halfway from O to P. Consider a small rectangular segment of the surface having one corner located at P, and of infinitesimal lengths dl_1 and dl_2 on each side.

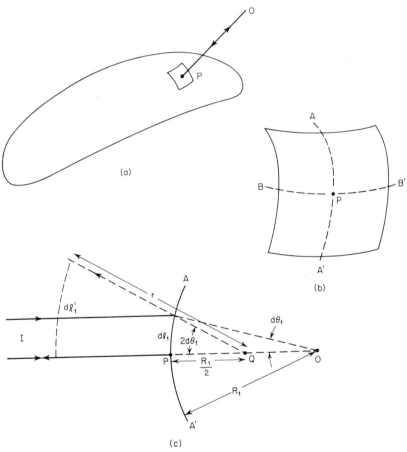

fig. 9.5 *Target strength of a convex target of large radius of curvature: (a) A target with sound normally incident along OP. (b) Plan view of the vicinity of P with AA' and BB' in the principal normal sections. (c) Cross section through AA' of radius of curvature R_1.*

If a plane wave of intensity I_i is incident on the surface, the power dP intercepted by the segment will be

$$dP = I_i\, dl_1\, dl_2$$

But from the cross section of Fig. 9.5c, it is apparent that

$$dl_1 = R_1\, d\theta_1$$

and similarly in the perpendicular plane containing BB' we would have

$$dl_2 = R_2\, d\theta_2$$

so that the power intercepted will be

$$dP = I_i R_1 R_2\, d\theta_1\, d\theta_2$$

On reflection from the sphere, this power is distributed, at range r from the acoustic center C, over an area

$$dA = dl'_1\, dl'_2 = 2r\, d\theta_1\, 2r\, d\theta_2$$

Hence the intensity at r will be

$$I_r = \frac{dP}{dA} = \frac{I_i R_1 R_2}{4r^2}$$

and the target strength will be

$$\text{TS} \equiv 10 \log \left.\frac{I_r}{I_i}\right|_{r=1} = 10 \log \frac{R_1 R_2}{4}$$

9.3 Target Strength of a Small Sphere

We consider now the target strength of a small sphere, in which the return of sound back toward the source is a process of scattering instead of reflection.

The theory of sound scattering by a small, fixed, rigid sphere was first worked out by Rayleigh (5). By *small* is meant a sphere whose ratio of circumference to wavelength is much less than unity ($ka = 2\pi a/\lambda \ll 1$); by *fixed*, a sphere that does not partake of the acoustic motion of particles of the fluid in which the sphere is embedded; by *rigid*, a sphere that is nondeformable by the incident acoustic waves and into which the sound field does not penetrate. Under these conditions, Rayleigh showed that the ratio of the scattered intensity I_r at a large distance r to the intensity I_i of the incident phase wave is

$$\frac{I_r}{I_i} = \frac{\pi^2 T^2}{r^2 \lambda^4}\left(1 + \frac{3}{2}\mu\right)^2$$

where T = volume of sphere ($\tfrac{4}{3}\pi a^3$)
λ = wavelength
μ = cosine of angle between scattering direction and reverse direction of incident wave

For backscattering $\mu = +1$. On reducing to $r = 1$ and on taking 10 times the logarithm, we obtain

$$\text{TS} \equiv 10 \log \left.\frac{I_r}{I_i}\right|_{r=1} = 10 \log \frac{\pi^2 T^2}{\lambda^4} \left(\frac{5}{2}\right)^2 = 10 \log \left[(1{,}082) \frac{a^6}{\lambda^4} \right]$$

where a and λ are in units of yards. Thus, the target strength of a small sphere varies as the sixth power of the radius and inversely as the fourth power of the wavelength.

If we define the backscattering cross section of the sphere as

$$\sigma \equiv 4\pi \left.\frac{I_r}{I_i}\right|_{r=1}$$

the ratio of backscattering cross section to geometric cross section becomes

$$\frac{\sigma}{\pi a^2} = 2.8 (ka)^4$$

Figure 9.6 is a plot of this normalized ratio against the nondimensional quantity $ka = 2\pi a/\lambda$. It is seen that this quantity varies as the fourth power of ka, or as the fourth power of the frequency, for ka less than about 0.5, and is unity for ka greater than 5.0. Oscillations occur in the intermediate region ($ka \approx 1$).

Rayleigh also dealt with the scattering by small spheres, not fixed and rigid, but possessing a compressibility κ' and a density ρ' in a fluid of compressibility κ and density ρ. Fixed, rigid spheres are incompressible ($\kappa'/\kappa \ll 1$) and very dense ($\rho'/\rho \gg 1$) compared to the surrounding fluid; spheres having bulk moduli and densities comparable with those of the fluid oscillate to and fro in

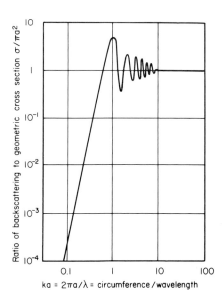

fig. 9.6 *Ratio of acoustic to geometric cross sections of a fixed rigid sphere. The oscillations in the range $1 < ka < 10$ are due to interference by the creeping wave. For small ka (low frequencies) the ratio varies as the fourth power of ka; for large ka the ratio is unity.*

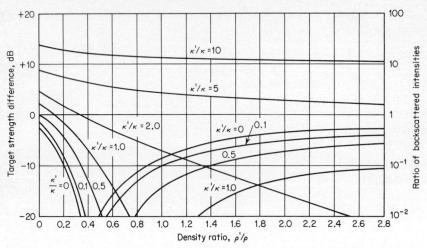

fig. 9.7 *Corrections to the target strength of a fixed rigid sphere for compressibility and density. The sphere is of compressibility ratio κ'/κ and density ratio ρ'/ρ relative to the surrounding fluid.*

the incident sound field. They also pulsate, or change their volume, in the compressions and rarefactions of the incident wave. These motions of the sphere modify the scattered wave. Rayleigh showed (6) that the term $(1 + \frac{3}{2}\mu)^2$ in the expression for the ratio of intensities for a fixed, rigid sphere becomes

$$\left[1 - \frac{\kappa'}{\kappa} + \frac{3(\rho'/\rho - 1)}{1 + 2\rho'/\rho}\mu\right]^2$$

in terms of the ratios of compressibility κ'/κ and density ρ'/ρ. Figure 9.7 is a plot of the quantity

$$10 \log \left[1 - \frac{\kappa'}{\kappa} + \frac{3(\rho'/\rho - 1)}{1 + 2\rho'/\rho}\mu\right]^2 \bigg/ \left(1 + \frac{3}{2}\mu\right)^2$$

This is the "correction" for compressibility and density to be applied to the target strength of a fixed, rigid sphere. This correction is seen from Fig. 9.7 to be large in most instances. For a sand grain in water, for example, for which $\rho'/\rho = 2.6$ and $\kappa'/\kappa \approx 0.1$, the target strength is seen to be 4 dB less than for the classic Rayleigh case of a fixed, rigid sphere. On the other hand, highly compressible spheres such as gas bubbles in water have enormously greater target strengths, even in the absence of resonance effects. For example, for a nonresonant air bubble in water at 1 atm, κ'/κ is approximately 19,000, and its backscattering is some 77 dB greater than that for classic Rayleigh scattering by a sphere of the same size.

A thorough theoretical treatment of sound scattering from a fluid sphere is given in a paper by Anderson (7).

9.4 Complications for a Large Smooth Solid Sphere

When a sound wave impinges on a large, smooth solid sphere in water, the sphere does not remain inert, but reacts to the impinging wave in different ways. First, sound can enter the sphere as a compressional wave and be reflected from the back side, so as to produce a secondary echo occurring slightly later than the specularly reflected echo from the front of the sphere. Second, flexural waves can be excited on the surface of the sphere at the place where their wavelength and that of sound in water "match." These waves of deformation travel with their own velocity around the sphere and contribute to the echo in the backward direction. Finally, "creeping" waves predicted by wave theory have been observed experimentally (8, 9). These are diffracted waves that originate at the edge of the geometrical shadow and travel around the sphere with the velocity of sound in water. Both surface and creeping waves reradiate in all directions as they travel around the sphere and in so doing become attenuated. The creeping or diffracted waves are the cause of the oscillations in the backscattering cross-section plot of Fig. 9.6 in the region $1 < ka < 10$.

These waves are illustrated in Fig. 9.8. Although they are somewhat academic, they illustrate some of the complications that can occur in the echo from real sonar targets in water.

The penetration of sound into a sphere can be used to enhance its target strength. Marks and Mikeska (10) filled a thin-wall stainless steel sphere with a mixture of the liquids freon and CCl_4, having a sound velocity 0.56 that of water, and found target strengths higher by 20 dB over the theoretical value

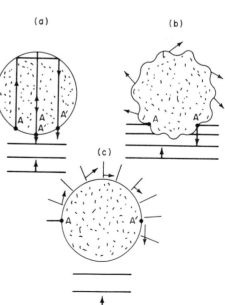

fig. 9.8 Contributions to the echo from a large, smooth solid sphere: (a) Back-reflected compressional waves take one or two bounces off the back side, entering at A and leaving at A'. (b) A flexural surface wave originates at A and leaves at A' reradiating as it travels aorund the sphere. (c) The "creeping" diffracted wave originates at the edge of the shadow boundary and similarly reradiates in all directions.

10 log $a^2/4$ for an impenetrable sphere. In addition, the beamwidth of the returned sound was narrow—about the same as that of a circular piston of the same diameter as the sphere. The filling fluid served to focus the incident sound to the rear of the sphere and thereby produced stronger echoes than would occur if the sphere were either hollow or of solid steel.

9.5 Target Strength of Simple Forms

The target strength of a number of geometric shapes and forms has been found theoretically, in most cases for applications to radar. Table 9.1 presents a list of a number of mathematical forms for which the target strength has been determined, together with the appropriate literature references. These idealized expressions should be viewed as no more than crude approximations for targets of complex internal construction for which penetration and scattering are suspected to occur. Moreover, the mobility and nonrigidity of sonar targets (unlike radar targets) cause them to have target strengths different from what they would have if they were fixed and rigid as theory requires. Yet the expressions will often be found useful for the prediction of the target strength of new and unusual objects for which no measured data are available and which, it is believed, can be approximated well enough by an ideal geometric shape. This use of these expressions will be illustrated later on for mines and torpedoes.

More complex targets can be modeled by breaking them up into elemental parts and replacing each part of one of the various simple forms. For example, a submarine can be modeled by a series of cylinders, wedges, plates, etc., each corresponding to some component of the hull structure. For a long-pulse sonar, the target strength can be found by adding up the contributions of the various simple forms into which the target has been divided. This technique appears to have been successfully employed in radar for estimating the target strength of targets as complex as an approaching jet bomber (11).

9.6 Bistatic Target Strength

Because of the fact that in sonar, as in radar, the monostatic geometry has usually been employed, much less is known about the target strength of underwater objects in the "bistatic" arrangement, that is, when widely separated sources and receivers are used. A rule of thumb assumption that is commonly made is that the target is isotropic: that is, that its target strength is the same in all directions, except possibly in the "glint" directions where specular reflection might exist.

In radar, a method has been found for estimating the bistatic target strength when the monostatic value is known (12). The bistatic theorem states that for large smooth objects, the bistatic target strength is equal to the monostatic target strength, taken at the bisector of "bistatic angle" between the

table 9.1 Target Strength of Simple Forms

Form	Target strength $t = 10 \log t$	Symbols	Direction of incidence	Conditions	References
Any convex surface	$\dfrac{a_1 a_2}{4}$	a_1, a_2 = principal radii of curvature r = range $k = 2\pi/\text{wavelength}$	Normal to surface	$ka_1, ka_2 \gg 1$ $r > a$	1
Sphere Large	$\dfrac{a^2}{4}$	a = radius of sphere	Any	$ka \gg 1$ $r > a$	1
Small	$61.7 \dfrac{V^2}{\lambda^4}$	V = volume of sphere λ = wavelength	Any	$ka \ll 1$ $kr \gg 1$	2
Cylinder Infinitely long Thick	$\dfrac{ar}{2}$	a = radius of cylinder	Normal to axis of cylinder	$ka \gg 1$ $r > a$	1
Thin	$\dfrac{9\pi^4 a^4}{\lambda^2} r$	a = radius of cylinder	Normal to axis of cylinder	$ka \ll 1$	3
Finite	$aL^2/2\lambda$	L = length of cylinder a = radius of cylinder	Normal to axis of cylinder	$ka \gg 1$ $r > L^2/\lambda$	4
	$aL^2/2\lambda (\sin \beta/\beta)^2 \cos^2 \theta$	a = radius of cylinder $\beta = kL \sin \theta$	At angle θ with normal		
Plate Infinite (plane surface)	$\dfrac{r^2}{4}$		Normal to plane		

table 9.1 Target Strength of Simple Forms (Continued)

Form	Target strength $t = 10 \log t$	Symbols	Direction of incidence	Conditions	References
Finite Any shape	$\left(\dfrac{A}{\lambda}\right)^2$	A = area of plate L = greatest linear dimension of plate l = smallest linear dimension of plate	Normal to plate	$r > \dfrac{L^2}{\lambda}$ $kl \gg 1$	5
Rectangular	$\left(\dfrac{ab}{\lambda}\right)^2 \left(\dfrac{\sin \beta}{\beta}\right)^2 \cos^2 \theta$	a, b = side of rectangle $\beta = ka \sin \theta$	At angle θ to normal in plane containing side a	$r > \dfrac{a^2}{\lambda}$ $kb \gg 1$ $a > b$	4
Circular	$\left(\dfrac{\pi a^2}{\lambda}\right)^2 \left(\dfrac{2 J_1(\beta)}{\beta}\right)^2 \cos^2 \theta$	a = radius of plate $\beta = 2ka \sin \theta$	At angle θ to normal	$r > \dfrac{a^2}{\lambda}$ $ka \gg 1$	4
Ellipsoid	$\left(\dfrac{bc}{2a}\right)^2$	a, b, c = semimajor axes of ellipsoid	Parallel to axis of a	$ka, kb, kc \gg 1$ $r \gg a, b, c$	6
Average over all aspects Circular disk	$\dfrac{a^2}{8}$	a = radius of disk	Average over all directions	$ka \gg 1$ $r > \dfrac{(2a)^2}{\lambda}$	5
Conical tip	$\left(\dfrac{\lambda}{8\pi}\right)^2 \tan^4 \psi \left(1 - \dfrac{\sin^2 \theta}{\cos^2 \psi}\right)^{-3}$	ψ = half angle of cone	At angle θ with axis of cone	$\theta < \psi$	7

table 9.1 Target Strength of Simple Forms *(Continued)*

Form	Target strength $= 10 \log t$	Symbols	Direction of incidence	Conditions	References
Any smooth convex object	$\dfrac{S}{16\pi}$	S = total surface area of object	Average over all directions	All dimensions and radii of curvature large compared with λ	4, 7
Triangular corner reflector	$\dfrac{L^4}{3\lambda^2}(1 - 0.00076\theta^2)^2$	L = length of edge of reflector	At angle θ to axis of symmetry	Dimensions large compared with λ	5
Any elongated body of revolution	$\dfrac{16\pi^2 V^2}{\lambda^4}$	V = body volume	Along axis of revolution	All dimensions *small* compared to λ	8
Circular plate	$\left(\dfrac{4}{3\pi}\right)^2 k^4 a^6$	a = radius $k = 2\pi/\lambda$	Perpendicular to plate	$ka \ll 1$	8
Infinite plane strip	$\dfrac{1}{4\pi k}\left[\dfrac{\cos\theta \sin(2ka\sin\theta)}{\sin\theta}\right]^2$	$2a$ = width of strip θ = angle to normal	At angle θ	$ka \gg 1$	8
	$\dfrac{ka^2}{\pi}$		Perpendicular to strip	$ka \gg 1$ $\theta = 0$	8

REFERENCES FOR TABLE 9.1
1. Physics of Sound in the Sea, pt. III, *Nat. Def. Res. Comm. Div. 6 Sum. Tech. Rep.* **8:**358–362 (1946). (Note: Eqs. 49, 50, 53, and 56 in this reference are in error.)
2. Rayleigh, Lord: "Theory of Sound," vol. II, p. 277, Dover Publications, Inc., New York, 1945.
3. Rayleigh, Lord: "Theory of Sound," vol. II, p. 311, Dover Publications, Inc., New York, 1945.
4. Kerr, D. E. (ed.): "Propagation of Short Radio Waves," M.I.T. Radiation Laboratory Series, vol. 13, pp. 445–469, McGraw-Hill Book Company, New York, 1951.
5. Propagation of Radio Waves, Committee on Propagation, *Nat. Def. Res. Comm. Sum. Rep.*, **3:**182 (1946).
6. Willis, H. F.: Unpublished (British) report, 1941.
7. Spencer, R. C.: Backscattering from Conducting Surfaces, RDB Committee on Electronics, Symposium on Radar Reflection Studies, September 1950.
8. Ruck, G. T., and others: "Radar Cross-Section Handbook," vols. I and II, Plenum Press, New York, 1970.

direction to source and receiver. This means, referring to Fig. 9.9, that the bistatic target strength with the source and receiver in the directions OS and OR from the target is the same as the monostatic target strength in the direction OP along the bisector of the angle SOR. This theorem, originating in physical optics, is said to be approximately true if the bistatic angle is considerably less than 180°. In the forward direction, where the bistatic angle equals 180°, physical optics also shows that the target strength of any large smooth object of projected area A equals $10 \log A^2/\lambda^2$, where λ is the acoustic wavelength, which must be small compared with all target dimensions. This is the same as the target strength of a flat plate of area A in the backward direction (Table 9.1).

Unfortunately, in the absence of measured data, it is not known how well these theoretical approximations apply, if they do at all, to the complex underwater targets of sonar.

9.7 Target-Strength Measurement Methods

The obvious and direct method of measuring the target strength of an underwater object is to place a hydrophone at a distance 1 yd from the object and to measure the ratio of the reflected (or scattered) intensity to the incident intensity. This method is impractical for a number of reasons. For many objects, it is difficult, if not impossible, to locate and place a hydrophone at this 1-yd point; even if it could be done, it would be difficult to separate the reflected sound from the incident sound at such a short distance. Moreover, the results would be invalid, in many cases, for use at longer ranges, since the target strength of objects like cylinders and submarines is different at short ranges than at long.

A more practical method is to use a reference target of known target strength, placed at the same range as the unknown, and to compare the levels of the echoes from the reference target and the target to be measured. The

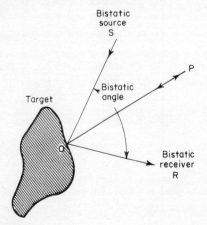

fig. 9.9 *The bistatic geometry. The direction OP is the bisector of the bistatic angle.*

fig. 9.10 *Target-strength measurement using a reference sphere. A sphere of diameter 12 ft is placed at the same range as the target being measured. Using a directional active sonar, the intensities of the echoes are compared.*

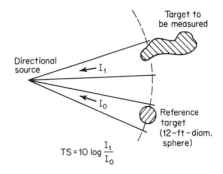

method is illustrated in Fig. 9.10. If the reference target is a sphere of diameter 4 yd, its target strength is 0 dB. Then the target strength of the target to be measured will be 10 times the logarithm of the ratio of the echo intensity of the unknown target to that of the 0-dB reference target. Suitable corrections can easily be made for differences in range and size of the reference target. This straightforward comparison method is particularly suited to measurements on small objects at short ranges. For large objects, such as submarines at longer ranges, the handling and the size requirements of the reference sphere make the method impractical. In all cases, the construction and dimensions of the reference target must be carefully controlled to make sure that its target strength approximates that of the ideal shape. Because of its ease of handling and its higher effective target strength, a calibrated transponder is a substitute for such a passive reference target.

Most target-strength measurements have been made by what may be called the conventional method, in which measurements of the peak or average intensity of the irregular echo envelope are made at some long range and then reduced to what they would be at 1 yd. This reduction requires, in effect, a knowledge of the transmission loss appropriate to the time and place the echoes were measured, together with the source level of the sound source producing the echoes. The method is illustrated in Fig. 9.11. A number of echo-level measurements of a sonar target are made at some range r. The average level of the echoes is reduced to 1 yd by adding the known or estimated transmission loss; the difference between the reduced 1-yd level and the known source level is the target strength desired. The method employs the active-sonar equation written in the form

$$EL = SL - 2TL + TS$$

where EL is the level of the echo. The equation is solved for the unknown, TS. This conventional method has the disadvantage of requiring an accurate knowledge of the transmission loss, which in turn requires either simple propagation conditions, or a special series of field measurements made for the purpose. Much of the scatter and diversity in existing target-strength data are doubtless attributable to erroneous transmission-loss assumptions and to

308 / *principles of underwater sound*

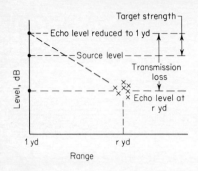

fig. 9.11 *Target-strength measurement using echo levels reduced to 1 yd. The range, source level, and transmission loss must be known.*

$EL = SL + 2TL + TS$

indiscriminate use of peak and average levels of the echo. Nevertheless, the method is basically simple and straightforward and requires no special equipment or instrumentation.

A method requiring no knowledge of the transmission loss, but needing special instrumentation, was used by Urick and Pieper (13). A measurement hydrophone and a transponder located about 1 yd from it were installed on the target submarine. On the measuring vessel (a surface ship) a hydrophone was suspended near the sound source producing the echoes to be measured. The relative levels of echo and transponder pulses were recorded aboard the surface ship; the relative levels of the incoming pulse and the transponder pulse were recorded aboard the submarine. As will be evident from Fig. 9.12, the target strength of the submarine is simply the difference in level between the two level differences recorded on the two vessels. The transponder serves, in effect, to "calibrate out" the underwater transmission path between the two vessels. No absolute calibration of the transducers used is needed, and the range separating the vessels need not be known.

9.8 Target Strength of Submarines

The target strength of submarines, of all underwater targets, has had the earliest historical attention and has been relatively well known from work

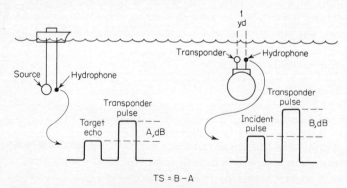

fig. 9.12 *A transponder method of target-strength measurement. No calibrations or knowledge of transmission loss are necessary.*

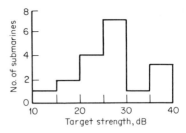

fig. 9.13 *Histogram showing the spread of the reported target strenths of 18 submarines at beam aspect, taken from the literature.*

done during World War II. This work has been well summarized in the NDRC Summary Technical Reports (14, 15). The present treatment of the subject will center about the variations of target strength with aspect, frequency, ping duration, depth, and range. The submarines considered will be limited to fleet-type diesel-powered submarines.

It should be emphasized at the outset that submarine target strengths are perhaps most noteworthy for their variability. Not only do individual echoes vary greatly from echo to echo on a single submarine, but also average values from submarine to submarine, as measured by different workers at different times and reduced to target strength, are vastly different.

Figure 9.13 shows the distribution of target strengths at beam aspect of diesel-powered submarines as reported in 18 separate reports by different observers using different sonars. As a result, the following discussion of the subject will be concerned with trends and broad effects, with the exception that individual measurements will show great differences from the mean or average values.

Variation with aspect   Figure 9.14 shows two examples of the variation of target strength with aspect around a submarine. Example A is a typical World War II determination (16) at 24 kHz with each point representing an average of about forty individual echoes in a 15° sector of aspect angle. Example B is the result of a postwar measurement (13) in which about five echoes were averaged in each 5° sector. Plots of this kind are based on measurements at sea of echoes from a submarine which runs in a tight circle at a distance from the source-receiver ship. Alternatively, the latter may circle a submarine running on a straight course at a slow speed.

The two examples of Fig. 9.14 illustrate the variability of target-strength measurements mentioned above. There is first a point-to-point variability with aspect that is due to the variability in level of individual echoes. There is also an average difference of about 10 dB between the two sets of measurements, with A being more typical of other determinations.

Nevertheless, when viewed in a broad fashion, the aspect dependence of target strength may be considered to have the pattern shown in Fig. 9.15. This "butterfly" pattern has the following characteristics:

310 / *principles of underwater sound*

1. "Wings" at beam aspect, extending up to about 25 dB and caused by specular reflection from the hull.

2. Dips at bow and stern aspects caused by shadowing of the hull and wake.

3. Lobes at about 20° from bow and stern, extending 1 or 2 dB above the general level of the pattern, perhaps caused by internal reflections in the tank structure of the submarine. These lobes do not appear in the aspect plots of nuclear-powered submarines not having ballast and fuel tanks outside the pressure hull.

4. A circular shape at other aspects due to a multiplicity of scattered returns from the complex structure of the submarine and its appendages.

All the characteristics of this idealized pattern are seldom seen in individual measurements. They may be absent entirely or lost in the scatter of the data. For example, Fig. 9.16 illustrates aspect plots of two submarines measured more recently at a frequency near 20 kHz with a pulse length of 80 ms while the submarines circled at a range of roughly 1,500 yd. Each plotted point

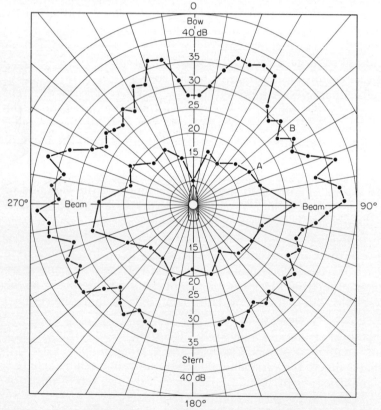

fig. 9.14 *Two determinations of the target strength of a submarine at various aspects. Example A (Ref. 15) is more typical of other determinations than B (Ref. 13).*

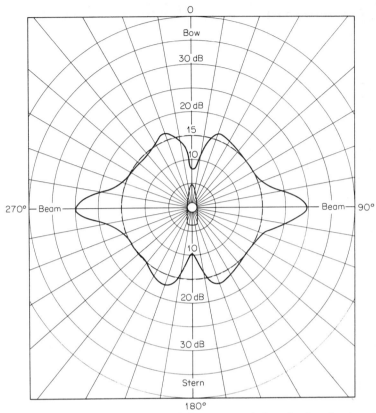

fig. 9.15 The "butterfly" pattern of the aspect variation of submarine target strength.

represents a single echo. Particularly striking is the variability between echoes during the circling maneuver, as well as the difference of about 5 dB between the target strengths of the two submarines when averaged over all aspects. One is hard-pressed to see in these data any sign of the "butterfly" pattern just described. Aspect patterns showing wide echo-to-echo variability are well known in radar (17). The tremendous variability between echoes is caused by a changing phase relationship between the different scatterers and reflectors on the target; as the target aspect changes, the various contributions add up with different phase relationships so as to produce a variable echo amplitude. Echo fluctuations are observed even from a target at an apparently constant aspect, as a result of the small changes of heading that must be made by the helmsman, or by the pilot of an aircraft, attempting to steer a steady course.

Variation with frequency Target-strength data were obtained during World War II (18) at 12, 24, and 60 kHz in an attempt to determine the effect of frequency on target strength. This attempt was unsuccessful, and it was concluded at that time that, if any frequency dependence existed, it was lost in the

312 / principles of underwater sound

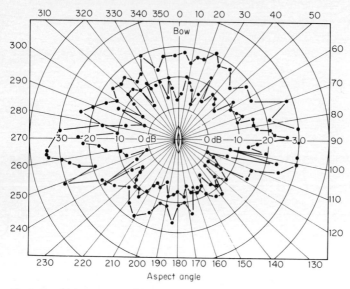

fig. 9.16 *Measurements of target strength versus aspect for two submarines. Each dot represents a single echo as the target circled at a distance.*

uncertainties of the data. The cause of this apparent frequency independence must be the many processes and sources that govern the return of sound from a submarine.

Variation with depth Except for the contribution of sound returned from the submarine's wake, no effect of depth on target strength has been established or expected. Depth effects on the echo are more likely the result of changing sound propagation with changing depth in the sea, instead of a depth effect on the submarine itself.

Variation with range For two reasons, the target strength of submarines is likely to be less at short ranges than at long ranges. One is the failure of a directional sonar, if one is used for the target-strength measurement, to insonify the entire target. The other is the fact that the echo of some geometrical forms does not fall off with range like the level from a point source. The solid curve in Fig. 9.17 shows the variation of echo intensity with range for a cylinder of length L at "beam" aspect. The intensity falls off like $1/r$ at short ranges ("cylindrical spreading") and like $1/r^2$ at long ranges ("spherical spreading") with a transition range approximately equal to L^2/λ. An echo at a short range r_1, when reduced to 1 yd by applying the transmission loss appropriate for a point source, would yield an apparent target strength TS_1 that would be less than the value TS_2 obtained with a long-range echo. Also, the target strength of an infinitely long cylinder, for which all points at a finite distance are in the near field or Fresnel region of the cylinder, is seen in Table 9.1 to increase linearly with range. The former effect—that of failure to insonify all the target—would apply to conditions of aspect and frequency where scattering by numerous scatterers on or in the target is the dominant echo-producing process; the latter effect applies to specular reflection. Both effects cause the short-range target strength of submarines to be less than the long-range target strength.

Variation with pulse length The target strength of an extended target, such as a submarine, may be expected to fall off with decreasing pulse length, inasmuch as a short pulse must fail to insonify the entire target. Thus, Fig. 9.18 illustrates that the target strength of an extended target should increase with the pulse length (in logarithmic units) until the pulse length is just long enough for all points on the target to contribute to the echo simultaneously at some instant of time. This occurs when the ping duration τ_0 is such that

$$\tau_0 = \frac{2l}{c}$$

where l = extension in range of target
c = velocity of sound

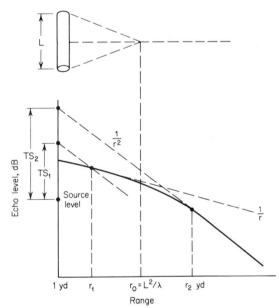

fig. 9.17 *Target strength versus range for a beam-aspect cylinder. At ranges less than r_0, the echo level varies with range like $1/r$ rather than $1/r^2$. This causes a lower target strength at short ranges than at long ranges ($TS_1 < TS_2$) if a $1/r^2$ transmission loss is used for the reduction to 1 yd.*

314 / principles of underwater sound

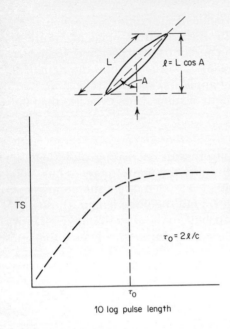

fig. 9.18 *Variation with pulse length. The target strength increases until the pulse length is great enough for the entire target to contribute to the echo at some one instant of time. This occurs at pulse length $\tau_0 = 2l/c$.*

If the target is of length L, then at aspect angle A,

$$l = L \cos A$$

This theoretical effect of reduced ping duration is not noticeable in target-strength measurements at beam aspect, where the extension in range is small and where specular reflection is the principal echo formation process, nor in measurements of peak target strength where individual target highlights are being measured.

9.9. Target Strength of Surface Ships

Because surface ships are not important targets for active sonars, comparatively little data are available on their target strengths. Three series of measurements at frequencies from 20 to 30 kHz were made during World War II on a total of 17 naval and merchant ships (19). They yielded beam-aspect values of 37, 21, and 16 dB for the three different groups of measurements and values of 13, 17, and 14 dB for corresponding off-beam target strengths. Standard deviations between 5 and 16 dB are quoted for these values. The small amount of available data show similar effects of aspect and range, as for submarines.

9.10 Target Strength of Mines

Modern mines are quasi-cylindrical objects a few feet long and 1 to 2 ft in diameter, flat or rounded on one end, and containing protuberances, depressions, and fins superposed on their generally cylindrical shape. Such targets should be expected to have a high target strength at "beam" aspect, as well as at aspects where some flat portion of the shape is normal to the direction of incidence; at other aspects, a relatively low target strength should be had. Measured target strengths range from about +10 dB within a few degrees of beam aspect, to much smaller values at intermediate aspects, with occasional lobes in the pattern attributable to reflections from flat facets of the mine shape. By Table 9.1, the target strength of a cylinder of length L, radius a, at wavelength λ is at long ranges

$$TS = 10 \log \frac{aL^2}{2\lambda}$$

If a be taken as 0.2 yd, $L = 1.5$ yd, and $\lambda = 0.03$ yd, corresponding to a frequency of 56 kHz, then

$$TS = 10 \log \frac{0.2 \times (1.5)^2}{2 \times 0.03} = 9 \text{ dB}$$

in approximate agreement with measurements on mines at beam aspect. The principal effects of frequency, aspect, range, and pulse length described for submarines apply for mines generally as well.

9.11 Target Strength of Torpedoes

A torpedo is, like most mines, basically cylindrical in shape with a flat or rounded nose. When specular reflection can be presumed to be the principal echo formation process, the target strength of a torpedo can be approximated by the theoretical formulas. An approaching torpedo of diameter 20 in. (radius = 0.28 yd) and having a hemispherical nose with a diameter equal to that of the torpedo, would have a target strength equal to $10 \log [(0.28)^2/4] = -17$ dB. At beam aspect its target strength could be approximated by the cylindrical formula illustrated for mines.

9.12 Target Strength of Fish

Fish are the targets of fish-finding sonars. Their target strengths are of interest for the optimum design of such sonars and, ideally, might be used for acoustic fish classification—that is, for estimating the number, type, and size of fish when fish finders are employed at sea.

Extensive data on the target strengths of fish of different species have been reported by Love (20, 21), both for fish in dorsal aspect (looking down at the fish) and in side aspect (looking broadside). Echoes from live fish, anesthetized to keep them motionless, were measured in a test tank, and the measurements were reduced to target strength in the conventional way (Sec. 9.7). Eight frequencies over the range 12 to 200 kHz were used; the fish specimens ranged in length between 1.9 and 8.8 inches. The results showed a strong dependence on the size of the fish, as measured by its length, and only a weak dependence on frequency or wavelength. When combined with older data reported by others, it was found that the measurements at dorsal aspect could be fitted by the empirical equation

$$TS = 19.1 \log_{10} L - 0.9 \log_{10} f - 54.3$$

where L is the fish length in inches, and f is the frequency of kilohertz. With L in centimeters, this equation becomes

$$TS = 19.1 \log_{10} L - 0.9 \log_{10} f - 62.0$$

The range of validity was $0.7 < L/\lambda < 90$; over this range an individual fish had a target strength differing on the average by about 5 dB from that given by the equation. Figure 9.19 shows TS plotted against L for frequencies 10 and 100 kHz according to these expressions. At side aspect, the measured values were on the average 1 dB higher at $L/\lambda = 1$ and 8 dB higher at $L/\lambda = 100$ than those at dorsal aspect. Additional measurements at lower frequencies over the range 4 to 20 kHz have been reported by McCartney and Stubbs (22), with generally similar results. These investigators point out that, at least in their frequency range, the swim bladder of the fish is the major cause of the backscattered return. Fish without swim bladders, such as mackerel, have a target strength some 10 dB lower than those that do, such as cod (23).

Although it is biologically incorrect to speak of whales as "fish," we may note

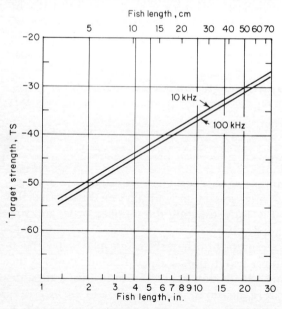

fig. 9.19 *Target strength of fish as a function of fish length for two frequencies, according to expressions given in text.*

here, if only as a curiosity, that the target strength of one species of whale in the open sea has been reported (24). Echoes were obtained from a number of humpback whales as they cavorted around the Argus Island oceanographic tower off Bermuda. At 20 kHz, two whales about 15 yd long were found to have a target strength of +7 at side aspect and −4 dB at head aspect; at 10 kHz, one smaller whale 10 yd long had a target strength of +2 dB at side aspect. These values are roughly what would be obtained by extrapolations in Fig. 9.19. Strange to say, they are not greatly different from what would be obtained from the expression TS = $10 \log R_1 R_2/4$ for the target strength of a large curved object of the same dimensions (Sec. 9.2).

9.13 Target Strength of Small Organisms

The target strength of various forms of zooplankton, such as squid, crabs, euphausid shrimp, and copepods has been reported by a number of investigators. The results have been reviewed by Penrose and Kaye (25) who found, strangely enough, that the expression of Love given above fitted the measurements fairly well. Thus copepods 3 mm long were found to have a target strength of about −80 dB, crabs 30 mm in size, −60 dB; squid 100 mm in size, −45 dB. Little or no frequency dependence was found, although Greenlaw (26), using preserved specimens in the frequency range 200 to 1,000 kHz, observed a strong ($20 \log f$) increase with frequency. The spread of measured data is large and the mechanism by which a small marine organism—which may or may not have a hard shell or contain a bubble of gas—interacts with sound is still not clear.

9.14 Echo Formation Processes

Complex underwater targets return sound back to the source by a number of processes. These processes will all occur, in general, for a complex target like a submarine, but only one or two will be dominant under any particular conditions of frequency and aspect angle.

Specular reflection The most simple and best understood echo formation process is *specular reflection*. It is illustrated by the return of sound from large spheres or convex surfaces, as described earlier in this chapter. In this process, the target remains stationary and does not move in the sound field and thereby generates a reflected wave having a *particle velocity* just sufficient, in wave theory, to cancel that of the incident wave over the surface of the object. Alternatively, the reflecting surface may be *soft* instead of *hard* or stationary, so as to form a reflection in which the *pressures* of the two waves will cancel. A specular reflection has a waveform that is a duplicate of the incident waveform and can be perfectly correlated with it. For submarines and mines, specular reflection appears to be the dominant process at beam aspects, where a much-enhanced return is observed and where a short incident pulse has a faithful replica as an echo.

Scattering by surface irregularities Irregularities, such as protuberances, corners, and edges, that have small radii of curvature compared with a wavelength, return sound by scattering rather than reflection. Most real objects possess many such irregularities on their surface, and the scattered return is composed of contributions from a large number of such scattering centers. When only a few scatterers on the underwater target are dominant, they form *highlights,* which may be observable in the echo envelope. Submarines, especially the older types, possess numerous scattering irregularities on the hull, such as the bow and stern planes, railings, periscopes; of these protuberances, the conning tower is likely to be a strong reflector or scatterer of sound.

Penetration of sound into the target Underwater targets seldom remain rigid under the impact of an incident sound wave, but move or deform in a complex manner. This reaction may be thought of as penetration of sound into the target and a complex deformation of the target by the exciting sound wave.

In solid metal spheres in water, Hampton and McKinney (27) found considerable experimental evidence that acoustic energy in the frequency range 50 to 150 kHz penetrated into solid spheres a few inches in diameter and thereby produced a complex echo envelope and a target strength which varied by as much as 30 dB with frequency. A theoretical study by Hickling (28) of hollow metal spheres in water demonstrated that part of the echo originates from a kind of flexural wave moving around the shell. Such surface waves were also observed experimentally by Barnard and McKinney (29) in a study of solid and air-filled cylinders in water. We have already described these kinds of deformation waves in Sec. 9.4.

Submarines are structurally much more complicated than the simple forms just mentioned. Figure 9.20 shows cross-sectional views of a fleet-type submarine of basically World War II construction. The hull is essentially a quasi-cylindrical pressure hull surrounded by a tank structure of thin steel. Penetration of sound into the tank structure is relatively easy, and a return of sound by corner reflectors should be expected. These internal reflections may be the cause of the enhanced target strength at about 20° off bow and stern in the butterfly pattern of Fig. 9.15. In addition, scattering and resonance effects in the tank structure are likely to be sources of the echo at off-beam aspects. Finally, the pressure hull itself may be expected to contribute a scattered return at some aspect angles.

Whenever a regular repeated structure or appendage exists on a target, an enhanced echo is likely to occur at the frequency and angle for which the scattered returns reinforce one another in a coherent manner. This is a diffraction-grating effect; it will occur when the projected distance between structural components is a half-wavelength or a multiple thereof. A regularly structured target appears to act like an array of radiating elements, and produces the same beam pattern in its echo as would a line of equally spaced elements at twice the spacing. This is an alternative explanation of the 20°

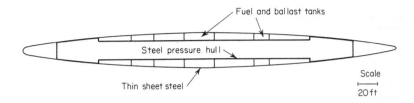

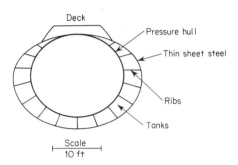

fig. 9.20 Diagrammatic fore-and-aft and athwartship cross sections of a World War II U.S. fleet submarine.

"ears" of the butterfly pattern just mentioned; the tank structure, the framing of the hull, or some other regularly repeated structural component may provide the regularity needed for selective reinforcement of the echo.

Resonant effects   Certain incident frequencies may correspond to various resonance frequencies of the underwater target. Such frequencies will excite different modes of oscillation or vibration of the target and give rise, in principle, to an enhanced target strength. Figure 9.21 is a pictorial description of a number of possible oscillatory modes of a compressible, flexible cigar-shaped object. When the sonar frequency is low enough, such modes may be excited by the incident sound wave and contribute to the target strength to an extent depending on the aspect and the Q or damping constant of the oscillatory mode. However, it should be pointed out that those modes which are good radiators of sound have a high damping constant because of radiation loading by the surrounding fluid, and as a result they would not necessarily give rise to a much greater target strength. Similarly, slightly damped modes of high Q might be poor reradiators of sound and thereby be poor contributors to the target strength. Such slightly damped modes would require a long sonar pulse of closely matched frequency for their excitation. It is even possible that resonant vibratory modes may be accompanied by a *lower* target strength if high internal losses in the target cause absorption of the incident energy, or if the mode is such that sound is reradiated, at resonance, into directions other than into the specular direction.

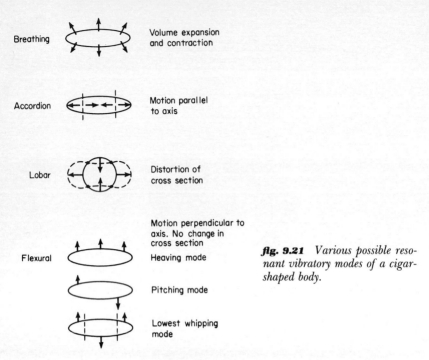

fig. 9.21 Various possible resonant vibratory modes of a cigar-shaped body.

A resonance of an entirely different type is the flexural resonance of the plates of which an underwater target like a submarine is composed. At certain angles a steel plate has been shown by the work of Finney (30) to be transparent to sound because of coupling between shear waves in the plate and sound waves in the surrounding water.

9.15 *Reduction of Target Strength*

It is sometimes desired to do something to a target, other than make major structural and compositional changes to it, for the purpose of reducing its target strength. This kind of acoustic camouflage can be achieved in a variety of ways. These are listed in Table 9.2.

table 9.2 Methods of Target Strength Reduction

Wavelength Large:* Low frequencies
  Volume reduction
Wavelength Small:* High frequencies
  Body shaping
  Anechoic coatings
  Viscous absorbers
  Gradual-transition coatings
  Cancellation coatings
  Quarter-wave layer
  Active cancellation

* Compared to all dimensions of the body.

At low frequencies, where the wavelength is large compared to all dimensions of the object, the only recourse is to reduce the *volume* of the target. This follows because, as we have seen, the target strength of a small sphere (Sec. 9.3) or of any small smooth object (Table 9.1) depends only on its volume at a fixed frequency. It follows that no alteration of body shape or application of a coating will be effective on a body that is small compared to a wavelength.

However, at high frequencies and short wavelengths, a variety of techniques are applicable, at least in principle. One is to *change the shape* of the body, if it is possible to do so, in such a way as to make the two principal radii of curvature ($R_1 R_2$, Sec. 9.2) everywhere small, avoiding in particular the infinities characteristic of flat plates and cylinders. The body shape should be smooth, without protuberances, holes, and cavities that act as scatterers of sound. In radar, a shape that has received much attention for missile applications is a cone with hemispherical base; this presents a particularly low target strength when viewed in the direction toward the conical tip. In radar, the radar cross section of the body need be minimized only in one (the forward) direction, whereas in sonar, a reduction usually must be achieved in many or all directions. For this reason, other shapes will be useful for sonar applications.

Anechoic coatings are materials that are cemented or attached to the body in order to reduce its acoustic return. The most important of these are various *viscous absorbing coatings* which attenuate the sound reaching, and returning from, the target by the process of viscous conversion to heat. Metal-loaded rubbers are an example of this kind of coating (31); here the tiny air cavities accompanying the metal particles cause a shear deformation of the rubber and a loss of sound to heat. Much larger air cavities causing shear deformation in rubber were employed in the Alberich coating used by the German Navy in World War II (32). This coating comprised a sheet of rubber 4 mm thick containing cylindrical holes 2 and 5 mm in diameter, with an overlying thin sheet of solid rubber to keep water out of the holes. This particular coating was effective in the octave 9 to 18 kHz; a number of German submarines were coated with it during the latter part of the war. A *gradual transition coating* consists of wedges or cones of lossy material with their apex pointing toward the incident sound. An example of this is a mixture of sawdust material called "insulkrete," described by Darner (33) for lining a tank to make it anechoic. This particular type of coating is massive and relatively fragile in construction for use on sonar targets.

A *cancellation coating* consists of alternate layers of acoustically hard and soft materials that reflect sound with opposite phase angles, so that no sound is returned from the target. Unfortunately, the cancellation occurs only at normal incidence, and there is little or no effect in other directions. A *quarter-wave layer* is a coating $\lambda/4$ in thickness having an acoustic impedance (ρc) equal to the geometric mean between the materials on either side, as, for example water and steel (34). Theory shows that under these conditions the layer is a perfect impedance match between the two materials, and no sound is reflected. But, because it is effective only at a single frequency (plus odd multiples thereof) and then only at normal incidence, the quarter-wave coating is of no value for practical sonar applications. Finally, we can mention the principle of *active cancellation*, wherein the incoming sound is monitored on the target and an identical signal is generated 180° out of phase to it by a small sound source. This technique has been used for reducing reflections in a tube for transducer calibrations (35), but requires expensive instrumentation; in any case, it would have little value for reducing reflections from the large complex targets of sonar.

9.16 Echo Characteristics

Echoes from most underwater objects differ from the incident pulse in a number of ways other than intensity, as described by the parameter target strength. The reflecting object imparts its own characteristics to the echo; it interacts with the incident sound wave to produce an echo that is, in general, different in wave shape and other characteristics from the incident pulse. These differences are useful to the sonar engineer in two ways: they may be employed as an aid in *detection,* as in filtering with narrow-band filters to enhance an echo buried in reverberation; they may be utilized to assist in target *classification* to distinguish one type of target from another, as in distinguishing a submarine from a school of fish.

Some distinguishing characteristics of echoes are:

Doppler shift Echoes from a moving target are shifted in frequency by the familiar doppler effect (36) by an amount equal to

$$\Delta f = \frac{2v}{c} f$$

where v = relative velocity or range rate between source and target
c = velocity of sound (in same units as v)
f = operating frequency of transmitter

In practical terms, and for a sound velocity of 4,900 ft/s,

$$\boxed{\Delta f = \pm 0.69 \text{ Hz/(knot)(kHz)}}$$

where the range rate is in units of knots and the sonar frequency is in kilohertz. An echo from an approaching target with a 10-knot relative velocity and a frequency of 10 kHz would thus be shifted higher in frequency by 69 Hz. The ± sign indicates that an approaching target produces an echo of higher frequency ("up-doppler"), and a receding target one of lower frequency ("down-doppler").

Extended duration Echoes are lengthened by the extension in range of the target. Whenever an underwater target is such that sound is returned by scatterers and reflectors distributed all along the target, the entire area or volume of the target contributes to the echo. For a target of length L at an aspect angle θ, the incident pulse is lengthened in duration by the time interval

$$\frac{2L \cos \theta}{c}$$

for the monostatic case, and by

$$\frac{L}{c}(\cos \theta_i + \cos \theta_r)$$

for the bistatic case with incidence at aspect angle θ_i, and an echo return at angle θ_r. This time elongation of the echo is most noticeable when short sonar pulses are employed. It occurs only for complex targets composed of numerous distributed scatterers and is negligible when specular reflection is the most important process of echo formation. Most sonar echoes are elongated, however, and so provide a clue concerning the size, and therefore the nature, of the echoing object.

Some actual measured data on the duration of echoes from a submarine target as a function of aspect angle are shown in Fig. 9.22. In this plot, each

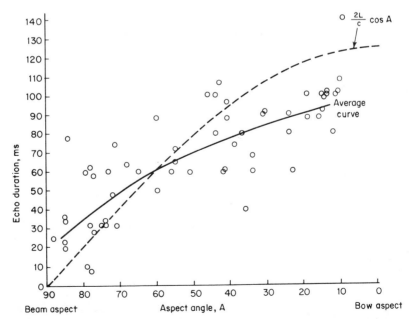

fig. 9.22 *Echo duration at different aspect angles of a submarine target. Each point represents a single echo from an explosive source.*

point represents the time duration of an explosive echo at the particular aspect angle at which it occurred. We observe that the theoretical relationship $t = 2(L/c) \cos A$ is only a rough approximation to the observations summarized by the average curve. The echoes are longer than they should be near beam aspect, probably because of multipath transmission (more particularly, the surface reflected path) from source to target and back again; the echoes are shorter than they should be near bow aspect, probably because of shadowing of the stern portions of the hull by the bow.

Irregular envelope The echo envelope is irregular, especially where specular reflection is not important. This irregularity arises from acoustic interference between the scatterers of the target. For some targets, individual highlights in the echo may be identified as arising from individual strong echoing portions of the target; for example, the conning tower of a submarine may yield a recognizable strong return of its own as part of the echo. Most often, however, the sonar echo from a sinusoidal ping is an irregular blob, without distinguishing features, that varies in envelope shape from echo to echo as the changing phase relationships among the scatterers and the propagation paths to and from the target take effect. Figure 9.23 is a sequence of three echoes from a short-range oblique-aspect submarine taken at intervals one second apart using a sonar with a pinglength of 15 ms. From this sequence we should note the rapid variability of the complex structure of the echo envelope and its extended duration relative to the 15-ms outgoing pulse.

table 9.3 Nominal Values of Target Strength

Target	Aspect	TS, dB
Submarines	Beam	+25
	Bow-stern	+10
	Intermediate	+15
Surface ships	Beam	+25 (highly uncertain)
	Off-beam	+15 (highly uncertain)
Mines	Beam	+10
	Off-beam	+10 to −25
Torpedoes	Bow	−20
Fish of length L, in.	Dorsal view	$19 \log L - 54$ (approx.)
Unsuited swimmers	Any	−15
Seamounts	Any	+30 to +60

Modulation effects At stern aspects on propeller-driven targets, the propeller may amplitude-modulate the echo in the same way that the propeller of an aircraft is known (37) to modulate a radar echo. The propeller thus produces a cyclic variation in the scattering cross section of the target. Another possible form of modulation can arise from interaction between the echo from a moving vessel and the echo from its wake. The difference in frequency between the two may appear as beats, or amplitude variations, in the envelope of the combined echo from the hull and wake at certain aspect angles.

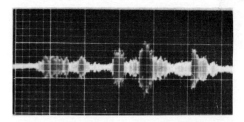
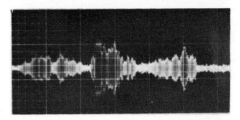

fig. 9.23 *Photographs of three submarine echoes taken one second apart. The width of the horizontal scale is 120 ms; the sonar pinglength was 15 ms.*

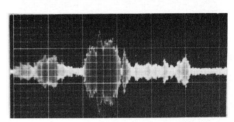

9.17 Summary of Numerical Values

Table 9.3 is a summary of target-strength values for the underwater targets described in this chapter, plus two others. As previously mentioned, these are subject to considerable variation in individual measurements on targets of the same type, and the target-strength values given are to be regarded only as nominal values useful for first-cut problem solving.

REFERENCES

1. Kerr, D. E. (ed.): "Propagation of Short Radio Waves," M.I.T. Radiation Laboratory Series, vol. 13, pp. 445–481, McGraw-Hill Book Company, New York, 1951.
2. Freedman, A.: Recent Approaches to Echo-Structure Theory, *J. Acoust. Soc. Am.,* **36:**2000(A) (1964). Also, A Mechanism of Acoustic Echo Formation, *Acoustica,* **12:**10 (1962).
3. Neubauer, W. G.: A Summation Formula for Use in Determining the Reflection of Irregular Bodies, *J. Acoust. Soc. Am.,* **35:**279 (1963).
4. Steinberger, R. L.: Theoretical Analysis of Echo Formation in the Fluctuation Environment of the Sea, *U.S. Nav. Res. Lab. Rep.* 5449, 1960.
5. Rayleigh, Lord: "Theory of Sound," vol. 2, p. 277, eqs. 26 and 27, Dover Publications, Inc., New York, 1945.
6. Rayleigh, Lord: "Theory of Sound," vol. 2, p. 24, eq. 13, Dover Publications, Inc., New York, 1945.

7. Anderson, V. C.: Sound Scattering from a Fluid Sphere, *J. Acoust. Soc. Am.*, **22**:426–431 (1950).
8. Harbold, M. L., and B. N. Steinberg: Direct Experimental Verification of Creeping Waves, *J. Acoust. Soc. Am.*, **45**:592 (1969).
9. Neubauer, W. G.: Pulsed Circumferential Waves on Aluminum Cylinders, *J. Acoust. Soc. Am.*, **45**:1134 (1969).
10. Marks, B. M., and E. E. Mikeska: Reflections from Liquid-filled Spherical Reflectors, *J. Acoust. Soc. Am.*, **59**:813 (1976).
11. Crispin, J. W., and K. M. Siegel (eds.): "Methods of Radar Cross-Section Analysis," Electrical Science Series, Academic Press, Inc., New York, 1968.
12. Ruck, G. T., and others: "Radar Cross-Section Handbook," vol. I, p. 11, Plenum Press, New York, 1970.
13. Urick, R. J., and A. G. Pieper: "Determination of the Target Strength of a Submarine by a New Method," U.S. Office of Naval Research Informal Report, April 1952.
14. Principles of Underwater Sound, *Nat. Def. Res. Comm. Div. 6 Sum. Tech. Rep.* 7, chap. 8, 1946.
15. Physics of Sound in the Sea: Reflection of Sound from Submarines and Surface Vessels, *Nat. Def. Res. Comm. Div. 6 Sum. Tech. Rep.* 8, 1946.
16. Physics of Sound in the Sea: Reflection of Sound from Submarines and Surface Vessels, *Nat. Def. Res. Comm. Div. 6 Sum. Tech. Rep.* 8, fig. 3, p. 391, 1946.
17. Corriher, H. A.: Radar Reflectivity of Aircraft, Georgia Institute of Technology, Final Report on Project A-1025, 1970.
18. Physics of Sound in the Sea: Reflection of Sound from Submarines and Surface Vessels, *Nat. Def. Res. Comm. Div. 6 Sum. Tech. Rep.* 8, pp. 408–410, 1946.
19. Physics of Sound in the Sea: Reflection of Sound from Submarines and Surface Vessels, *Nat. Def. Res. Comm. Div. 6 Sum. Tech. Rep.* 8, chap. 24, 1946.
20. Love, R. H.: Dorsal Aspect Target Strength of Individual Fish, *J. Acoust. Soc. Am.*, **49**:816 (1971).
21. Love, R. H.: Maximum Side-Aspect Target Strength of Individual Fish, *J. Acoust. Soc. Am.*, **46**:746 (1969).
22. McCartney, B. S., and A. R. Stubbs: Measurements of the Target Strength of Fish in Dorsal Aspect, *Proc. Int. Symp. on Biol. Sound Scattering in the Ocean, Maury Center for Ocean Sci. Rep.* MC 005, p. 180, 1970.
23. Foote, K. G.: Importance of the Swim Bladder in Acoustic Scattering by Fish, *J. Acoust. Soc. Am.*, **67**:2084 (1980).
24. Love, R. H.: Target Strengths of Humpback Whales, *J. Acoust. Soc. Am.*, **54**:1312 (1973).
25. Penrose, J. D., and G. T. Kaye: Acoustic Target Strengths of Marine Organisms, *J. Acoust. Soc. Am.*, **65**:374 (1979).
26. Greenlaw, C. F.: Backscattering Spectra of Preserved Zooplankton, *J. Acoust. Soc. Am.*, **62**:44 (1977).
27. Hampton, L. D., and C. M. McKinney: Experimental Study of the Scattering of Acoustic Energy from Solid Metal Spheres in Water, *J. Acoust. Soc. Am.*, **33**:664 (1961).
28. Hickling, R.: Analysis of Echoes from a Hollow Metallic Sphere in Water, *J. Acoust. Soc. Am.*, **36**:1124 (1964).
29. Barnard, G. R., and C. M. McKinney: Scattering of Acoustic Energy by Solid and Air-filled Cylinders in Water, *J. Acoust. Soc. Am.*, **33**:226 (1961).
30. Finney, W. J.: Reflection of Sound from Submerged Plates, *J. Acoust. Soc. Am.*, **20**:626 (1948).
31. Cramer, W. S.: Acoustic Properties of Metal Loaded Rubber, *U.S. Nav. Ord. Lab. Rep.* NAVORD 3756, 1954.
32. Richardson, E. G.: "Technical Aspects of Sound," vol. II, chap. 7, American Elsevier Publishing Company, Inc., New York, 1957.
33. Darner, C. L.: An Anechoic Tank for Underwater Sound Measurements, *J. Acoust. Soc. Am.*, **26**:221 (1954).

34. Cramer, W. S.: Theoretical Study of Underwater Sound Absorbing Layers, *U.S. Nav. Ord. Lab. Rep.* NAVORD 2803, 1953.
35. Beatty, L. G., and others: Sonar Transducer Calibration in a High Pressure Tube, *J. Acoust. Soc. Am.*, **39:**48 (1966).
36. Kinsler, L. E., and A. R. Frey: "Fundamentals of Acoustics," p. 453, John Wiley & Sons, Inc., New York, 1950.
37. Lawson, J. L., and G. E. Uhlenbeck: "Threshold Signals," M.I.T. Radiation Laboratory Series, vol. 24, p. 288, McGraw-Hill Book Company, New York, 1950.

ten

radiated noise of ships, submarines, and torpedoes: radiated-noise levels

Ships, submarines, and torpedoes are excellent sources of underwater sound. Being themselves machines of great complexity, they require numerous rotational and reciprocating machinery components for their propulsion, control, and habitability. This machinery generates vibration that appears as underwater sound at a distant hydrophone after transmission through the hull and through the sea. Of particular importance is the machinery component called the propeller, which serves to keep the vehicle in motion, and in doing so, generates sound through processes of its own.

Radiated noise is of particular importance for passive sonars, which are designed to exploit the peculiarities of this form of noise and to distinguish it from the background of self-noise or ambient noise in which it is normally observed.

In this chapter the principal characteristics of radiated noise will be discussed, and some data on the level of this noise for various vehicles will be presented. The discussion will be restricted to conventional designs and conventional propulsion systems, omitting any discussion of the noise of nuclear-propelled vessels or unusual propulsion methods. The word "vessel" will be used to refer to surface ships, submarines, or torpedoes indiscriminately when it is unnecessary to distinguish among them.

10.1 Source Level and Noise Spectra

The parameter *source level* for radiated noise in the sonar equations is the intensity, in decibel units, of the noise radiated to a distance by an underwater source, when measured at an arbitrary distance and reduced to 1 yd from the *acoustic center* of the source. By acoustic center is meant the point inside or outside the vessel from which the sound, as measured at a distance, *appears* to be radiated. Practically speaking, radiated-noise measurements must of necessity be made at a distance from the radiating vessel, typically 50 to 200 yd, and must be reduced to 1 yd by applying an appropriate spreading or distance correction. Source levels are specified in a 1-Hz band, and will be referred to a reference level of 1 μPa. The source levels for radiated noise are accordingly *spectrum* levels relative to the 1-μPa reference level.

When 1 meter is used as the reference distance instead of 1 yd, a correction of -0.78 dB must be applied to source levels referred to 1 yd. This same correction is required for all the other range-dependent sonar parameters when using 1 meter instead of 1 yd as the unit of range.

Noise spectra are of two basically different types. One type is *broadband noise* having a continuous spectrum. By continuous is meant that the level is a continuous function of frequency. The other basic type of noise is *tonal noise* having a discontinuous spectrum. This form of noise consists of tones or sinusoidal components having a spectrum containing *line components* occurring at discrete frequencies. These two spectral types are illustrated diagrammatically in Fig. 10.1. The radiated noise of vessels consists of a mixture of these two types of noise over much of the frequency range and may be characterized as having a continuous spectrum containing superposed line components.

10.2 Methods of Measurement

The noise radiated by vessels is nearly always measured by running the vessel past a stationary distant measurement hydrophone. Various types of hydrophones and hydrophone arrays have been employed for this purpose. The simplest arrangement uses a single hydrophone hung from a small measurement vessel. More elaborate configurations involve an array of hydrophones, either strung in a line along the bottom in shallow water, or hung vertically in

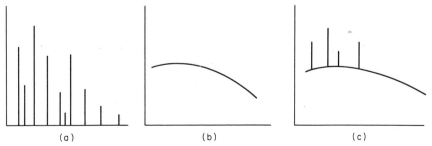

fig. 10.1 Diagrams of (a) line-component spectrum, (b) continuous spectrum, and (c) composite spectrum obtained by superposing (a) and (b).

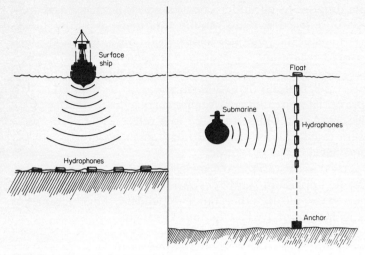

fig. 10.2 *Arrangements of hydrophones used at sound ranges for measuring radiated noise.*

deep water, as illustrated in Fig. 10.2. The former arrangement is suitable for measuring the noise of surface ships, and the latter is useful for measuring the noise of submarines or torpedoes running at deep depths. In both cases the vessel under test is arranged to run at a constant speed and course so as to pass the measurement hydrophones at a known distance. Suitable techniques are used to determine the range of the vessel while its radiated-noise output is being measured. During the run, broadband tape recordings are made, and later subjected to analysis in different frequency bands.

Figure 10.3 shows the instrumentation employed at the Atlantic Undersea Test and Evaluation Center (AUTEC) for the measurement of the radiated noise of submarines. An elaborate system is used for tracking the submarine during its passage past the measurement hydrophones. This noise-measurement range is located in the "Tongue of the Ocean" east of the Bahama Islands.*

Although radiated noise is commonly expressed as *spectrum levels*, that is, in 1-Hz bands, frequency analyses are often made in wider bands. The results are reduced to a band of 1 Hz by applying a bandwidth reduction factor equal to 10 times the logarithm of the bandwidth used. That is to say, if BL is the noise level measured in a band w hertz wide, the spectrum level in a 1-Hz band is BL $-$ 10 log w (Sec. 1.5). This reduction process is valid for continuous "white" noise having a flat spectrum; it can be shown to be also valid for noise having a continuous spectrum falling off at the rate of -6 dB/octave, if the center frequency of the band is taken to be the geometric mean of the two

---

* Five papers on AUTEC were presented at the 74th meeting of the Acoustical Society of America in 1967. Abstracts appear in *J. Acoust. Soc. Am.*, **42**:1187 (1967).

ends of the frequency band. But for line-component noise containing one or several strong lines within the measurement bandwidth, this reduction process is not valid and yields spectrum levels lower than the level of the line component dominating the spectrum. In short, the nature of the spectrum must in principle be known before the reduction is made. Many old reported data are almost useless at low frequencies because an excessively broad frequency band had been employed in the original analysis.

Similarly, a correction for distance is required to reduce the measurements to the 1-yd reference distance. The ubiquitous spherical-spreading law is generally applied for this purpose. Several investigations have indicated that spherical or inverse-square spreading is a good rule of thumb for expressing the variation of ship noise with range at close distances, even for low frequencies in shallow water (1). In any case, in spite of the apparent artificiality of the reduction processes in many instances, the original levels can be recovered from the published reduced values if the reduction process, bandwidth, and measurement distance are stated.

Studies of the sources of noise under various operating conditions, as distinct from measurements of noise level alone, have employed various clever schemes to pinpoint the dominant sources of noise. For example, when ships and torpedoes are run very close to the measuring hydrophone, the principal source of noise—whether machinery noise originating amidships or propeller noise originating near the stern—can be identified by the correspondence between the peak of the noise and the closest part of the ship at that instant. Overside surveys (2) of surface ships and submarines, in which a hydrophone

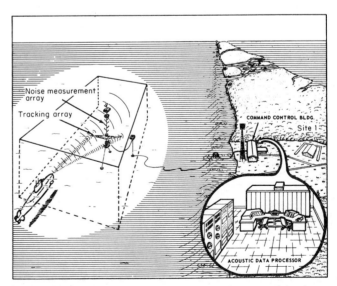

fig. 10.3 *Noise measurement range at the Atlantic Undersea Test and Evaluation Center (AUTEC).*

is lowered over the side of the moored vessel, are sometimes made to measure the noise radiated into the water by different pieces of ship's machinery. Various modifications to a running vessel have been made to study their effect on the noise output; such modifications include towing a surface ship without its propeller and operating a torpedo in a "captive" condition alongside a measurement platform. The effect of a bubble screen has been simulated by turning a ship so as to cross its own wake and noting the effect on the self-noise and radiated noise. All these methods have contributed to our present knowledge of the sources of radiated noise.

10.3 Sources of Radiated Noise

The sources of noise on ships, submarines, and torpedoes can be grouped into the three major classes listed in Table 10.1. *Machinery noise* comprises that part of the total noise of the vessel caused by the ship's machinery. *Propeller noise* is a hybrid form of noise having features and an origin common to both machinery and hydrodynamic noise. It is convenient to consider propeller noise separately because of its importance. *Hydrodynamic noise* is radiated noise originating in the irregular flow of water past the vessel moving through it and causing noise by a variety of hydrodynamic processes.

table 10.1 Source of Radiated Noise (Diesel-Electric Propulsion)

Machinery noise:
 Propulsion machinery (diesel engines, main motors, reduction gears)
 Auxiliary machinery (generators, pumps, air-conditioning equipment)
Propeller noise:
 Cavitation at or near the propeller
 Propeller-induced resonant hull excitation
Hydrodynamic noise:
 Radiated flow noise
 Resonant excitation of cavities, plates, and appendages
 Cavitation at struts and appendages

Machinery noise Machinery noise originates as mechanical vibration of the many and diverse parts of a moving vessel. This vibration is coupled to the sea via the hull of the vessel. Various paths, such as the mounting of the machine, connect the vibrating member to the hull. Machine vibration can originate in the following ways:

 1. Rotating unbalanced parts, such as out-of-round shafts or motor armatures

 2. Repetitive discontinuities, such as gear teeth, armature slots, turbine blades

 3. Reciprocating parts, such as the explosions in cylinders of reciprocating engines

4. Cavitation and turbulence in the fluid flow in pumps, pipes, valves, and condenser discharges
5. Mechanical friction, as in bearings and journals

The first three of these sources produce a *line-component spectrum* in which the noise is dominated by tonal components at the fundamental frequency and harmonics of the vibration-producing process; the other two give rise to noise having a *continuous spectrum* containing superposed line components when structural members are excited into resonant vibration. The machinery noise of a vessel may therefore be visualized as possessing a low-level continuous spectrum containing strong line components that originate in one or more of the repetitive vibration-producing processes listed above.

A diagrammatic view of the sources of machinery noise aboard a diesel-electric vessel is shown in Fig 10.4. Each piece of machinery produces periodic vibrational forces at the indicated fundamental frequency and thereby generates a series of line components at this frequency and at its harmonics. However, at a distance in the sea, the sound produced by these vibrational forces depends not only on their magnitude, but also on how such forces are transmitted to the hull and coupled to the water. A notable example is the resonant excitation of large sections of the hull by machinery vibration—called "hull drone"—such as that produced by the rotation of the massive propeller shaft, wherein certain frequencies of the excitation spectrum are reinforced by a short of sounding-board effect. The manner of mounting of the machine and the resulting vibration of the hull are determining factors in the radiation of sound. Another source of variability is in the propagation of different fre-

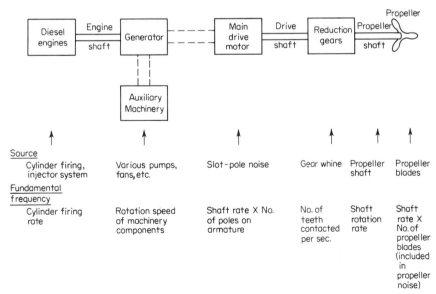

fig. 10.4 *Machinery components and noise sources on a diesel-electric vessel.*

quencies to a distant point in the sea. Because of these various effects, the harmonic structure of radiated noise is complex, and the line-component series generated by even a single source of noise is irregular and variable. When many noise sources are present, as in a vessel under way, the machinery-noise spectrum contains line components of greatly different level and origin, and is, consequently, subject to variations of level and frequency with changing conditions of the vessel.

Propeller noise   Even though the propeller is a part of the propulsion machinery of a vessel, the noise it generates has both a different origin and a different frequency spectrum from machinery noise. As just described, machinery noise originates *inside* the vessel and reaches the water by various processes of transmission and conduction through the hull. Propeller noise, on the other hand, originates *outside* the hull as a consequence of the propeller action and by virtue of the vessel's movement through the water. The location of the sources of noise along the hull is different as well. When a vessel passes close to a nearby hydrophone, it is observed that noises ascribable to the vessel's machinery reach a peak level *before* those originating at the propellers, in keeping with the place of origin aboard the noise-producing vessel.

The source of propeller noise is principally the noise of cavitation induced by the rotating propellers. When a propeller rotates in water, regions of low or negative pressure are created at the tips and on the surfaces of the propeller blades. If these negative (tensile) pressures become high enough, physical rupture of the water takes place and cavities in the form of minute bubbles begin to appear. These cavitation-produced bubbles collapse a short time later—either in the turbulent stream or up against the propeller itself—and in so doing emit a sharp pulse of sound. The noise produced by a great many of such collapsing bubbles is a loud "hiss" that usually dominates the high-frequency end of the spectrum of ship noise when it occurs. The production and collapse of cavities formed by the action of the propeller is called propeller cavitation.

Propeller cavitation may be subdivided into *tip-vortex cavitation*, in which the cavities are formed at the tips of the propeller blades and are intimately associated with the vortex stream left behind the rotating propeller, and *blade-surface cavitation*, where the generating area lies at front or back sides of the propeller blades. Of these two types of cavitation, the former has been found by laboratory measurements with model propellers and by analyses of field data (3) to be the more important noise source with propellers of conventional design.

Because cavitation noise consists of a large number of random small bursts caused by bubble collapse, it has a continuous spectrum. At high frequencies, its spectrum level *decreases* with frequency at the rate of about 6 dB/octave, or about 20 dB/decade. At low frequencies, the spectrum level of cavitation noise *increases* with frequency, although this reverse slope tends to be obscured in measured data by other sources of noise. There is, therefore, a peak in the spectrum of cavitation noise which, for ships and submarines, usually occurs within the frequency decade 100 to 1,000 Hz. The location of the peak in the spectrum shifts to lower frequencies at higher speeds and (in the case of submarines) at smaller depths. Figure 10.5 shows diagrammatic cavitation-noise spectra for three combinations of speed and depth for a hypothetical submarine. The behavior of the spectral peak is associated with the generation of larger cavitation bubbles at the greater speeds and the lesser depths and with the resulting production of a greater amount of low-frequency sound. On an actual submarine, the variation of propeller noise with speed at a constant depth and with depth at a constant speed is illustrated by the measured curves of Fig. 10.6, which were obtained with a hydrophone placed 4 ft from the propellers on the two submarines, "Hake" and "Hoe."

It has long been known that as the speed of the ship increases, there is a speed at which propeller cavitation begins and the high-frequency radiated noise of the vessel suddenly and dramatically increases. This speed has been called the *critical speed* of the vessel. The submarines measures during World War II were observed to have critical speeds between 3 and 5 knots when operating at periscope depth and to have well-developed propeller cavitation at 6 knots and beyond. At speeds well beyond the critical speed, the noise of cavitation increases more slowly with speed. The noise-speed curve of a cavitating vessel accordingly has a shape like the letter S, with an increase of 20 to 50 dB at high frequencies when the speed is a few knots beyond the critical speed and with a slower rise, at a rate of 1.5 to 2.0 dB/knot, beyond. Measured data illustrating the S-shape characteristics of the noise-speed curve for submarines are shown in Fig. 10.7. The flattening of the curve at high speeds may be surmised to be due to a self-quenching or an internal absorption effect of the cloud of cavitation bubbles. Surface ships do not exhibit the S-shape feature in their noise-speed curves, but show, instead, a more gradual, and nondescript, increase of level with speed.

Cavitation noise is suppressed, and the critical speed is increased, by sub-

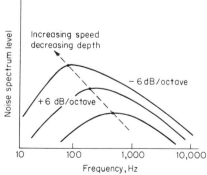

fig. 10.5 *Variation of the spectrum of cavitation noise with speed and depth.*

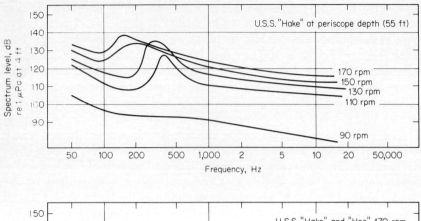

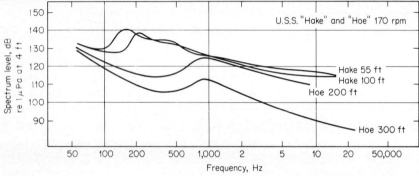

fig. 10.6 Propeller noise as measured with a hydrophone located 4 ft from the tips of the propeller blades. World War II data on the submarines "Hake" and "Hoe." (Ref. 4.)

merging to a greater depth. This effect has long been well known to submariners as a means to avoid detection. By simple hydrodynamic theory, it is known that the intensity of tip-vortex cavitation is governed by the *cavitation index*, defined by

$$K_T = \frac{p_0 - p_v}{(\frac{1}{2})\rho v_T^2}$$

where p_0 = static pressure at propellers
p_v = vapor pressure of water
ρ = density of water
v_T = tip velocity of propeller blades

When K_T lies between 0.6 and 2.0, tip cavitation begins; when K_T is less than 0.2, the occurrence of cavitation is certain; when K_T is greater than 6.0, cavitation is unlikely. If it is assumed that cavitation begins at some fixed value of K_T and if p_v is neglected, we conclude that the critical speed must vary as the square root of the static pressure at the depth of the propellers. This is illustrated by the measurements on "Hake" and "Hoe" previously referred to.

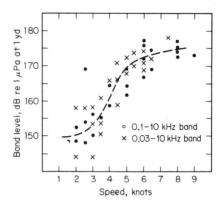

fig. 10.7 Broadband measurements of the radiated noise of a number of World War II submarines operating at periscope depth. 200-yd data reduced to 1 yd. (Ref. 4.)

For these submarines, Fig. 10.8 shows noise levels in the 10- to 30-kHz band observed with a hydrophone mounted 4 ft from the propeller tips and normalized by dividing by the square root of the hydrostatic pressure at the operating depth.

Although the cavitation noise of submarines is suppressed by depth, the reduction of noise with depth does not occur uniformly. When strong cavitation at high speeds occurs, the radiated noise of submarines is observed to first *increase* as the submarine dives, before the onset of the normal suppression of noise with depth. Figure 10.9 illustrates the effect of depth on the cavitation noise of a German type XXI submarine. At 8 knots, for example, this particular submarine had to submerge to 150 ft before experiencing the beginning of lower noise levels. This has been called the *anomalous depth effect*, although the term is now a misnomer, since it can be accounted for by theoretical analysis of cavitation-noise formation.

Many factors other than speed and depth affect propeller noise. A damaged propeller makes more noise than an undamaged one. More noise is made during turns and accelerations in speed than during uniform cruising. Truly anomalous conditions are sometimes found. "Singing" propellers generate strong tones between 100 and 1,000 Hz as a result of resonant excitation of the propeller by vortex shedding. The sound made by a singing propeller is

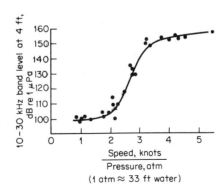

fig. 10.8 Normalized curves of cavitation noise. (Ref. 4.)

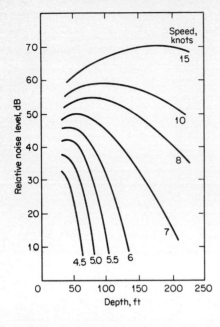

fig. 10.9 The anomalous depth effect, illustrated by measured wartime data at audio frequencies on a type XXI German submarine. Relative levels only.

very intense and can be heard underwater at distances of many miles. When it occurs, it can readily be cured, either by using a propeller made of a high-damping alloy or more simply, by changing the shape of the tips of the propeller blades (6).

Propeller noise is not radiated uniformly in all directions, but has a characteristic directional pattern in the horizontal plane around the radiating vessel. Less noise is radiated in the fore-and-aft directions than abeam, probably because of screening by the hull in the forward direction and by the wake at the rear. The dips in the pattern generally occur within 30° of the fore-and-aft direction, with the bow dip a few decibels deeper than the dip at the stern. A directivity pattern in the 2.5- to 5-kHz band for a freighter traveling at 8 knots is given in Fig. 10.10, where the contours show the locations, relative to the ship, of equal sound intensity on the bottom in 40 ft of water.

Propeller noise has been known for many years to be amplitude-modulated and to contain "propeller beats," or periodic increases of amplitude, occurring at the rotation speed of the propeller shaft, or at the propeller blade frequency equal to the shaft frequency multiplied by the number of blades. Propeller beats have long been used by listening observers for target identification and for estimating target speed. They have been observed (9) to be present in the noise of torpedoes as well as in the noise of ships and submarines. Propeller beats are most pronounced at speeds just beyond the onset of cavitation and diminish in the great roar of steady cavitation noise at high speeds.

Propeller noise, with its origin in the flow of water about the propeller, creates tonal components in addition to the continuous spectrum of cavitation noise. One tonal component is the "singing" tone of a vibrating singing propeller just mentioned. More normally, at the low-frequency end of the spectrum, propeller noise contains discrete spectral "blade-rate" components occurring at multiples of the rate at which any irregularity in the flow pattern into or about the propeller is intercepted by the propeller blades. The frequency of the blade-rate series of line components is given by the formula

$$f_m = mns$$

where f_m is the frequency, in hertz, of the mth harmonic of the blade-rate series of lines, n is the number of blades on the propeller, and s is the propeller rotation speed in number of turns per second. In one Russian experiment (10), these "blade" lines were observed with and without the emission of air bubbles around the propeller. At speeds when cavitation was well developed, they were found to be intimately associated with the cavitation process; in addition, line components not falling in the expected series were observed as well. Blade-rate line components were long ago observed (11) to be the dominant source of the noise of submarines in the 1- to 100-Hz region of the spectrum. It should be noted in passing that line components generated by the rotating propeller shaft—a form of machinery noise—fall in the same harmonic series as the blade-rate lines. Propeller noise, like machinery noise, can excite and be reinforced by the vibrational response of mechanical structures in the vicinity of the propeller.

The propellers of surface ships cavitate strongly at normal operating speeds. As a result, their low-frequency radiated noise spectrum is dominated by the blade-rate series of line components at the propeller blade frequency and its harmonics. Two examples of narrow-band spectra of surface ships—one a freighter, the other a supertanker—are shown in Fig. 10.11. The blade-rate series, with a fundamental frequency of about 8 Hz for both ships, is the principal feature of their radiation below about 50 Hz. The blade-rate series of

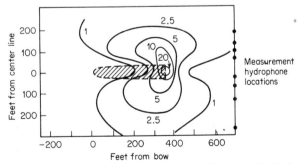

fig. 10.10 *Equal pressure contours on the bottom in 40 ft of water of a freighter at a speed of 8 knots. Contour values are pressures, in dynes per square centimeter in a 1-Hz band, at a point on the bottom, measured in the octave band, 2,500 to 5,000 Hz. (Ref. 8.)*

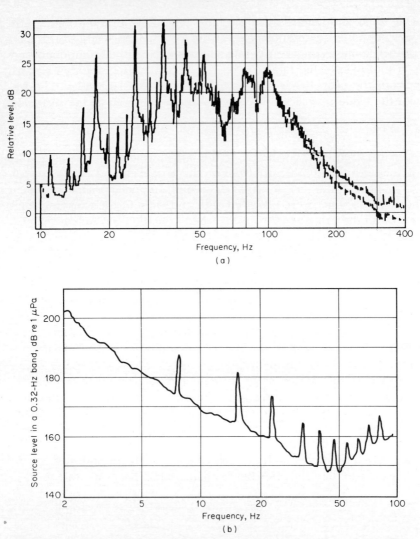

fig. 10.11 Narrow-band spectra of two merchant ships. (a) A bulk cargo ship, deadweight tons 12,200, speed 14.7 knots, analysis bandwidth 0.1 Hz. (b) A supertanker, "World Dignity," deadweight tons 271,000, speed 16 knots, shaft rate 1.44 rps, number of propeller blades 5, analysis bandwidth 0.32 Hz. (Ref. 7.)

line components are evenly spaced at 8-Hz intervals, although they do not appear to be so because of the logarithmic frequency scale.

The emission of air around the propeller is an effective practical way of reducing propeller noise. When cavitation occurs, air bubbles emitted in the neighborhood of the cavitating propeller replace the water-vapor bubbles created by the physical rupture of the water. The bubbles of air collapse with less force and thereby soften the effect of collapse caused by cavitation.

Hydrodynamic noise Hydrodynamic noise originates in the irregular and fluctuating flow of fluid past the moving vessel. The pressure fluctuations associated with the irregular flow may be radiated directly as sound to a distance, or, more importantly, may excite portions of the vessel into vibration. The noise created by the turbulent boundary layer is sometimes called "flow noise."

The excitation and reradiation of sound by various structures of the vessel are an important source of hydrodynamic noise. One kind of such noise is propeller singing, mentioned above. In addition, the flow of fluid may have an "aeolian harp" effect on other structures of the vessel, such as struts, and excite them into a vibrational resonance. Like Helmholtz resonators, cavities may be excited by the fluid flow across their openings, in the manner that a bottle can be made to "sing" by blowing over its opening. These resonant occurrences can sometimes be easily diagnosed and curative measures applied.

The form of hydrodynamic noise called *flow noise* is a normal characteristic of flow of a viscous fluid and occurs in connection with smooth bodies without protuberances or cavities. Flow noise can be radiated directly or indirectly as flow-induced vibrations of plates or portions of the body. The former—direct radiation to a distance—is an inefficient process not likely to be important at the low Mach numbers (ratio of speed of the vessel to the speed of sound in water) reached by vessels moving through water, since it is of quadrupole origin and is therefore not efficiently radiated to a distance. The latter process—flow excitation of a nonrigid body—depends on (1) the properties of the pressure fluctuation in the turbulent boundary layer, (2) the local response of the structure to these fluctuations, and (3) the radiation of sound by the vibrating portion. The response of plates to a random pressure field has been studied theoretically by Dyer (12). Flow noise is a more important contribution to self-noise than to radiated noise, and will be discussed more fully as a part of self-noise.

Other kinds of hydrodynamic noise are the roar of the breaking bow and stern waves of a moving vessel and the noise originating at the intake and exhaust of the main circulating water system.

Under normal circumstances, hydrodynamic noise is likely to be only a minor contributor to radiated noise, and is apt to be masked by machinery and propeller noises. However, under exceptional conditions, such as when a structural member or cavity is excited into a resonant source of line-component noise, hydrodynamic noise becomes a dominant noise source in the region of the spectrum in which it occurs.

10.4 Summary of the Sources of Radiated Noise

Of the three major classes of noise just described, machinery noise and propeller noise dominate the spectra of radiated noise under most conditions. The relative importance of the two depend upon frequency, speed, and depth.

This is illustrated by Fig. 10.12 which shows the characteristics of the spectrum of submarine noise at two speeds. Figure 10.12*a* is a diagrammatic spectrum at a speed when propeller cavitation has just begun to appear. The low-frequency end of the spectrum is dominated by machinery lines, together with the blade-rate lines of the propeller. These lines die away irregularly with increasing frequency and become submerged in the continuous spectrum of propeller noise. Sometimes, as indicated by the dotted line, an isolated high-frequency line or group of lines appear amid the continuous background of propeller noise. These high-frequency lines result from a singing propeller or from particularly noisy reduction gears, if the vessel is so equipped.

At a higher speed (Fig. 10.12*b*), the spectrum of propeller noise increases and shifts to lower frequencies. At the same time, some of the line components increase in both level and frequency, whereas others, notably those due to auxiliary machinery running at constant speed, remain unaffected by an increase in ship speed. Thus, at the higher speeds, the continuous spectrum of propeller cavitation overwhelms many of the line components and increases its dominance over the spectrum. A *decrease* in depth at a constant speed has, as indicated above, the same general effect on the propeller-noise spectrum as an *increase* in speed at constant depth.

For a given speed and depth, therefore, a "crossover" frequency may be said to exist, below which the spectrum is dominated by the line components of the ship's machinery plus its propeller, and above which the spectrum is in large part the continuous noise of the cavitating propeller. For ships and submarines, this frequency lies roughly between 100 and 1,000 Hz, depending on the individual ship and its speed and depth; for torpedoes, the crossover frequency is higher and the line components extend to higher frequencies because of the generally higher speeds of operation of torpedo machinery.

For illustrating the nature of ship spectra, a frequency-time analyzer, such as that used for speech analysis, is particularly convenient. This kind of analyzer, called a sound spectrograph, was first described by Koenig, Dunn, and

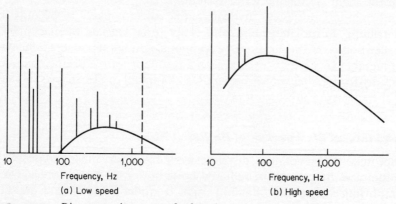

fig. 10.12 *Diagrammatic spectra of submarine noise at two speeds.*

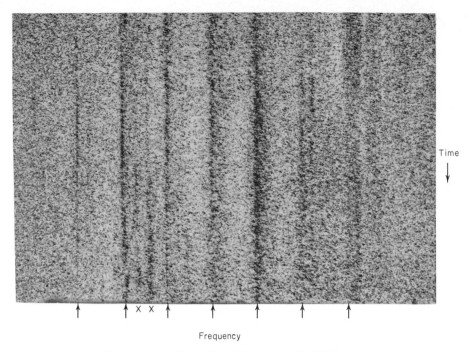

fig. 10.13 *Sound spectrogram of a surface ship at a speed of 11 knots.*

Lacy (13) and is widely used for the analysis of speech (14). It gives a plot of frequency against time and shows the intensity of the sound in the analysis bandwidth by a darkening of the record. Figure 10.13 is an example of a sound spectrogram of the noise of a large surface ship as it passed over a deep hydrophone. The frequency scale extends from 0 to 150 Hz, and the duration of the recording was approximately ½ hour. The harmonic series of line components marked by the arrows are blade-rate lines. The lines marked X are of unknown origin.

An interesting recent observation is that the noise spectrum of fishing boats, when trolling for fish at a speed of 10 to 12 knots, apparently influences the size of their catch of fish. Boats having spectral peaks above 1,500 Hz in their noise radiation were found to have an appreciably smaller catch of albacore than similar boats fishing at the same time and location, but having no strong high-frequency radiation, as if the high-frequency tones were scaring away, rather than luring, this particular species of fish (15).

10.5 Total Radiated Acoustic Power

It is of interest to determine how much total acoustic power is radiated by a moving vessel and how it compares with the power used by the vessel for propulsion through the water. This can easily be done by integration of the spectrum. When the spectrum is continuous and has a slope of -6 dB/octave, the intensity dI in a small frequency band df centered at frequency f can be written as

$$dI = \frac{A}{f^2} df$$

where $1/f^2$ represents the -6 dB/octave slope of the spectrum and A is a constant.

By integrating between any two frequencies f_1 and f_2, the total intensity between f_1 and f_2 becomes

$$I_{(f_1,f_2)} = \int_{f_1}^{f_2} \frac{A}{f^2} df = A\left(\frac{1}{f_1} - \frac{1}{f_2}\right)$$

Letting $f_2 \to \infty$, we find that the total intensity for all frequencies above f_1 is

$$I_{\text{tot}} = \frac{A}{f_1}$$

Recognizing that A/f_1^2 is the intensity in a 1-Hz band centered at frequency f_1, we note that

$$I_{\text{tot}} = (I_{f_1})(f_1)$$

where $10 \log I_{f_1}$ is the spectrum level at f_1.

As an example (Fig. 10.14), we take a spectrum having a spectrum level of 160 dB at 200 Hz with slope of $1/f^2$ above this frequency. The total level of the radiated noise above 200 Hz becomes

$$10 \log I_{\text{tot}} = 10 \log I_{f_1} + 10 \log f_1$$
$$= 160 + 23 = 183 \text{ dB}$$

The total acoustic power represented by this source level, if assumed to be radiated nondirectionally, amounts to 12 watts. It may be observed that if the continuous spectrum extended below 200 Hz to zero Hz at the constant level of 160 dB, as shown by the dashed line in Fig. 10.14, the total radiated power

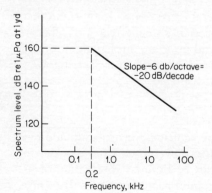

fig. 10.14 *Idealized spectrum integrated to obtain the total acoustic power radiated by a vessel.*

at frequencies *below* 200 Hz would also be 12 watts, and the total radiation over the entire spectrum would become 24 watts.

If the vessel is regarded as a mechanical source of sound, this radiated power may be compared with the shaft horsepower developed by the radiating vessel. Let us assume that the selected spectrum corresponds to that of a destroyer at a speed of 20 knots. At this speed, an average destroyer is known to generate a shaft horsepower of approximately 14,000, or about 10^7 watts. Comparing this power with that radiated as sound in the continuous part of the spectrum, we observe that the efficiency of the vessel as a sound producer is only of the order of 10^{-6}. Surface ships are therefore extremely inefficient radiators of sound in terms of the developed shaft power of their propulsion system, even when allowance is made for the power represented by the tonal components of the radiation.

10.6 Radiated-Noise Levels

During World War II, the United States and Great Britain, motivated by the design needs of acoustic mines, made a great many measurements of the radiated noise of surface ships at a number of acoustic ranges. At locations such as Wolf Trap, Virginia; Treasure Island, California; Thames River, New London, Connecticut; Puget Sound, Washington; and Waipio Point, Honolulu, Hawaii, literally thousands of "runs" on hundreds of ships of all types were made. Far fewer wartime measurements were made on submarines and torpedoes. Although much of this old data is obsolete because many of the vessels measured are no longer in existence, the general run of sound levels and their variation with speed, frequency, and depth will still be pertinent to many presently existing vessels, if only as a guide for approximation.

In the following sections, a few extracts have been selected from this vast wartime literature to illustrate the main quantitative aspects of the subject. Particular attention is called to two excellent summaries on the radiated-noise levels of submarines (4) and surface ships (5) and to one of the NDRC Summary Technical Reports (17) on the radiated noise of torpedoes.

Surface ships Table 10.2 shows typical radiated source levels for various classes of ships current during World War II, as reduced to 1 yd from an original reference distance of 20 yd by the conventional assumption of spherical spreading.

In graphical form, Fig. 10.15 gives in the upper set of curves average spectrum levels at 5 kHz as a function of speed for a number of classes of surface ships. The lower curve is a relative spectrum for use in obtaining values at other frequencies. The standard deviation (in decibels) of individual measurements from the line drawn is indicated for each class of ships.

Empirical expressions also were devised to fit the mass of data. In terms of the propeller-tip speed V in feet per second, the displacement tonnage T of

table 10.2 *Typical Average Source Levels for Several Classes of Ships in dB vs. 1 μPa in a 1-Hz Band at 1 yd.* (Ref. 5)*

Frequency	Freighter, 10 knots	Passenger, 15 knots	Battleship, 20 knots	Cruiser, 20 knots	Destroyer, 20 knots	Corvette, 15 knots
100 Hz	152	162	176	169	163	157
300 Hz	142	152	166	159	153	147
1 kHz	131	141	155	148	142	136
3 kHz	121	131	145	138	132	126
5 kHz	117	127	141	134	128	122
10 kHz	111	121	135	128	122	116
25 kHz	103	113	127	120	114	108

* Originally reported at 20 yd.

the ship, the frequency F in kilohertz, and distance D in yards, the source level for the average radiated noise of large ships was found to be given by

$$SL = 51 \log V + 15 \log T - 20 \log F + 20 \log D - 13.5$$

This formula, based on 157 runs of 77 ships of 11 different classes (mostly freighters, tankers, and large warships), was found to fit individual measurements to a standard deviation of 5.4 dB. It is applicable only at frequencies above 1 kHz where propeller cavitation is the principal source of noise. A more convenient formula, in terms of the speed of the ship, for use when information concerning the propeller-tip speed is lacking, was found to be

$$SL = 60 \log K + 9 \log T - 20 \log F + 20 \log D + 35$$

In this formula K is the forward speed of the ship in knots. This expression was found to fit the measured levels of passenger ships, transports, and warships at a frequency of 5 kHz to a standard deviation of 5.5 dB, but to be unreliable for freighters and tankers.

Figure 10.16 shows the radiated noise of a postwar destroyer as a function of speed for three frequencies. World War II levels as read from Fig. 10.15 for 500 and 5,000 Hz are also plotted. The residual levels at speeds less than 10 knots are caused by auxiliary and other non-speed-dependent machinery aboard the vessel.

At the low-frequency end of the spectrum, Fig. 10.17 illustrates other measured data (1) on four destroyers measured in the frequency bands 2 to 17 Hz and 7 to 35 Hz. The levels shown are band levels giving the total intensity measured in each band. The levels of any individual line components in the spectrum would be an indefinite number of decibels higher than the level indicated, depending on the number and relative strengths of the lines occurring in each band.

The distribution of radiated levels in the octave 75 to 150 Hz, based on a large number of ship runs at a number of wartime acoustic ranges (18), is shown in Fig. 10.18. In terms of the spectrum level at 1 yd, the ordinate in this figure gives the percentage of runs in which the measured level exceeded that

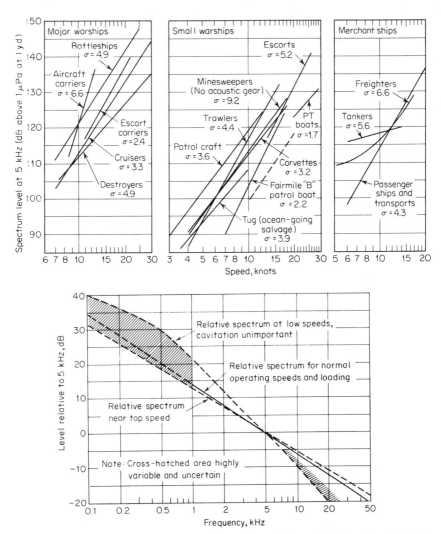

fig. 10.15 *Average radiated spectrum levels for several classes of surface ships. (Ref. 5.)*

shown on the horizontal scale. The levels have been reduced to spectrum levels at 1 yd from octave-band data given at a distance of 100 ft beneath the ship.

Regarding such averaged, reduced data, it should be borne in mind that individual ships occasionally deviate greatly from the average levels. Moreover, at low frequencies, the use of spectrum levels is questionable because of the likely tonal content of the radiated sound. Finally, the data are entirely based on measurements on the bottom in shallow water averaging 20 yd in depth, and do not necessarily indicate the levels that would be observed in deep water.

348 / *principles of underwater sound*

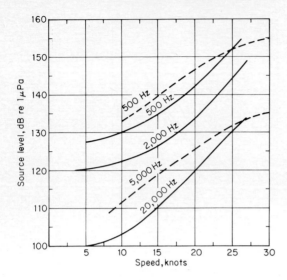

fig. 10.16 Solid curves, noise level versus speed at 0.5, 2, and 20 kHz for a postwar destroyer. Dashed curves, noise level versus speed for World War II destroyers from Fig. 10.15.

A quantitative model for the radiation at the blade-rate fundamental frequency has been presented by Gray and Greeley (16). Based on this model, expressions for the source level at the design speed of the blade-rate radiation of merchant ships were found to be

$$SL_D = 6 + 70 \log L \pm 11 \text{ dB}$$
$$= 92 + 94 \log D \pm 8 \text{ dB}$$

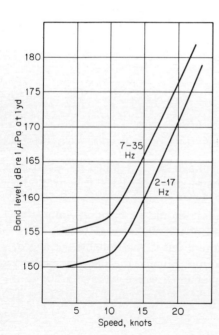

fig. 10.17 Average levels on four destroyers in two frequency bands as measured on the bottom in 105 ft of water at the Puget Sound acoustic range. Values reduced from 105 ft to 1 yd. (Ref. 1.)

fig. 10.18 *Cumulative distribution curves of the radiated noise of surface ships in the octave 75 to 150 Hz. (Ref. 18.) Horizontal scale is spectrum level obtained from band levels by subtracting 10 log bandwidth and reduced from 100 ft to 1 yd by assuming spherical spreading.*

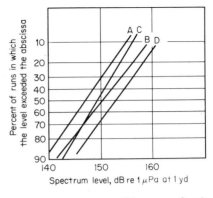

		Ships	Speed
A	(88 runs)	< 5,000 tons	< 10 knots
B	(132 runs)	< 5,000 tons	≥ 10 knots
C	(72 runs)	> 5,000 tons	< 10 knots
D	(129 runs)	> 5,000 tons	≥ 10 knots

where L and D are the ship length and propeller blade diameter, both in meters. The uncertainty values of 11 and 8 dB are the estimated standard deviation of a prediction based on these expressions, relative to calculated values based on the model and the known characteristics—such as ship length, draft, and propeller size—of merchant ships. The source level SL_D is the dipole source level at 1 meter. From it, the free-field source level can be found from

$$SL = SL_D + 6 + 20 \log kd$$

where k is the wave number $2\pi/\lambda = 2\pi f/c$, and where d is the operating depth of the propeller. For merchant ships the blade-rate fundamental frequency f falls largely in the range 6.7 to 10.0 Hz, and lies near 8 Hz for many ships. However, no data are available to compare the predictions based on this hydrodynamic model with noise measurements on actual ships under way at sea.

Submarines The available data on submarines are much less extensive and involve only a few submarines under limited operating conditions. Figure 10.19 shows average spectra for three World War II submarines at periscope depth (about 55 ft to the keel) and at the surface. Spectra appreciably higher, though similar in shape, were obtained for the single British submarine, HMS "Graph," as illustrated in Fig. 10.20.

The effect of depth on the radiated levels of submarines is illustrated in Fig. 10.21, which shows the reduction in noise involved in submerging from periscope depth to 200 ft. The effect of depth may also be determined from Fig. 10.8, given previously, in which the spectrum level was plotted as a function of speed divided by the square root and the hydrostatic pressure. These depth

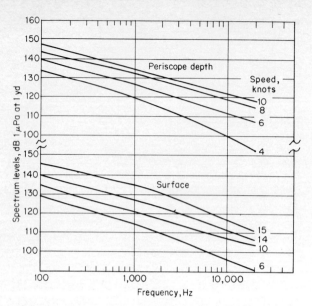

fig. 10.19 Smoothed spectra of three submaries (USS S-48, "Hake," and "Runner") on electric drive. Levels, originally reported at 200 yd, reduced to 1 yd. (Ref. 4.)

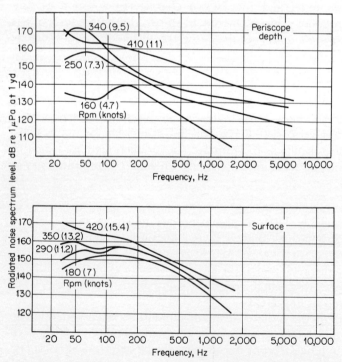

fig. 10.20 Radiated-noise spectra of the British submarine HMS "Graph" at periscope depth and on the surface. 200-yd reported data reduced to 1 yd. (Ref. 4.)

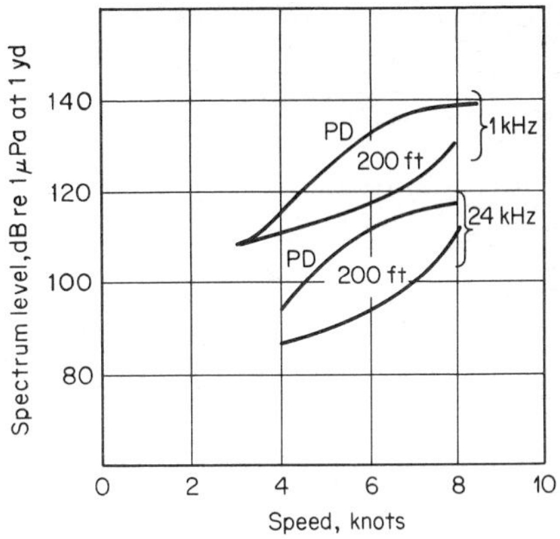

fig. 10.21 *Representative submarine-radiated-noise spectrum levels at two frequencies at periscope depth (PD) and at a depth of 200 ft on electric drive. (Ref. 19.)*

effects occur only for the noise produced by the cavitating propeller; when machinery noise predominates, as at very low speeds or very low frequencies, little or no quieting on submergence to deep depths is expectable.

Torpedoes Figure 10.22 illustrates measured radiated-noise spectra over a wide frequency range for a variety of torpedoes running at different speeds, and Fig. 10.23 is a compilation of torpedo noise levels at 25 kHz as a function of speed. Although the torpedoes listed had a variety of propulsion systems, the noise radiated at kilohertz frequencies must be presumed to be dominated by propeller cavitation.

10.7 Cautionary Remark

The potential user of the numerical data presented in this chapter will discover that it represents, for the most part, World War II measurements on obsolete vessels. Although it may be useful for first-cut problem solving and for showing qualitative effects, it must be stressed that the source level data will be almost useless for practical work. The subject of the radiated noise of

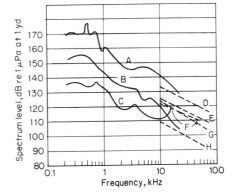

fig. 10.22 *Noise spectra of various World War II torpedoes running at shallow depths. [(A–C), Ref. 20; (D–H), Ref. 17.] (A) Highest measured values for several U.S. torpedoes. (B) Japanese Mark 91, 30 knots. (C) U.S. Mark 13, 30 knots. (D) U.S. Mark 14, 45 knots. (E) British Mark VIII, 37 knots. (F) U.S. Mark 18, 30 knots. (G) U.S. Mark 13, 33 knots. (H) British Mark VIII, 20 knots.*

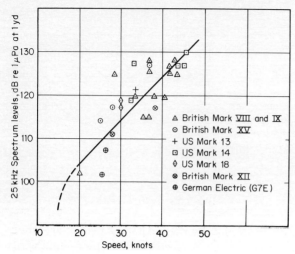

fig. 10.23 *25-kHz spectrum levels of various torpedoes as a function of speed. The straight line has a slope of 1 dB/knot increase of speed. (Ref. 17.)*

vessels is, and has long been, almost completely classified on security grounds, and the reader must resort to the classified literature for the information and data needed on modern existing vessels.

REFERENCES

1. Pomerantz, J., and G. F. Swanson: Underwater Pressure Fields of Small Naval Vessels in the 2–17 and 7–35 cps Bands at the Puget Sound Acoustic Ranges, *U.S. Nav. Ord. Lab. Rep.* 1022, 1945.*
2. Ship Acoustical Surveys, *U.S. Navy Bur. Ships Rep.* NAVSHIPS 250–371, no date.
3. Strasberg, M., and W. J. Sette: Measurements of Propeller Noise on Three Submarines of the SS212 Class, *David Taylor Model Basin Rep.* R-205, 1944.*
4. Knudsen, V. O., R. S. Allford, and J. W. Emling: Survey of Underwater Sound No. 2: Sounds from Submarines, *Nat. Def. Res. Comm. Div. 6 sec.* 6.1–NDRC-1306, 1943.*
5. Dow, M. T., J. W. Emling, and V. O. Knudsen: Survey of Underwater Sound No. 4: Sounds from Surface Ships, *Nat. Def. Res. Comm. Div. 6 sec.* 6.1–NDRC-2124, 1945.*
6. Ross, D.: "Mechanics of Underwater Noise," Sec. 9.5, Pergamon Press, New York, 1976.
7. Cybulski, C. J.: Probable Origin of Measured Supertanker Noise Spectra, *Conf. Proc. OCEANS '77*, October 1977.
8. Pomerantz, J.: An Analysis of Data in the 2500–10,000 cps Region Obtained at the Wolf Trap Range, *U.S. Nav. Ord. Lab. Rep.* 733, 1943.*
9. Irish, G.: "Low Frequency Modulation of the Supersonic Pressure Field of Torpedoes," U.S. Naval Ordnance Laboratory unpublished memorandum, 1944.*
10. Aleksandrov, I. A.: Physical Nature of the Rotation Noise of Ship Propellers in the Presence of Cavitation, *Sov. Phys. Acoust.*, **8**(1):23 (July–September 1962).
11. Jaques, A. T.: Subsonic Pressure Variations Produced by Submarines, *U.S. Nav. Ord. Lab. Rep.* 744, 1943.*
12. Dyer, I.: Response of Plates to a Decaying and Convecting Random Pressure Field, *J. Acoust. Soc. Am.*, **31**:922 (1959).

13. Koenig, W., H. K. Dunn, and L. Y. Lacy: The Sound Spectrograph, *J. Acoust. Soc. Am.,* **18:**19 (1946).
14. Potter, R. K., G. A. Kopp, and H. C. Green: "Visible Speech," D. Van Nostrand Company, Inc., Princeton, N.J., 1947.
15. Erickson, G. J.: Some Frequencies of Underwater Noise Produced by Fishing Boats Affecting Albacore Catch, *J. Acoust. Soc. Am.,* **66:**216 (1979).
16. Gray, L. M., and D. S. Greeley: Source Level Model for Propeller Blade-Rate Radiation for the World's Merchant Fleet, *J. Acoust. Soc. Am.,* **67:**516 (1980).
17. Acoustic Torpedoes, *Nat. Def. Res. Comm. Div. 6 Sum. Tech. Rep.,* vol. 22, 1946.*
18. Pomerantz, J.: "Cumulative Percentage Distribution of Ship's Sound Pressure in the 75–150 cps Octave," Naval Ordnance Laboratory unpublished memorandum, 1944.
19. Prediction of Sonic and Supersonic Listening Ranges, *Nat. Def. Res. Comm. Div. 6 sec.* 6.1–ser. 1131–1884, 1944.*
20. Listening Systems, *Nat. Def. Res. Comm. Div. 6 Sum. Tech. Rept.,* vol. 14, 1946.*

* References indicated by an asterisk were originally issued during World War II as classified reports. These reports have been declassified and are available from the U.S. Department of Commerce, National Technical Information Service, Springfield, Va. 22151.

eleven

self-noise of ships, submarines, and torpedoes: self-noise levels

Self-noise differs from radiated noise in that the measurement hydrophone is located on board the noise-making vessel and travels with it, instead of being fixed in the sea at a location some distance away. Although the fundamental causes of noise are the same, the relative importance of the various noise sources is different. Moreover, in self-noise, the paths by which the noise reaches the hydrophone are many and varied and play a dominant role in affecting the magnitude and kind of noise received by the hydrophone on the moving vessel.

In the sonar equations, radiated noise occurs as the parameter *source level* SL, where it is the level of the source of sound used by passive sonar systems. By contrast, self-noise is a particular kind of background noise occurring in sonars installed on a noisy vehicle; in the sonar equations, self-noise occurs quantitatively as the *noise level* NL. Self-noise exists in fixed hydrophones as well, whenever the manner of mounting or suspension creates noise of its own.

Self-noise is one of many different kinds of undesired sound in sonar and originates in a variety of ways. Figure 11.1 illustrates the interrelationships of the various kinds of sonar backgrounds. Self-noise refers to those noise sources between the dashed lines.

Self-noise depends greatly upon the directivity of the hydrophone, its mounting, and its location on the vehicle. On surface ships, the sonar transducer is located in a streamlined dome projecting below the keel of the ship. On older submarines, the principal passive

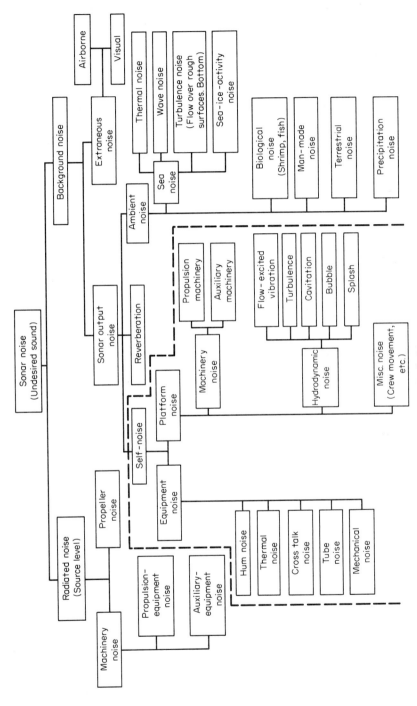

fig. 11.1 *Interrelationships of various sonar noise sources. (Ref. 1.)*

356 / *principles of underwater sound*

sonars were the JT and JP sonars, having horizontal line transducers placed topside toward the bow of the vessel; on modern submarines, a large portion of the bow of the vessel is taken over by the sonar transducer. Acoustic homing torpedoes often have their transducers in the nose of the torpedo and face forward in the direction of travel. On all these vessels, the sonar transducer is placed as far forward on the vessel as is practicable, so as to be removed as much as possible from the propulsion machinery and the propeller noises of the vessel. Figure 11.2 is a pictorial view of the location of the sonar transducer on these three kinds of vessels.

11.1 Self-Noise Measurements and Reduction

Self-noise measurements on these vehicles have been made in the past with sonar hydrophones of varying sizes and shapes and hence of differing directivity. In order to make these various measurements compatible with one another, and at the same time make them useful for other directional sonars, it is convenient to express self-noise levels as *equivalent isotropic levels*. The equivalent isotropic self-noise level is the level that would be indicated by a nondirectional hydrophone of sensitivity equal to that of the directional transducer with which the self-noise measurements were made. If the noise level measured with a directional hydrophone is NL′, the equivalent isotropic level is

$$NL = NL' + DI$$

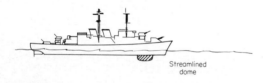
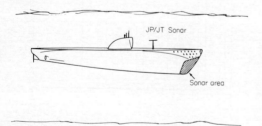
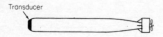

fig. 11.2 *Locations of the sonar transducer aboard surface ships, submarines, and torpedoes.*

In converting to equivalent isotropic levels, therefore, the measured levels NL' are corrected for directivity by applying the directivity index of the measurement hydrophone. The isotropic level NL is the level required for use in the sonar equations previously written.

Although this convention serves to bring self-noise measurements made on the same kind of vessel with different transducers into some degree of agreement, some hesitancy must be felt when applying self-noise data obtained with one sonar to a sonar of different design, directivity, mounting, and location on the same vehicle. Self-noise, like ambient noise, is seldom isotropic in its directional and coherence characteristics, and the DI is not always a satisfactory measure of the discrimination of transducers against it. Often noise coming from a single direction will predominate, and the DI will be almost meaningless as a measure of the discrimination against noise. In torpedoes, for example, it has been found that the front-to-back ratio of the transducer beam pattern—defined as the difference in response between the forward and backward directions—is a more useful measure of the discrimination against self-noise than the directivity index. In general, it is necessary to understand the sources and paths of the prevailing kind of noise before transferring self-noise data from one sonar to another. When this knowledge is lacking, the equivalent isotropic self-noise level must be used in the sonar equations, with recognition of its crude nature and of the likelihood that large errors may occur in some circumstances.

11.2 Sources and Paths of Self-Noise

The three major classes of noise—*machinery noise, propeller noise,* and *hydrodynamic noise*—apply for self-noise as well as for radiated noise. The sound and vibration generated by each kind of noise reach the sonar hydrophone through a variety of different acoustic paths.

Figure 11.3 shows a surface ship and the paths in the ship and through the sea by which sound generated at the propeller and in the machinery spaces of the ship can reach the sonar hydrophone. Path *A* is an all-hull path by which the vibration produced by the machinery, the propeller shaft, and the propeller itself reach the vicinity of the sonar array at a forward location. Here it may be reradiated by the hull or, more importantly, cause vibration of the wall of the streamlined dome and the mounting of the hydrophone array. Path *B* is an all-water path leading directly from the ship's propellers to the hydrophone. Path *C* shows propeller noise backscattered by volume scatterers located in the volume of the sea. In general, these scatterers are the same as those causing volume reverberation. Path *D* is the bottom-reflected or scattered path by which propeller noise can reach the vicinity of the hydrophone. This path is likely to be a major contributor to self-noise on surface ships operating in shallow water.

For submarines and torpedoes, the upward analog of Path *D*—reflection and scattering from the sea surface—is, similarly, an important acoustic path

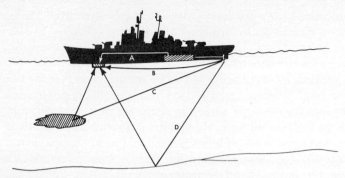

fig. 11.3 *Paths of self-noise on a surface ship.*

when the vehicle is running at a shallow depth. An example of this kind of path is shown in Fig. 11.4. This is an artist's conception of a torpedo with three areas of forward and side scattering on the surface by which sound from the propellers can reach hydrophones located near the nose. Of all the vessels of interest, the self-noise of torpedoes long ago received a considerable amount of analytical attention, probably because of the relative ease of performing experiments upon them. For example, torpedoes were "run" with and without propellers, in water and in air, and with various modifications, all during the World War II years (2).

Both machinery noise and propeller noise are prominent contributors to self-noise. The self-noise contribution of the vessel's machinery occurs principally at low frequencies as tonal components in the overall noise. Unlike other kinds of noise, machinery noise tends to be relatively independent of speed,

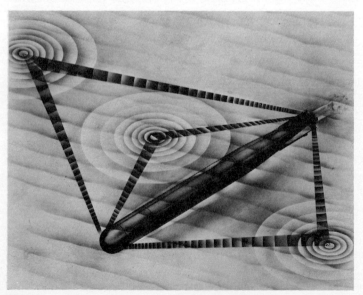

fig. 11.4 *Surface-reflected paths of torpedo self-noise.*

since much of it originates in the constant-speed auxiliary machinery of the vessel. Hence, at slow speeds, where other kinds of noise are of low level, the noise of the vessel's auxiliary machinery is often a troublesome source of self-noise. In the wartime JP equipment on submarines operating below 5 knots, it was observed that listening was affected by power operation of the bow and stern planes, the steering machinery, and certain rotating equipment aboard the submarine. At higher speeds, propeller noise becomes the dominant contributor to self-noise under conditions of high frequencies, shallow water depths, and stern bearings. At high speeds also, the many and diverse forms of hydrodynamic noise become important.

Hydrodynamic noise includes all those sources of noise resulting from the flow of water past the hydrophone and its support and the outer hull structure of the vessel. It includes the turbulent pressures produced upon the hydrophone face in the turbulent boundary layer of the flow (flow noise), rattles and vibration induced by the flow in the hull plating, cavitation around appendages, and the noise radiated to a distance by distant vortices in the flow. Hydrodynamic noise increases strongly with speed, and because the origin of this noise lies close to the hydrophone, it is the principal source of noise at high speeds whenever the noise of propeller cavitation—itself a form of hydrodynamic noise—is insignificant.

The relative importance of the sources of self-noise can be illustrated by showing their areas of dominance on a plot having vessel speed and frequency as the two coordinates (Fig. 11.5). At very low speeds, the hydrophone "sees" the ambient noise of the sea itself. With increasing speed, machinery noise tends to dominate the low-frequency end of the spectrum, and a combination of propeller and hydrodynamic noise becomes important at high frequencies. The scales and the crosshatched regions separating the noise sources in Fig. 11.5 vary with the kind of vessel and with the directivity, location, and mounting arrangement of the measuring hydrophone.

Hydrodynamic noise occurs in stationary hydrophones as well as in those on moving vessels. Around the case of an acoustic mine on a rocky bottom, for example, a swift tidal current was found (3) to produce pressures of 1,000

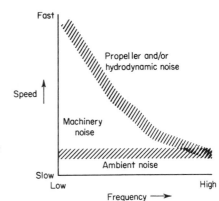

fig. 11.5 *Regions of dominance of the sources of self-noise.*

dyn/cm² in the 1- to 4-Hz band and 50 dyn/cm² in the 5- to 32-Hz band on the crystal hydrophone of the mine.

Electrical noise is occasionally bothersome in sonar sets as a form of self-noise. However, its existence indicates a pathological condition whose cure is normally obvious. Except under exceptionally quiet conditions, as at times in the Arctic under a uniform ice cover, electrical noise is not a serious problem in well-designed sonars.

11.3 Flow Noise

A particular kind of hydrodynamic noise has been called *flow noise*. Because it is amenable to theoretical and experimental study and has an urgent application to the reduction of cabin noise inside aircraft, it has received relatively abundant attention in the literature. Flow noise, in actual practice, may be said to be what is "left over" after all other sources of hydrodynamic noise aboard the vessel have been accounted for or removed.

Flow noise consists of the pressures impinging upon the hydrophone face created by turbulent flow in the turbulent boundary layer about the hydrophone. Figure 11.6 illustrates a rigid, flat boundary containing a flush-mounted pressure hydrophone, above which a viscous fluid is flowing. Between the free stream and the boundary lies a turbulent boundary layer within which fluctuating pressures are created and are transmitted to the hydrophone in the boundary. Although these turbulent pressures are not true sound, in that they are not propagated to a distance, they form what has been termed "pseudosound" (4) and give rise to a fluctuating noise voltage at the output of the pressure hydrophone. In the following sections, some of the salient characteristics of flow noise will be briefly summarized. The interested reader is referred to the quoted literature for more detailed information; a good tutorial paper on the subject has been published by Haddle and Skudrzyk (5).

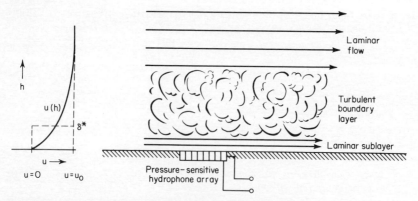

fig. 11.6 *Flow structure in a fluid moving over a stationary surface. The diagram at the left shows the variation of flow velocity with height above the boundary.*

Total fluctuating pressure The rms pressure p_{rms} on the boundary due to the turbulent flow is related to the free-stream dynamic pressure by

$$\frac{p_{rms}}{\frac{1}{2}\rho u_0^2} = 3 \times 10^{-3}\alpha$$

where ρ = fluid density
u_0 = free-stream flow velocity

α is a constant, called the Kraichman constant after a pioneering investigator of turbulent flow, ranging from 0.6 to 4 in different measurements using different data (6), but centering roughly about unity. In any one series of measurements, α is found to be a constant over a wide range of flow velocities.

Spectrum of flow noise The power spectrum of flow noise, or the distribution in frequency of the mean-squared pressure, is found to be flat at low frequencies and to slope strongly downward at high frequencies at the rate of f^{-3}, or as -9 dB/octave, as shown by the flow-noise spectra observed with a flush-mounted hydrophone ½ in. in diameter inside a rotating cylinder (6). The transition frequency f_0 between the flat and the sloping portions of the spectrum is given by

$$f_0 = \frac{u_0}{\delta}$$

where δ is the thickness of the boundary layer. Since the flow velocity is a continuous function of distance away from the wall (Fig. 11.6), the boundary-layer thickness is more exactly expressed by the "displacement thickness" δ^* such that

$$\delta^* u_0 = \int_0^\infty u(h)\, dh$$

where $u(h)$ is the flow velocity at a distance h normal to the wall. The actual boundary layer is thus replaced by an equivalent layer (in the above sense) of zero velocity (that is, the layer is attached to the wall) and of thickness δ^*. It is found that $\delta = 5\delta^*$, approximately. In experiments of Skudrzyk and Haddle (6), δ^* was found to be 0.153 in. at a speed u_0 of 20 ft/s and 0.135 in. at 60 ft/s.

Variation with speed At frequencies less than f_0, the spectrum level of flow noise varies as the cube of the speed u_0; at frequencies appreciably greater than f_0, it varies as the sixth power of the speed. The latter rate of increase amounts to 18 dB per speed doubled. This is equivalent to a straight-line rate-of-rise of 1.8 dB/knot in the speed range 10 to 20 knots, in approximate agreement with observations of self-noise on large naval surface ships over this range of speed.

A practical way to measure flow noise without the contaminating effects of machinery noise is to use a sinking streamlined body as a test vehicle or,

362 / *principles of underwater sound*

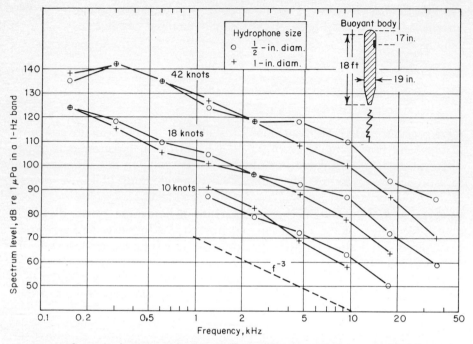

fig. 11.7 *Flow-noise levels on a buoyant body at different speeds for two hydrophone diameters. (Ref. 5.)*

alternatively, to use a buoyant body rising under its own buoyancy from a depth in the sea. Examples of measurements by Haddle and Skudrzyk (5) on a buoyant streamlined body are shown in Fig. 11.7. The data shown here were obtained with two hydrophones of different sizes placed 17 in. behind the nose of the body 18 ft long. The results roughly confirm the 18-dB increase per speed doubled just described, as well as the f^{-3} variation with frequency. It will be noted that doubling the hydrophone diameter from ½ to 1 in. *decreased* the flow-noise level by some 5 to 10 dB near the high-frequency end of the spectrum.

Effect of surface roughness Flow noise across a rough surface has been studied by Skudrzyk and Haddle (6) using a rotating cylinder having different degrees of surface roughness obtained by cementing grit of various sizes to the outside. It was found that at 24 kHz the noise produced by the roughnesses became equal to the flow noise across the perfectly smooth surface when the height of the roughness was such that

$$h = \frac{0.06}{u'}$$

where u' = flow velocity, knots
h = roughness height, in.

Surfaces need not be optically smooth to be considered "smooth" for flow noise, but must be free of roughnesses high enough to extend above the laminar boundary layer (Fig. 11.6) and affect the turbulent flow.

Coherence of turbulent pressures Measurements of the longitudinal and transverse correlation of the pressures due to the flow have been made with small hydrophones in the walls of tubes and pipes (7, 8). In a frequency band w hertz wide centered at frequency f, the crosscorrelation coefficient of the pressure measured at two points along the walls at distance d apart has been found (7) to be

$$\rho(d,w) = \rho(s) \frac{\sin(\pi w d/u_c)}{\pi w d/u_c} \cos 2\pi s$$

where s is the (nondimensional) Strouhal number defined as

$$s = \frac{fd}{u_c}$$

in which u_c is the "convection velocity" equal to the velocity at which turbulent patches are carried past the hydrophone by the flow. u_c is somewhat smaller than the free-stream velocity u_0 and varies from $0.6u_0$ to $1.0u_0$, depending on the frequency f. The correlation function $\rho(s)$ has been found experimentally to be

$$\rho_L(s) = e^{-0.7|s|}$$

for longitudinal separations d parallel to the flow (7), and

$$\rho_T(s) = e^{-5|s|}$$

for transverse separations at right angles to the flow (8).

Discrimination against flow noise A pressure hydrophone of finite size, that is, an array, will discriminate against flow noise to an extent determined by the spatial correlation coefficients ρ_L and ρ_T described above. The magnitude of this discrimination β is defined as

$$\beta = \frac{R'}{R}$$

where R' is the mean-square voltage output of an array placed in a flow noise field, and R is the mean-square voltage output of a very small pressure hydrophone placed in the same noise field and having a sensitivity equal to the plane-wave axial sensitivity of the array. The quantity $10 \log(1/\beta)$ is equivalent to the *array gain for flow noise;* the ratio β measures the reduction of flow noise experienced by an array of pressure-sensitive elements relative to the noise pickup of a single small element. The magnitude of the discrimination factor β has been worked out by Corcos (8) for circular and square arrays, on the assumption that the correlation function in oblique directions to the flow is given by the product of the two principal components ρ_L and ρ_T. White (9) has extended this work by means of a unified theory and has carried out computations for rectangular arrays with long side parallel to and perpendicular to the direction of flow. Figure 11.8 shows the quantity β as defined above for a rectangular array (after White) and for a circular array (after Corcos). For a large square hydrophone of side equal to L, Corcos shows that

$$\beta = \frac{0.659}{\gamma^2}$$

where $\gamma = 2\pi f L/u_c$, and for a circular hydrophone of radius r,

$$\beta = \frac{0.207}{\gamma^2}$$

where $\gamma = 2\pi f r/u_c$. In both cases, γ must be much greater than unity.

Comparison with an isotropic sound field If now we define a *convection wavelength* λ_c such that

$$\lambda_c = \frac{u_c}{f}$$

in analogy with acoustic wavelength

$$\lambda_s = \frac{c}{f}$$

where c is the velocity of sound, the expression for the discrimination factor of a circular hydrophone becomes

$$\beta = 0.207 \left(\frac{\lambda_c}{2\pi r}\right)^2$$

For isotropic noise, the corresponding expression is (Table 3.2)

$$\beta_{iso} = \left(\frac{\lambda_s}{2\pi r}\right)^2$$

where $10 \log (1/\beta_{iso})$ is the ordinary directivity index of the circular array. Comparing the two expressions, we observe that $\beta = 0.207 \beta_{iso}$ *when the appropriate wavelength is used for the two types of noise*. But λ_c is far smaller than λ_s for vehicles traveling in the sea; the ratio λ_c/λ_s is approximately equal to the Mach number of the vessel, or the ratio of its speed to the speed of sound in the sea. Since M is a small quantity, it follows that the discrimination against flow noise for a large array of a given size is much greater than it is for isotropic noise. In terms of the Mach number M,

$$\frac{\beta}{\beta_{iso}} = 0.207 M^2$$

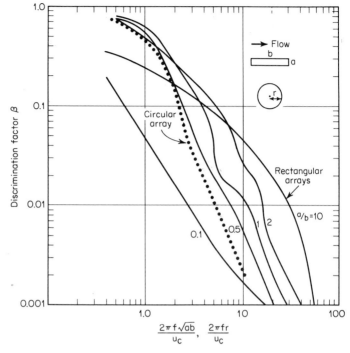

fig. 11.8 *Discrimination of rectangular and circular arrays against flow noise. The sides a and b of the rectangular array are oriented perpendicular and parallel to the direction of flow, respectively.* [After White (9) and Corcos (8).]

As an example, at a speed of 2.0 knots, $M = 7 \times 10^{-4}$, and the quantity $10 \log (\beta/\beta_{iso})$ becomes -70 dB. In actual practice on a moving vessel, however, such great benefits of large arrays for reducing noise are not likely to be observed because of the existence of sources of noise other than the turbulent-flow pressures on the rigid wall. If the wall is not rigid, resonant wall vibration as well as the radiated noise of distant turbulence are likely to overwhelm the ideal flow-noise pressures picked up by a large array.

11.4 Flow-Noise Reduction

For a well-streamlined body with a minimum of vanes, fins, and appendages, various techniques can be used to reduce the sensitivity to flow noise of a hydrophone located on it. Some of these are:

1. *Make the hydrophone larger.* The effect of a larger hydrophone size is to cause the turbulent flow-noise pressures to average out as a result of their small correlation distance. In wind and water tunnels, reductions of 40 dB or more can be obtained in this way, but on moving bodies such great reductions

cannot be achieved, as mentioned above, because of effects such as shell vibration and the flow-excited radiated noise of fins and other appendages.

2. *Move the hydrophone forward.* A forward location is quieter than one toward the tail of the body because of a thinner boundary layer and a greater distance from the noise-producing surfaces at the rear of the body. The quietest place of all is right at the nose itself in the region of the "stagnation point" where the flow separates; here the boundary layer is absent, and flow noise is received only by radiation and diffraction around the front of the body. Figure 11.9 shows diagrammatically the output of a hydrophone at a forward location and one toward the stern of a streamlined body in the turbulent boundary layer. A hydrophone located at the nose of a blunt-nose body picks up the radiated noise of the turbulence by diffraction around the nose, as shown by the experimental work of Lauchle in a water tunnel (10).

3. *Remove the hydrophone from the turbulent boundary layer.* This can be done by placing it in a cavity or in a recess in the wall. The effect here is to allow the positive and negative pressures to cancel out, in much the same way as with a hydrophone of larger size.

4. *Eject polymer fluids.* The ejection of very small amounts of polymers consisting of long-chain unbranched molecules of high molecular weight (about 10^6) has been found effective in the reducing fluid drag in pipes and on moving bodies. The drag-reduction process appears to be one of thickening the laminar sublayer (Fig. 11.6) and thereby reducing self-noise by separating the hydrophone from the turbulent boundary layer, as in technique 3 above. The concentration of polymer found to be effective is very low (100 parts per million or less), and, on a moving body, ejection can be achieved through holes in the nose. Long-chain polymers are viscoelastic fluids that cause the solution to be non-Newtonian: that is, they cause the shear stress to be no longer linearly proportional to the rate of shear. Just how their effect on

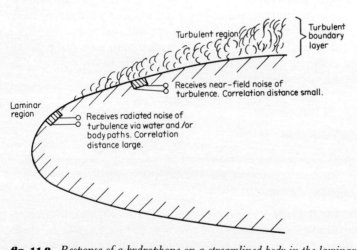

fig. 11.9 *Response of a hydrophone on a streamlined body in the laminar and turbulent regions of the flow.*

the boundary layer occurs is uncertain; one idea is that the long-chain molecules tend to align themselves parallel to the wall and thereby inhibit the formation of stress differences normal to the wall that result in the formation of turbulence. On a self-propelled body like a torpedo, polymer ejection has an added beneficial effect on self-noise in that, by reducing drag, the propulsive power needed to achieve a given speed is less; in other words, the machinery and propeller contributions to self-noise are lower. The hydrodynamic effects of polymer ejection have received much attention in the literature, and an excellent review paper by Hoyt (11) containing many references has been published.

11.5 *Domes*

It was observed many years ago that large reductions of what is now known as hydrodynamic noise could be achieved on surface ships by surrounding the sonar transducer by a streamlined housing. Such housings are called *sonar domes*. They reduce self-noise by minimizing turbulent flow, by delaying the onset of cavitation, and by transferring the source of flow noise to a distance from the transducer. Sonar domes were originally spherical in shape but were soon streamlined into a teardrop shape to prevent the occurrence of cavitation at high speeds. Some examples of domes used during World War II are shown in Fig. 11.10. The domes in this figure are of all-metal construction and have a thin stainless steel window to permit the ready exit and entrance of sound out of and into the transducer inside. Modern domes are constructed of rubber reinforced with thin steel ribs. Many have baffles, such as those seen in Fig. 11.10*a* and 11.10*c,* to reduce machinery and propeller noises coming from the rear.

The acoustic and mechanical requirements of dome design are severe. The dome must be acoustically transparent, so as to introduce only a small transmission loss and produce no large side lobes in the directivity pattern of the enclosed transducer. The latter requirement means the absence of internal specular reflection from the dome walls. At the same time the dome should be sufficiently streamlined to delay the onset of cavitation on its surface beyond the highest speed reached by the vessel and should be of sufficient mechanical strength to resist the hydrodynamic stresses upon it when under way. These requirements are to a large extent mutually incompatible.

Expressions have been obtained theoretically (12, 13) and generally verified experimentally for the transmission loss and the specular reflection produced by a dome of a given material and wall thickness on a transducer of given frequency, directivity, and position in the dome. Both the transmission loss and reflectivity of the dome increase with frequency and with the thickness and density of the dome walls. Internal reflections in domes can be greatly reduced by increasing the horizontal, and particularly the vertical, curvature of the dome walls. Hence, for both acoustic and hydrodynamic reasons, sonar domes employ materials as thin and light as possible formed into curved

368 / **principles of underwater sound**

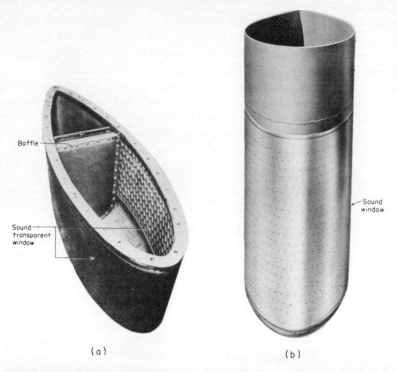

fig. 11.10 *Photographs of streamlined domes. (a) QBF dome. (b) QGA dome. (c) A modern surface ship dome; note the curved baffle in this dome. (a, b, from Ref. 14; c, courtesy B. F. Goodrich Co., Akron, Ohio.)*

streamlined shapes. The acoustic windows of sonar domes used during World War II, like those of Fig. 11.10, ranged from 0.020 to 0.060 in. in thickness (14).

For low self-noise, sonar domes must be kept undamaged and free of marine fouling. A dome with a rough exterior surface will produce a higher flow noise, as well as noise caused by local cavitation at "hot spots" on its surface at the higher speeds of the vessel on which it is mounted.

For many years the sonar domes of surface ships were located aft of the bow, and were retractable for use in shallow water and for ease of dry docking. However, the recent trend toward lower frequencies, higher powers, and larger sizes of modern sonars has required a shift in location and an abandonment of the rectractable feature of sonar domes. The dome of the A/N-SQS 26 sonar is located just at the bow and is bulbous in shape to accommodate the tremendous size of the transducer (Fig. 11.11). It is about 50 ft long, 10 ft high, and 18 ft wide, and weighs some 40 tons. Compared to older installations, the location and size of this dome has both benefits and disadvantages. It is as far removed from the propellers as is possible with a hull installation,

fig. 11.11 *The A/N-SQS 26 pressurized sonar dome with rubber window on the USS "Willis A. Lee" (DL-4) in drydock. (Photo courtesy of B. F. Goodrich Company.)*

and so experiences a lower level of propeller noise. The bulbous shape of the bow reduces—rather than increases—the drag of hull in its motion through the water, and is said to reduce pitching of the ship as well. On the other hand, the draft of the vessel is increased, and the drydocking operation is made more complicated. The alternative to such a large dome is to use the sides and length of the vessel to provide the necessary surface area for efficient power radiation and directivity, but the problems of element phasing and interaction in this kind of array are formidable.

11.6 Self-Noise of Cable-Suspended and Bottomed Hydrophones

Hydrophones that are hung from the end of a cable are likely to suffer from a peculiar kind of noise called *strumming* or *flutter* noise. This form of self-noise occurs in a current of water, and is the result of cable vibration induced by the eddies or vortices shed by the cable. This is the "aeolian harp" effect, or the singing of telephone wires in a wind, that has long been known in air acoustics (15). The frequency of vortex shedding is given by the simple expression $f = Sv/d$, where S is the dimensionless "Strouhal number," v is the water current speed, and d is the cable diameter in the units of length of v. The Strouhal number happens to be a constant equal to 0.18 over much of the range of current speeds and cable sizes occurring in practice. Thus, with a 1-cm-diameter cable in a 1-knot current (51.5 cm/s), the strumming frequency will be 9 Hz. Strumming noise can be readily alleviated by a number of means, including using a faired cable, keeping the natural frequency of cable vibration well separated from the strumming frequency, isolating the hydrophone from the cable (as by such simple means as suspending it from the cable by rubber bands), and employing a hydrophone having an acceleration canceling design.

Another malady of cables is *triboelectric noise* or the spurious voltages resulting from friction ("tribo") between the cable conductor and the shield. Any motion of the cable tends to produce varying frictional charges in the cable dielectric; these appear as voltages at the end of the cable. They occur in voids or gaps between the conductor and the dielectric that act as variable air capacitors. Triboelectric noise exists whenever the cable is bent or altered in shape; it enhances the effect of cable strumming. It can be alleviated by coating the dielectric with graphite (Aquadag) or by binding the dielectric material tightly to the conductors. This kind of noise has been investigated empirically with various kinds of cables subjected to different kinds of deformation (16), and a number of low-noise cables are commercially available.

Hydrophones resting on the ocean floor are susceptible to the motional effects of water currents flowing past, and around, the hydrophone and its mounting. Strasberg (17) has discussed and made quantitative estimates of the apparent noise caused by the impingement of the turbulent and thermal microstructure carried along by a current and striking a hydrophone. Vortices and turbulences shed by the hydrophone and its structure are also a form of low frequency nonacoustic noise, as shown by laboratory measurements of McGrath et al. (18). These current-produced pseudonoises make difficult valid measurements of the infrasonic ambient background of the sea, especially in shallow water, and require careful design of the hydrophone structure, as well as a thermal shield around the hydrophone for quiet operation at frequencies below 20 Hz.

11.7 Self-Noise of Towed Sonars

A sonar housed in a streamlined body and towed at a depth behind a surface craft (see Fig. 1.2) is subject to a wide variety of self-noise sources. These range from "pure" flow noise—the pressure fluctuations of the boundary layer adjacent to the hydrophone—to vibrations of the towing cable.

Some direct field trials of the noise of hydrophones inside a towed streamlined body of revolution have been made by Nishi, Stockhausen, and Evensen (19). The body they used was 6 ft long and 1½ ft in maximum diameter. A number of small hydrophones 0.07 in. in diameter were placed on it for self-noise measurements at various places along its length. Towing was done by a hydrofoil craft at speeds up to 30 knots. It was found during the field trials that a hydrophone along the side of the body, where the turbulent boundary layer was well developed, picked up mainly flow noise at frequencies above 1 kHz; its levels were not greatly different from those of the buoyant body shown in Fig. 11.7, when allowance is made for hydrophone size. A hydrophone at the nose was found to be quieter than the one located along the side of the body. This sensor picked up the radiated noise of the towing craft, together with the wall vibrations caused by vibrations of the tow cable and the tail-fin structure of the body.

Another kind of towed sonar is a *flexible-line hydrophone array* towed at a considerable distance (up to several miles) behind a surface craft to reduce pickup of the noise from the towing vessel. This is the successor of the "eel" worked on the World War I (Sec. 1.1); more recently towed lines have been extensively used in waterborne seismic prospecting for oil (20). Typically, a seismic "streamer" consists of 24 or more hydrophone groups, each up to several hundred feet long, and having an overall length of about 8,000 feet (21, 22). The hydrophones are placed inside a thin-walled flexible plastic jacket, typically 2½ in. in diameter, filled with a fluid of low specific gravity so as to be nearly neutrally buoyant. In the shallow water of waterborne prospecting, a constant depth is maintained by dynamic controllers or "birds," which themselves are a source of noise for hydrophones near them. Fig. 11.12 shows smoothed measured spectra of a hydrophone group in a streamer towed at two speeds a distance of 1,000 feet behind the towing vessel in sea states 0 to 1. Added for comparison is an average curve of the ambient noise level for a wind speed of 6 knots at seven Pacific Ocean locations, due to Wenz (Fig. 7.8).

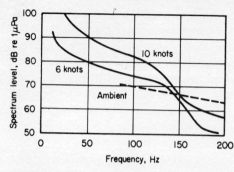

fig. 11.12 *Seismic streamer self-noise spectrum levels at two towing speeds, and the ambient-noise spectrum for a wind speed of 6 knots in shallow water as given by Wenz (Fig. 7.8). (After Ref. 22.)*

Although towed arrays do reduce greatly the radiated noise of the ship by means of their distance and directivity, they are subject to various motional effects and to various forms of hydrodynamic noise, including "pure" flow noise. Especially deleterious are the results of vibration induced by towing, which causes effects such as acceleration response in the array hydrophones, hydrostatic pressure changes due to the vertical motion of the array, changing pressures in the oil filling of the flexible hose, and vortex shedding by the vibrating tow cable. Nevertheless, flexible towed-line arrays make possible the rapid exploration for oil in coastal areas and, as passive sonars, provide a capability for listening from surface ships to low frequencies and in stern directions—something that hull-mounted sonars do not possess.

11.8 Self-Noise Levels

Figure 11.13 shows equivalent isotropic self-noise levels at 25 kHz on a number of World War II American and British destroyers. These data, taken from

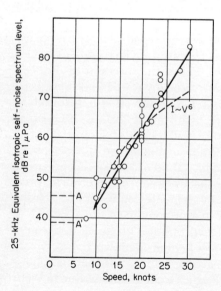

fig. 11.13 *Equivalent isotropic spectrum levels at 25 kHz on destroyers (Ref. 23). The levels A and A' are deep-sea ambient levels at sea states 3 and 6. The dashed curve shows the theoretical variation with speed of flow noise.*

a wartime study of Primakoff and Klein (23), were obtained with various sonar transducers of differing directivity, and have been corrected to equivalent isotropic levels, as discussed above, by allowing for (adding) the directivity index of the transducer. The curved line shows an increase of self-noise intensity as the sixth power of the speed, in agreement with theoretical expectations for the speed variation of flow noise. Although higher noise levels are observed at high speeds, probably because of dome cavitation and other sources of noise, the general agreement suggests the dominance of some form of flow noise in these data at moderate speeds.

Figure 11.14 is a similar plot for small warships of the PC and SC classes and for British DE-type ships and frigates. Here the increase of noise with speed is much more rapid and suggests the dominance of propeller cavitation noise in these smaller ships, on which the sonar dome, its distance from the propellers, and the amount of screening by the hull are all much smaller than on destroyers.

The levels of Figs. 11.13 and 11.14 apply for forward bearings, when the directional transducer in its dome is trained in a forward direction. When it is trained toward the stern of the ship, higher levels are observed, particularly on small ships, due to the noise of the cavitating propeller.

At the lower frequencies more typical of modern destroyers, the same sort of behavior with speed as that shown in Fig. 11.13 is observed. Figure 11.15 is representative of modern destroyer sonars operating at frequencies of 10 kHz and below. At very slow speeds, or when lying to, the sonar self-noise level is close to the ambient background level of the sea in the prevailing sea state. At speeds between 15 and 25 knots the self-noise level increases sharply with speed at the rate of about 1½ dB/knot, and represents the domi-

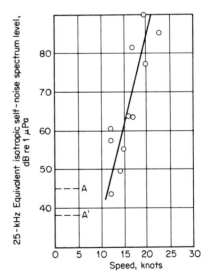

fig. 11.14 *Equivalent isotropic spectrum levels at 25 kHz on American PC and SC class ships and on British DE types and frigates. A and A' are ambient levels at sea states 3 and 6. (Ref. 23.)*

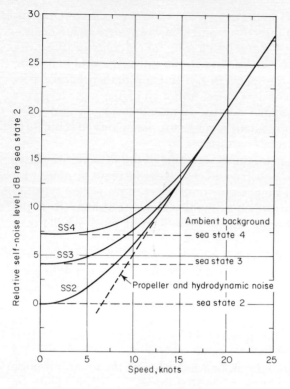

fig. 11.15 Self-noise levels versus speed on a modern destroyer.

nant contributions of hydrodynamic and propeller noise to the total noise level.

Self-noise levels on submarines are illustrated in Fig. 11.16a and b. These show average spectra ad the increase of noise with speed measured during World War II with the JP-1 listening equipment. The JP sonar had a 3-ft horizontal line hydrophone mounted on the deck of the submarine forward of the conning tower and could be trained manually so as to listen in different directions. The spectra of Fig. 11.16a are for noisy, average, and quiet installations at a speed of 2 knots. They approximate the levels of deep-water ambient noise at the high-frequency end, but rise more rapidly with decreasing frequency, probably as a result of machinery-noise contributions. The extremely rapid rise with speed suggests the influence of propeller cavitation as the speed increases, as does the fact that the self-noise of the JP-1 sonar decreased with increasing depth of submergence (24). In other submarine sonars, less exposed to cavitation noise, the effect of speed is less marked and approximates the rate of rise of 1½ to 2 dB/knot found on destroyer-like surface ships (Fig. 11.13).

Just as for destroyers, the self-noise of submarines observed with the JP-1 sonar was found to be independent of bearing on forward bearings but to increase sharply as the hydrophone was trained toward the stern. Examples of the "directivity pattern" of self-noise at several speeds are given in Fig. 11.17.

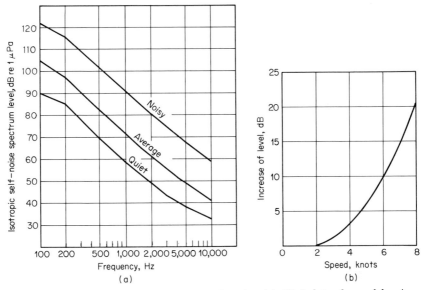

fig. 11.16 Average self-noise spectra on submaries: (a) JP-1 data, forward bearings, speed 2 knots, periscope depth. (b) Increase of noise level with speed relative to 2 knots. (Ref. 25, figs. 9 and 10.)

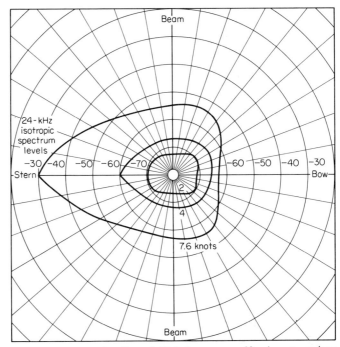

fig. 11.17 Directionality of high-frequency self-noise at various speeds. Hydrophone directivity index assumed to be 20 dB. (Ref. 24.)

REFERENCES

1. Halley, R.: Physics of Passive Sonar, *U.S. Navy Electron. Lab. Rep.* NAVPERS 92934, 1962.
2. Hunt, F. V.: Completion Report on Machinery Noise of the Electric Torpedo, *Nat. Def. Res. Comm. Div. 6 sec.* 6.1-ser. 287–2053, 1945.*
3. Childs, W. B., and A. T. Jaques: Water Noise around Mine Cases in Strong Tidal Currents, *U.S. Nav. Ord. Lab. Rep.* 889, 1944.*
4. Lighthill, M. J.: Sound Generated Aerodynamically—the 1961 Bakerian Lecture, *Proc. Roy. Soc. London*, A**267:**147 (1962).
5. Haddle, G. P., and E. J. Skudrzyk: The Physics of Flow Noise, *J. Acoust. Soc. Am.*, **46:**130 (1969).
6. Skudrzyk, E. J., and G. P. Haddle: Noise Production in a Turbulent Boundary Layer by Smooth and Rough Surfaces, *J. Acoust. Soc. Am.*, **32:**19 (1960).
7. Bakewell, H. P.: Longitudinal Space Time Correlation Function in Turbulent Airflow, *J. Acoust. Soc. Am.*, **35:**936 (1963).
8. Corcos, G. M.: Resolution of Pressure in Turbulence, *J. Acoust. Soc. Am.*, **35:**192 (1963).
9. White, F. M.: A Unified Theory of Turbulent Wall Pressure Fluctuations, *U.S. Navy Underwater Sound Lab. Rep.* 629, 1964.
10. Lauchle, G. C.: Noise Generated by Axisymmetric Turbulent Boundary-Layer Flow, *J. Acoust. Soc. Am.*, **61:**694 (1977).
11. Hoyt, J. W.: Effect of Additives on Fluid Friction, *J. Basic Eng.*, **94:**258 (1972).
12. Primakoff, H.: The Acoustic Properties of Domes, Parts I and II, Office of Scientific Research and Development, *Nat. Def. Res. Comm. Div. 6 sec.* 6.1-ser. 1130–1197, 1944.*
13. Basic Methods for the Calibration of Sonar Equipment, *Nat. Def. Res. Comm. Div. 6 Sum. Tech. Rep.*, vol. 10, chap. 9, 1946.
14. A Manual of Calibration Measurements of Sonar Equipment, *Nat. Def. Res. Comm. Div. 6 Sum. Tech. Rep.*, vol. 11, chap. 3, 1946.*
15. Rayleigh, Lord: "Theory of Sound," vol. II, p. 413, Dover Publications, Inc., New York.
16. Donovan, J. E.: Triboelectric Noise Generation in Some Cables Commonly Used with Underwater Electroacoustic Transducers, *J. Acoust. Soc. Am.*, **48:**714 (1970).
17. Strasberg, M.: Non-acoustic Noise Interference in Measurements of Infrasonic Ambient Noise, *J. Acoust. Soc. Am.*, **66:**1487 (1979).
18. McGrath, J. R., O. M. Griffin, and R. A. Finger: Infrasonic Flow-noise Measurements Using an H-J8 Omnidirectional Cylindrical Hydrophone, *J. Acoust. Soc. Am.*, **61:**390 (1977).
19. Nishi, R. Y., J. H. Stockhausen, and E. Evensen: Measurement of Noise on an Underwater Towed Body, *J. Acoust. Soc. Am.*, **48:**753 (1970).
20. Parma, E. M.: Marine Seismic Exploration, *Oceanol. Int.*, p. 32, January 1970.
21. Bedenbender, J. W., and others: Electroacoustic Characteristics of Marine Seismic Streamers, *Geophysics*, **35:**1054 (1970).
22. Schoenberger, M., and J. F. Mifsud: Hydrophone Streamer Noise, *Geophysics*, **39:**781 (1974).
23. Primakoff, H., and M. J. Klein: The Dependence of the Operational Efficacy of Echoranging Gear on Its Physical Characteristics, *Nat. Def. Res. Comm. Div. 6 sec.* 6.1-ser. 1130–2141, 1945.*
24. Principles of Underwater Sound, *Nat. Def. Res. Comm. Div. 6 Sum. Tech. Rep.*, vol. 7, fig. 5, p. 248, 1946.
25. Prediction of Sonic and Supersonic Listening Ranges, *Nat. Def. Res. Comm. Div. 6 sec.* 6.1-ser. 1131–1884, 1944.*

* References indicated by an asterisk were originally issued during World War II as classified reports. These reports have been declassified and are available from the U.S. Department of Commerce, National Technical Information Service, Springfield, Va. 22151.

twelve

detection of signals in noise and reverberation: detection threshold

Sonar signals, whether the echoes of active sonars or the target sounds of passive sonars, must almost always be observed amid a background of noise or reverberation. Before any other system function is performed, the sonar system must first *detect* the presence of the signal in this background; that is, the system must determine, or help a human observer to determine, whether or not the signal is occurring within some given interval of time. With a human observer, the process of detection calls for a decision on the part of the observer using certain criteria as to whether, in his judgment, the signal really occurred, or not, during the observation interval.

In this chapter, some simple expressions will be obtained for the input signal-to-noise ratio required for making this decision at some preassigned level of correctness of the decision as to "target present" or "target absent." At the preassigned level, the signal-to-noise ratio is called the *detection threshold,* and refers to the input terminals of the receiver-display-observer combination. It is the term in the sonar equations which satisfies the equality in the equation when the signal is just being detected. Although other functions besides detection will sometimes be of prime concern to the sonar, such as "classifying" or determining the identity of a target, detection requires the least signal-to-noise ratio of all and places the greatest demands on the design engineer. Moreover, the detection process is the one that has received greatest theoretical attention in the literature. The words "detection threshold" imply two of the most important aspects involved in extracting a signal from the background in which it is embedded: (1) the function of detection itself, and (2) the existence of a threshold somewhere near the output of the receiving system.

In seeking expressions for the detection threshold, the method of approach is to view the decision as the end product that must be related to the signal-to-noise ratio at the input of the processing system (1). Mathematical detail will be avoided in what follows. The reader interested in the statistical theory of signals in noise is referred particularly to a book by Helstrom (2), as well as to an excellent review paper by Bennett (3), and to numerous other books on the subject (4–11). At the same time, only peripheral attention will be given to sonar processing methods, since a number of papers have appeared on the theory of correlators and matched filters for application to active sonar (12–16).

12.1 Definition of Detection Threshold

Figure 12.1 illustrates in diagrammatic form the components of a sonar system that lie between the hydrophone array and the decision "target present" or "target absent." These components are (1) a *receiver,* designed to process the signal appearing across its terminals A-A' in the most advantageous manner, (2) a visual or an aural *display* of present and past signals and backgrounds, and (3) a human *observer* who, based on the display, makes the required *decision.*

Referring to Fig. 12.1, *detection threshold* is defined as the ratio, in decibel units, of the signal power (or mean-squared voltage) in the *receiver bandwidth*

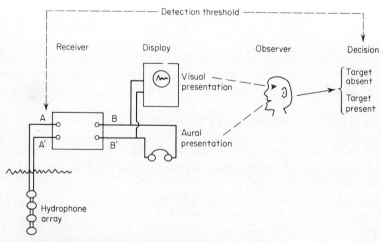

fig. 12.1 *Elements of a receiving system.*

fig. 12.2 Binary decision matrix.

		Decision:	
		Signal present	Signal absent
At input:	Signal present	Correct detection p(D)	Miss 1-p(D)
	Signal absent	False alarm p(FA)	Null decision 1-p(FA)

to the noise power (or mean-squared voltage), *in a 1-Hz band,** measured at the receiver terminals, required for detection at some preassigned level of correctness of the detection decision. If S is the signal power in the receiver bandwidth at A-A' of Fig. 12.1 and N is the noise power in a 1-Hz band at A-A', then

$$DT = 10 \log \frac{S}{N}$$

when the decision is made under certain probability criteria of correct decisions and errors.

When a signal is in fact present at the receiver input terminals, two decisions, absent or present, are possible. When a signal is in fact absent, the same two decisions can be made. In making this binary, forced-choice decision (the observer *must* make a "yes" or "no" choice), four possibilities occur, as shown by the decision matrix of Fig. 12.2. Two decisions are correct, and two incorrect; they appear as the diagonal elements of the matrix. The probability that if a signal is present, the *correct* decision, "signal present," is made is called the *detection probability* $p(D)$. The probability that if a signal is absent, the *incorrect* decision, "signal present," is made is called the *false-alarm probability*, $p(FA)$. The detection threshold depends upon these two independent probabilities in a way that will be described below.

An alternative, though less satisfactory, way of describing the performance of a processing system is by means of the term *processing gain*. The processing gain of a receiver is defined as the difference between the signal-to-noise ratio in decibels at the input and the signal-to-noise ratio in decibels at the output, when *both* are referred to or measured in the receiver bandwidth. In Fig. 12.1, the input and output terminals are A-A' and B-B'. Although useful for comparing receivers, processing gain describes the performance of only a part of the input to the decision chain and has only limited value as a working concept for the overall detection system.

* This bandwidth convention for DT permits us to obtain a single expression for both sinusoidal and broadband signals, and also reflects the custom of expressing noise levels as *spectrum levels*, that is, in bands 1 Hz wide.

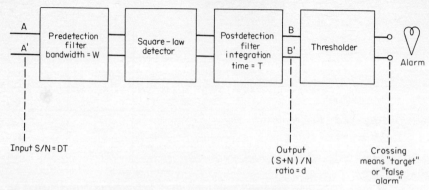

fig. 12.3 *Basic components of an energy detection system.*

Fig. 12.3 shows the basic components of a basic energy detector, consisting of a predetection bandpass filter of bandwidth w, a square-law detector, and a postdetector integrator or averager of integration time T. The system is called an "energy" detector because it converts the voltage at A-A' into power at the output of the squarer and then into energy at B-B' by the action of the integrator.

12.2 The Threshold Concept

The decision process requires the setting of a threshold such that when it is exceeded, the decision "target present" will be made. An example of such a decision is the closing of a relay and the sounding of an alarm when the threshold is exceeded. If the threshold is set too high, however, only strong targets will be detected. If it is set too low, too many "false alarms" will be sounded. At a high threshold setting, both the probability of target detection and the probability of a false alarm are low; at a low threshold, both probabilities become high. This effect of threshold setting is illustrated in Fig. 12.4. The three target signals having the envelope shown in (a), when added to the noise envelope (b), appear as signal plus noise in (c). At threshold T_1, only the third target is detected, and there are no false alarms within the time interval shown in the figure. At a lower threshold T_2, all three targets are detected, but a number of false alarms will occur as well. For a fixed output signal-to-noise ratio, various threshold settings correspond to different pairs of values of the two probabilities. On a plot of detection probability (defined as the probability that a signal, when present, will be detected) and false-alarm probability (defined as the probability that a threshold crossing is caused by noise), a curve will be traced out as the threshold setting is varied. This curve is one of a family of curves called *receiver-operating-characteristic* (ROC) curves. An example of the ROC curve that would result from Fig. 12.4 as the threshold is changed from T_1 to T_2 is shown in Fig. 12.5. Figure 12.6 shows the family of ROC curves on a linear scale as originally drawn by Peterson and

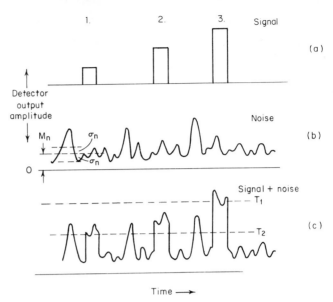

fig. 12.4 *Signal and noise at two threshold settings T_1 and T_2: (a) Shows three target signals which, when added to the noise (b), appear as signal plus noise in (c). M_n and σ_n are the mean and standard deviation of the noise.*

Birdsall (17); Fig. 12.7 is the equivalent family of straight lines on a probability scale extending down to lower values of $p(D)$ and $p(FA)$.

ROC curves are strictly determined by the probability-density functions of signal and noise at the receiver output terminals where the threshold setting is made. Consider Fig. 12.8, which shows curves of probability density $P(a)$ plotted against amplitude a for noise alone and signal plus noise. $P(a)$ is the probability that the amplitude of the envelope of the output lies within a small unit interval of amplitude centered at a. The mean noise amplitude is $M(N)$ and the mean signal-plus-noise amplitude is $M(S + N)$. Both noise and signal

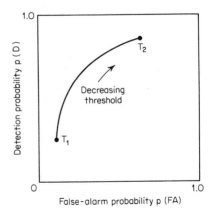

fig. 12.5 *A sample ROC curve that might be constructed from Fig. 12.4.*

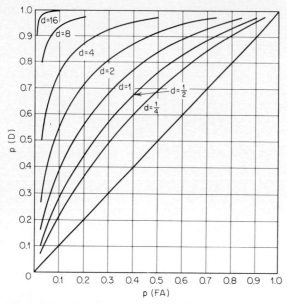

fig. 12.6 *Receiver-operating-characteristic (ROC) curves of detection probability p(D) against false-alarm probability p(FA), with detection index d as a parameter. (Ref. 17.)*

plus noise are assumed* to be gaussian with equal variance σ^2. In terms of these quantities, a ratio d may be defined as equal to the squared difference between the means divided by the variance of the probability-density functions, or

$$d = \frac{[M_{(S+N)} - M_N]^2}{\sigma^2}$$

In psychoacoustics and the theory of auditory masking, a quantity d', equal to the square root of d, is employed. The parameter d is called the *detection index*, and is equivalent to the signal-plus-noise to noise ratio of the envelope of the receiver output at the terminals where the threshold setting T is established. The area under the curve of signal plus noise to the right of T is the probability that an amplitude in excess of T is due to signal plus noise and is equal to the detection probability $p(D)$; that under the noise curve to the right of T is, similarly, the false-alarm probability $p(FA)$. These two probabilities vary as T is changed and depend upon the detection index d.

* By the central-limit theorem (18), the distribution of the sum of a large number of samples is gaussian, regardless of the distribution of the population from which the samples were drawn. In the present context, the sum refers to the envelopes of noise and signal plus noise and implies time integration in the processing such that large sample sizes (large bandwidth-time products) exist.

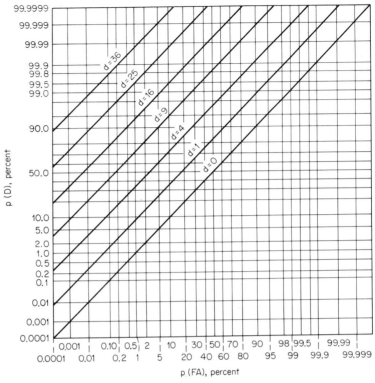

fig. 12.7 *ROC curves on probability coordinates.*

12.3 Input Signal-to-Noise Ratio for Detection

The pioneering work of Peterson and Birdsall (17) and Peterson, Birdsall, and Fox (19) related the ROC curves to the signal-to-noise ratio *at the receiver input* required for detection. It was shown that the *optimum receiver*, defined as one satisfying certain optimum criteria, is one that calculates, and presents at its output, the *likelihood ratio* for each receiver input. This is the ratio of the

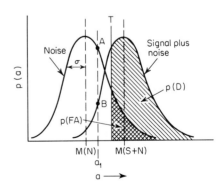

fig. 12.8 *Probability-density distributions of noise and signal plus noise.*

probability that a given input amplitude represents signal plus noise (signal present) to the probability that it represents noise alone (signal absent). If the probability-density curves of Fig. 12.8 be imagined to refer to the receiver input, then the ratio B/A is the likelihood ratio for an input sample of amplitude a_1. For a number of cases involving different degrees of knowledge about the signal in a gaussian noise background, Peterson and Birdsall computed the ROC curves and related the detection index d to the input signal-to-noise ratio for an optimum receiver. Two extreme cases of this knowledge may be called case I and case II. Case I applies for a signal known exactly, that is, one whose waveform as a function of time is completely known. For case I it was found that

$$d = \frac{2E}{N_0}$$

where d = detection index of ROC curves
E = total input signal energy in receiver band
N_0 = noise power in 1-Hz band

If S is the signal power and t is its duration, then $E = St$, and

$$d = 2t \frac{S}{N_0}$$

where S/N_0 is the input signal-to-noise ratio referred to a 1-Hz band of noise. For this case the detection threshold becomes

$$\boxed{\text{DT} = 10 \log \frac{S}{N_0} = 10 \log \frac{d}{2t}}$$

For this case of an exactly known signal, the optimum receiver is a *crosscorrelator*, in which signal plus noise is correlated, or multiplied, with a noise-free replica of the known signal; alternatively, the optimum receiver for white noise* is the *matched filter* (2, pp. 91–95) whose impulse response is the same as the waveform of the known signal reversed in time. A practical realization of the matched filter is the crosscorrelator just mentioned.

Case II is that of the completely unknown signal in a background of gaussian noise. Under the conditions of small signal-to-noise ratios ($S/N \ll 1$) and large sample sizes (large bandwidth-time product), Peterson and Birdsall found that the same ROC curves apply for case II that applied for case I, provided that d is taken such that

$$d = wt \left(\frac{S}{N}\right)^2$$

---

* Noise is said to be "white" when it has a "flat" spectrum, of constant spectral power density, over the bandwidth of interest.

where w = bandwidth
S = signal power in bandwidth w
N = noise power in bandwidth w

Solving for S/N and converting to a 1-Hz noise band, we obtain

$$\frac{S}{N_0} = \frac{Sw}{N} = \left(\frac{dw}{t}\right)^{1/2}$$

and the detection threshold becomes

$$\boxed{DT = 10 \log \frac{S}{N_0} = 5 \log \frac{dw}{t}}$$

For case II, the optimum type of processing involves an energy detector preceded by a filter. It can be shown that the optimum or likelihood-ratio detector is, for gaussian backgrounds, a *square-law detector,* having an output proportional to the square of the input. Other types of nonlinear devices, such as the full-wave and half-wave linear amplifiers, have a detection performance somewhat poorer than that of the square-law detector. Faran and Hills (20) have shown, for example, that the full-wave rectifier requires a detection threshold higher (poorer) by $|5 \log [1/(\pi - 2)]|$, or 0.3 dB, and the half-wave rectifier is poorer by the amount $|5 \log [1/(2\pi - 2)]|$, or 3.1 dB, than the square-law detector; higher-power-law detectors are poorer still.

So far we have not considered the effect on pulsed signals of a postdetector averager, or smoothing filter, that smooths out the fluctuations in the output caused by noise. When a smoothing filter of integration time T is used, the above expressions hold only when T equals the signal duration t. When $T > t$, too much smoothing is used; the pulse does not build up to its full value, and the detection threshold increases. When $T < t$, insufficient smoothing is used; the passband $1/T$ of the postdetector filter is excessively wide, and too much noise reaches the output. The effect of a mismatched output filter is to *increase* the detection threshold by the amount $|5 \log (T/t)|$, and the expression for DT becomes

$$\boxed{DT = 5 \log \frac{dw}{t} + \left|5 \log \frac{T}{t}\right|}$$

where the meaning of the vertical bars is that the mismatch correction always *increases* DT, regardless of whether $T > t$ or $T < t$.

When repeated signals are used for detection, as when a number of successive echoes in different samples of background are incoherently added before the detection decision is made, the effect is to lengthen the signal duration t. Hence, the effect of using n incoherently added signals is to decrease the detection threshold by $5 \log n$, or 1.5 dB for each doubling of the number of signals. When a human observer is used in the detection chain, however, this relationship is found to hold only approximately. For example, for visual detection of radar pips on an A-scan oscilloscope screen, it was found during World War II (21) that the detection index increased by 5 log n for small n, but was less than this for large values of n. This implies a deterioration in the integration ability of the eye for large n, as might be expected. However, it must be pointed out that no improvement appears to occur when the same signal *in the same background* is presented repeatedly to an alert human observer.

When detecting constant sinusoidal signals in ambient noise, the variability of the ocean noise background (Sec. 7.4) limits the gain in detection produced by integration to less than 5 log t. Hodgkiss and Anderson (22) have shown, by means of theory supported by ambient ocean noise data, that the nonstationarity of the noise can cause the gain to be as small as 0.2 log t for averaging times between 0.5 and 120 seconds. However, for very short or very long averaging times, the expected 5 log t variation was achieved.

12.4 Modifications to the ROC Curves

The conventional ROC curves (Fig. 12.7) apply for only certain idealized and limiting conditions of signal and noise. These conditions include a steady signal in stationary gaussian noise, large bandwidth-time products, and the requirement that only a single signal at a time be detected. When these conditions do *not* hold, modifications to the ROC's are necessary.

Fluctuating signals The tacit assumption has hitherto been made that at the output of the detector where amplitude thresholding is done, the occurrence of the signal does nothing more than shift the mean level of the noise, leaving its σ the same and causing σ_{S+N} to be the same as σ_N. This behavior is evident in the distributions of signal plus noise and noise that we have already seen in Fig. 12.8. But when the signal fluctuates, either rapidly within the signal duration or slowly between occurrences of the signal—as from echo to echo in echo ranging—σ_{S+N} becomes greater than σ_N and the ROC's must be modified to an extent dependent on the magnitude of the fluctuations. When gaussian fluctuations are present, the ROC's can be expressed by the equations (23)

$$p(D) = \frac{1}{2}\operatorname{erfc}\left[\frac{1}{\sqrt{2}} \cdot \frac{1}{k}\left(\frac{T}{\sigma_N} - d^{1/2}\right)\right]$$

$$p(FA) = \frac{1}{2}\operatorname{erfc}\left[\frac{1}{\sqrt{2}} \cdot \frac{T}{\sigma_N}\right]$$

where erfc is the complementary error function, T is the threshold setting, d is the detection index, and k is a fluctuation parameter defined by $k^2 =$

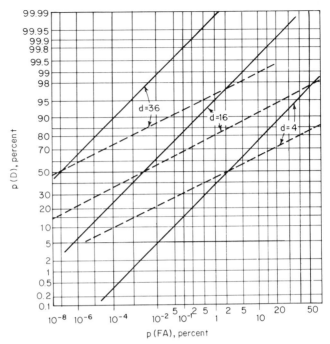

fig. 12.9 ROC curves without (solid lines) and with (dashed lines) signal fluctuation, for d = 4, 16, and 36. The fluctuation index k was taken to be 2. The ROC's for higher values of k would be flatter still. (Ref. 23.)

$\sigma_{S+N}^2/\sigma_N^2$. When fluctuations are absent, $k = 1$. Figure 12.9 shows for several values of d the conventional ROC's ($k = 1$) and the fluctuation—modified ROC's with $k = 4$. With gaussian fluctuations, the ROC's are straight lines on double-probability coordinates and have a slope that depends on k. For the same d, the steady ROC's and the fluctuating ROC's intersect at $p(D) = 0.5$. This happens as a result of the assumption of the same mean value for the signal in the two cases.

It will be evident on inspection of Fig. 12.9 that at a fixed value of $p(FA)$ the effect of fluctuations is to increase $p(D)$ when the signal is weak and to decrease $p(D)$ when the signal is strong; for example, at $p(FA) = 10^{-2}$ percent, $p(D)$ *increases* from 5 to 21 percent for $d = 4$, and *decreases* from 99 to 87 percent for $d = 36$. This effect—better detection during signal surges at low S/N ratios and poorer detection during signal fades at high S/N ratios— is observed regularly during detection trials at sea as a slower falloff of $p(D)$ with range than would be predicted by using the conventional ROC curves.

An alternate approach to estimating the effect of fluctuations of signal-to-noise ratios on detection is through the use of *transition curves,* or plots of probability of detection against *signal excess,* defined as the excess or deficiency of input signal-to-noise ratio relative to that needed for $p(D) = 0.5$.

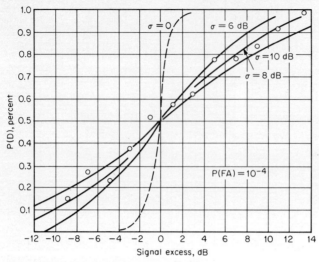

fig. 12.10 *Log-normal transition curves for $\sigma = 0, 6, 8$ and 10 dB for $p(FA) = 10^{-4}$. The circles are data points averaged from a number of sea-going detection trials.*

Letting DT_{50} be the detection threshold for $p(D) = 0.5$ at a selected value of $p(FA)$, we have by this definition

$$DT = DT_{50} + SE$$

where SE denotes the signal excess. Thus, $SE = 0$ for $p(D) = 0.5$; SE is positive for $p(D) > 0.5$, and negative for $p(D) < 0.5$. In this expression $DT_{50} = 5 \log d_{50} w$ where d_{50} is the value of d for the selected value of $p(FA)$ and for $p(D) = 0.5$.

Figure 12.10 shows transition curves [so called because they allow a "transition" to be made from 0.5 to any other value of $p(D)$] for log-normal fluctuations of signal-to-noise ratio for $\sigma = 0$ (no fluctuation), 6, 8, and 10 dB. The light circles are averages obtained by analysis of a number of operational real-world detection exercises using a variety of passive and active sonars. It is seen that the field data fit approximately the transition curve for $\sigma = 8$ dB. Del Santo and Bell (24) long ago measured a large number of values of SE in decibels and found them to be normally distributed with a standard deviation σ between 8 and 9 dB.

This normality of the decibels of SE is not at all obvious. At least one of the sonar parameters is known not to be log-normally distributed; for example, signals from a steady source were found to have a Rician rather than a log-normal distribution (Sec. 6.7). A reasonable explanation for log normality lies in the Central Limit Theorem (25) applied to the sum of the decibel values of the sonar parameters in the sonar equations; by this theorem the distribution of the sum will be normal in decibels, regardless of the individual variability of the parameters themselves. However, this will be only approximately true because the number of parameters is not large, nor are they always independently variable.

The above applies to unpredictable, stochastic fluctuations (Sec. 6.7). Some fluctuations are deterministic rather than stochastic, such as the signal from a ship passing a receiver, and special processors have been described to use this knowledge to improve detection (55, 56).

Small bandwidth-time products When the product of the predetector receiving bandwidth w and the time-constant t of the postdetector integrator is not infinite, a higher signal-to-noise ratio is required for detection at a given $p(D)$ and $p(FA)$ than if an infinite wt existed. With small wt products, the noise at the output of the processor is not gaussian, but is spiky, and the signal detectability deteriorates. The problem of detection with small wt products has been considered theoretically for sinusoidal and broadband signals by Nuttall and Magaraci (26) and by Nuttall and Hyde (27). Figure 12.11 shows the increase in DT required for detection of a band-limited gaussian signal in gaussian noise; slightly smaller values were found for sinusoidal signals. It is seen that a sizable correction to DT is required at high values of $p(D)$ and low values of

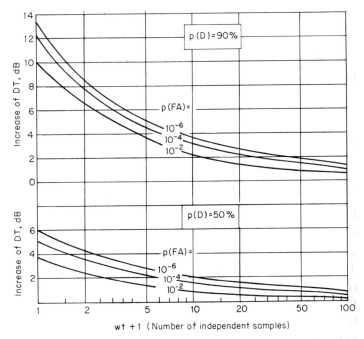

fig. 12.11 *Increase of DT for bandwidth-time products (wt) less than 100 for p(D) = 90% and 50% and several values of p(FA). The horizontal scale is the number of independent samples averaged by the output averages or integrator at the place where the thresholding is done, and is equal to wt + 1. (Ref. 27.)*

$p(FA)$ whenever the wt product employed in the processor is not infinitely large.

Multiple signals Let the signal to be detected be any one of M signals having the same energy. The signals must be "orthogonal" or statistically independent—more precisely, they must satisfy the requirement that

$$\int_0^T s_k(t) \cdot s_l(t) \, dt = 0$$

where $s_k(t)$ and $s_l(t)$ are any two of the signals expressed as a function of time, and T is the interval of time over which the signals can occur. Examples of multiple orthogonal signals occur in communication, where any one of M message signals may be sent, and in echo ranging, where the echo may fall into any one of M doppler bins, depending on target speed. Multiple processing of M channels in parallel, such as with multiple doppler channels or with multiple range bins, is commonly done with modern processing techniques. Nolte and Jaarsma (28), following Peterson, Birdsall, and Fox (19), considered the matter theoretically. For M independent signals, the ROC curves were found to be given approximately by the equations

$$p(D) = 1 - \Phi^{m-1}(\lambda) \cdot \Phi(\lambda - d^{1/2})$$
$$p(FA) = 1 - \Phi^m(\lambda)$$

where

$$\Phi(\lambda) = \frac{1}{(2\pi)^{1/2}} \int_{-\infty}^{\lambda} e^{-t^2/2} dt$$

Fig. 12.12 shows the modified ROC curves computed from these expressions for $d = 16$. We observe that the effect of multiple signals is to lower $p(D)$ at a fixed $p(FA)$; to put it another way, the price to be paid for the luxury of having multiple processing is a substantial increase in $p(FA)$ if $p(D)$ is to be maintained constant.

12.5 Estimating Detection Threshold

In estimating a detection threshold for a particular sonar, it is first necessary to decide upon an acceptable detection probability and an acceptable false-alarm probability for the system *under the conditions in which it is to be used*. These two probabilities are determined by the values of the two possible correct responses (Fig. 12.2) and the costs of the two incorrect responses. Decision theory (2, 29) provides a method for finding the likelihood ratio, and hence the threshold setting, that is optimum for the values and costs involved in the decision process. This optimization process will not be gone into here. In an actual sonar design problem, these quantities are exceedingly difficult to estimate, and will vary greatly with the tactics, strategy, and environmental factors applying for the conditions likely to be encountered. Most often, $p(D)$

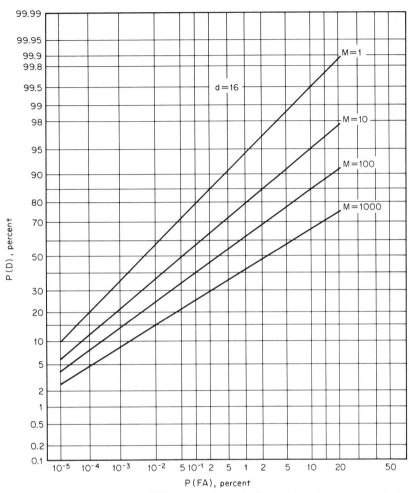

fig. 12.12 ROC curves for different numbers M of multiple orthogonal signals for $d = 16$. (Ref. 28.)

and $p(FA)$ are selected by a combination of experience and intuition and a clear understanding of the use to be made of the system under design. Moreover, in operational exercises with a sonar set where a human observer is involved in the decision process, the threshold criteria under which the human observer says "target present" or "target absent" are often ill-defined, and the detection threshold is, correspondingly, uncertain.

In selecting a value for the false-alarm probability, it should be remembered that $p(FA)$ occurring as a ROC-curve coordinate is the probability of occurrence of a false alarm in an *interval of time equal to the signal duration*. Practically speaking, for many sonars this probability is extremely small. Of greater interest is the probability of occurrence of a false alarm within some longer period. For a given value of $p(FA)$, the probability of getting exactly x false alarms in the time t' is given by the Poisson distribution

$$p'(FA) = \frac{e^{-\lambda}\lambda^x}{x!}$$

where $\lambda = p(FA)\, t'/t$

As an example, let $p(FA)$ be 10^{-4}, and let us find the probability of getting *at least* one false alarm in one hour with an active sonar having an echo duration of 0.1 second. This is the same as one minus the probability of occurrence of *zero* false alarms in one hour. Taking $x = 0$ in the above equation, and with $\lambda = 10^{-4}(3,600)/0.1 = 3.6$, then

$$1 - p'_0(FA) = 1 - \frac{e^{-3.6}(3.6)^0}{0!} = 1 - e^{-3.6} = 0.973$$

Operationally one would make an estimate of $p'(FA)$ using operational experience and then, for entering the ROC curves, calculate $p(FA)$ by trial and error or by the use of tables of the Poisson distribution.

12.6 Effect of Duration and Bandwidth

In the design of receivers and displays, the objective is to achieve the *minimum* detection threshold consistent with system requirements, among which are a detection probability and false-alarm probability compatible with the practical use of the sonar under design. The expressions for DT given above show that, for minimum DT, *it is desirable to use the longest signal duration permitted by system considerations*. For active sonars, this means the longest pulse length; for passive sonars, it means the longest observation time possible before the detection decision must be made. In active sonars, the maximum allowable pulse length is usually determined by the emergence of reverberation as the background that masks the echo; thereafter no increase in echo-to-reverberation ratio is had by an increased pulse length. The use of long pulses without a corresponding buildup of the coherent reverberation background is achieved in sonars using pseudo-random noise for transmission and clipped time-compressed correlators for reception (13); such sonars are useful against reverberation backgrounds or against targets of low doppler shift. In passive sonars, long observation times are limited by the memory of the processing system and the display, by the changing target sounds over a long period of time, and by the necessity of making an immediate detection decision.

Concerning bandwidth, it will be observed that *the detection threshold for the completely known signal (case I) is independent of bandwidth*. Although the processing gain, as defined above in terms of signal-to-noise ratios *in the receiver bandwidth*, is $10 \log 2wt$ and increases with bandwidth, the detection threshold does not. In practice, however, a filter ahead of the correlator is used to remove noise contributions outside the frequency band occupied by the signal. *For case II, it may be seen that minimum DT is obtained when the minimum possible bandwidth is used,* provided that the signal power in the receiver bandwidth remains constant. The latter condition obtains for sinusoidal signals for bandwidths wider than $1/\tau$. For broadband signals, even though the signal power in the receiver bandwidth tends to increase as the first power of the bandwidth, depending on the signal-power spectrum, the improvement in detection threshold is only as the square root of the bandwidth; hence for broadband signals, it is desirable to use as broad a frequency band as possible whenever both signal and noise have about the same spectral distribution. This would occur, for example, in detecting the broadband sound of a cavitating vessel against a background of ambient noise in the region of the spectrum where both have the same spectral shape.

Detection threshold, defined as a ratio of powers or intensities, is not entirely appropriate as a measure of performance of all kinds of processors. It is sometimes incomplete, since it ignores properties of signal and noise other than their power over a period of time. An example is the degradation of performance of correlation sonars caused by multiple propagation paths in the sea. Such multiple paths tend to destroy the correlation between received and transmitted signals, even though they may leave unaltered, or indeed may enhance, the signal intensity. In general, detection threshold must be defined in terms of the particular processing system employed, as well as the appropriate properties of signal and background. Although, as a power ratio, detection threshold is adequate for processors using signal and noise power or energy, such as a square-law detector, it is only an approximation—if that—for sophisticated processing systems using other characteristics of signal and background.

12.7 Example of a Computation

As examples of a computation of DT in echo ranging for the two cases considered above, we consider first the idealized case of a fixed point target at a known range in an ideal medium free from distortion or multiple-path propagation effects. In this idealized case, the echo from a sonar pulse of known wave shape will be a perfect replica of the transmitted pulse and will occur at a known instant of time. We take as parameters the following:

$p(D) = 0.5$ (50% probability of detection)
$p(FA) = 0.0001$ (1 chance in 10^4 of a false alarm occurring in the 0.1-second signal duration)
$t = 0.1$ second (signal duration)
$w = 500$ Hz (receiver bandwidth)

Referring to the ROC curves of Fig. 12.6, we find $d = 15$ for the specified detection and false-alarm probabilities. Since the wave shape of the echo is known, the optimum processor is a correlator in which signal plus noise is correlated against a noise-free replica of the echo. The detection threshold is then

(Case I) $\text{DT} = 10 \log \dfrac{d}{2t} = 10 \log 75 = +19 \text{ dB}$

At the other end of the scale of knowledge is the completely unknown echo, in which neither the wave shape (frequency), time of occurrence, or duration can be specified. Such a situation occurs in echo ranging with a short sinusoidal pulse against a target of unknown range and range rate, where a single frequency band 500 Hz wide is required for reasons of simplicity to accommodate the doppler shift of the target. Now

(Case II) $\text{DT} = 5 \log \dfrac{dw}{t} = 5 \log \dfrac{15 \times 500}{0.1} = +24\frac{1}{2} \text{ dB}$

If the output averager were not "matched" to the signal duration t, but had an averaging time T of, say, 0.05 second, the mismatch would amount to $|5 \log (T/t)| = |5 \log 0.05/0.1| = 1\frac{1}{2}$ dB, and DT would rise to 26 dB.

It should be noted that these cases represent the extremes of knowledge concerning the signal. When this knowledge is complete, the processing can be tailored to the signal, and the minimum DT for the probability levels and signal duration demanded by system requirements can be achieved. In the example, this minimum threshold amounts to 19 dB. The additional 7 dB needed for detecting the echo whose frequency and duration are only approximately known is the price that must be paid for an incomplete knowledge about the signal to be detected.

More elaborate processing can be done to approach the minimum DT. Examples are a bank of 50 adjacent narrow-band filters each of bandwidth $1/t = 10$ Hz, instead of the single broadband filter; another example is a deltic correlator (13) to provide continuous time correlation for use when the range is unknown. In such processing schemes, the maximum gain in DT can be found from the two extremes presented above, and the decibel improvement can be weighed against the increased cost and complexity required.

Actual processing schemes usually involve some degradation in DT relative to the theoretical values. We have already considered the effects on DT of small wt products and of multiple orthogonal signals by means of appropriate modifications of the ROC curves (Sec. 12.4). In correlation sonars, multipath propagation effects in the sea cause a *correlation loss* because of the fact that the received signal is not an exact duplicate of the stored replica. In sonars employing clipped processing, such as in a polarity coincidence correlator (30) or with DIMUS processing (32), a *clipping loss* exists because only phase information is retained in the signal, and amplitude information is rejected by the hard limiting or "clipping" that is done before correlating. Clipper correlators are convenient for engineering reasons, inasmuch as the dynamic range and linearity requirements in the processor are greatly reduced; moreover, digital beam forming by the use of shift registers to obtain the needed time delays is relatively easy to do. Fortunately, the clipping loss in correlation sonars amounts to only 1 to 3 dB, depending on the details of the processing employed and the type of background in which the signal occurs (31). This is usually a small price to pay for the advantages that this kind of processing provides.

12.8 Detection Threshold for Reverberation

All of the above pertains to a background of broadband gaussian noise. Reverberation, however, is neither broadband nor gaussian, but is a decaying, quivering background of scattering whose characteristics are determined by the nature of the scatterers as well as the characteristics of the sonar itself. In Sec. 8.5 and Sec. 8.6 a method was outlined for computing the reverberation level at any instant of time or of range. The level so computed is the total level in the frequency band of the reverberation.

Unfortunately, no experimental data exist on the detection of echoes in realistic seagoing reverberation backgrounds. However, if we boldly assume that the reverberation is noiselike with an envelope smoothed and kept constant by automatic gain control, the expressions for DT previously found may be a useful approximation. That is, as a rough rule of thumb, we may take the detection threshold for reverberation to be

$$DT_R = 5 \log \frac{dw}{t}$$

However, in the active sonar equation for reverberation

$$SL - 2TL + TS = RL + DT_R$$

the equivalent reverberation level is spread out over the reverberation bandwidth w', a bandwidth generally larger than the receiver bandwidth w.

The frequency spread w' of reverberation arises from a number of causes. One is the source pulse itself. For example, if the pulse is a flat-topped sinusoidal of duration t, then $w = 1/t$ approximately; if it is an FM pulse with frequency increasing linearly from f_1 to f_2, $w = f_2 - f_1$. Another cause of frequency spreading is the result of the relative motion of the scatterers producing the reverberation, such as the moving sea surface for which some reverberation spectra may be seen in Fig. 8.36. A third cause of spreading is the combination of beamwidth and orientation of the transducer and the speed, relative to the scatterers, of the platform on which it is mounted. Thus if the transducer has an equivalent ideal two-way beamwidth Φ (Sec. 8.5) and is trained at angle θ to the heading of the echo-ranging vessel, the reverberation bandwidth would be

$$w = \Delta f \left[\cos\left(\theta - \frac{\Phi}{2}\right) - \cos\left(\theta + \frac{\Phi}{2}\right) \right]$$

where Δf is the doppler shift relative to the vessel's speed through the water and the frequency employed (Sec. 9.15).

The signal processing used in many active sonars involves small bandwidth-time (wt) products. For example, if the reverberation bandwidth is the result only of the pulse length, wt is unity, and the deterioration in DT caused by small wt products (Fig. 12.11) becomes appreciable. A linear frequency-modulated pulse, called an FM-slide, in which the frequency changes during the pulse length, has the combined benefits of spreading the reverberation over a larger bandwidth while at the same time increasing the bandwidth-time product.

In summary, assuming the validity of the noise expression for DT, the sonar equation for reverberation becomes

$$\text{SL} - 2\text{TL} + \text{TS} = \text{RL} - 10 \log w' + 5 \log \frac{dw}{t}$$

where RL is the reverberation level in the reverberation bandwidth w' computed as outlined in Chap. 9.

An excellent review of sonar processing and display techniques is given in a review paper by Winder (33), while the radar literature is replete with the theory and methods of target detection in radar sea and ground clutter (34, 35). A discussion of detection threshold problems in reverberation is given in a paper by Kroenert (36).

12.9 Tabular Summary

Table 12.1 gives expressions for the detection threshold for one input and for two inputs from one and two hydrophones, respectively, together with the type of detector and the signal-to-noise ratios before and after the output averager, and for m hydrophones combined in various ways, as adapted from a previously referenced report by Faran and Hills (20). In this table the expressions given for DT, for hydrophone outputs added together before additional processing, include the term $5 \log 1/m^2$, equivalent to the directivity index of an m-element array in an incoherent noise field. Hence the expressions, in effect, refer to the two terms DT and DI in the sonar equations. For multiplicative processing (Sec. 3.12), it is not possible to separate the effects of the array and of the processing system on the overall system performance.

12.10 Auditory Detection

The ear historically has been the earliest detection device used in sonar. Before the use of amplifiers and the perfection of visual displays, the hearing of a human observer provided the only means of detecting sonar signals. In recent years, the human ear has tended to be replaced by other detection devices having a longer integration time as well as a longer memory that

table 12.1 Detection Threshold for Single and Multiple Units

$$\text{Filter} \xrightarrow{S(t) + N_1(t)} \text{Multiplier} \xrightarrow{(S/N)'_{out}} \begin{array}{c}\text{Averager}\\ \text{time}\\ \text{constant} = t\end{array} \xrightarrow{(S/N)_{out} = d}$$

$$\text{Filter} \xrightarrow{S(t) + N_2(t)}$$

Bandwidth w

Case	Type of detector	$(S/N)'_{out}$	$(S/N)_{out} = d$ $= (S/N)'\ 2wt$	Detection threshold $= 10 \log [S/N/w]_{in}$
Single hydrophone: Signal known completely $N_2(t) = 0$	Coherent detector	$(S/N)_{in}$	$2wt(S/N)_{in}$	$10 \log \dfrac{d}{2t}$
Signal unknown $N_1(t) \equiv N_2(t)$	Autocorrelator (square-law detector)	$\tfrac{1}{2}\left(\dfrac{S}{N}\right)^2_{in}$	$wt\left(\dfrac{S}{N}\right)^2_{in}$	$5 \log \dfrac{dw}{t}$
Two hydrophones: $N_1(t) \neq N_2(t)$, but of same variance Outputs multiplied	Crosscorrelator	$\left(\dfrac{S}{N}\right)^2_{in}$	$2wt\left(\dfrac{S}{N}\right)^2_{in}$	$5 \log \dfrac{dw}{2t}$
Outputs added first and then multiplied (squared)	Adder + square-law detector	$2\left(\dfrac{S}{N}\right)^2_{in}$	$4wt\left(\dfrac{S}{N}\right)^2_{in}$	$5 \log \dfrac{dw}{4t}$

Limitations

1. Small S/N, i.e., $\overline{S(t)^2/N(t)^2} \ll 1$.
2. Stationary white gaussian noise, though not much change may be expected for other reasonable noise statistics.
3. $N_1(t)$, $N_2(t)$, and $S(t)$ are all uncorrelated, that is, $\overline{N_1(t)N_2(t)} = 0$.
4. $2wt \gg 1$. The number of sample points represented by the averager is large.
5. An alert observer, if one is used.

Continuation to m hydrophones (compiled from Ref. 20)

Case	$(S/N)_{out} = d$	Detection threshold $= 10 \log [S/N/w]_{in}$
m hydrophones (incoherent noise): Added and square-law detected	$m^2 wt \left(\dfrac{S}{N}\right)^{2*}_{in}$	$5 \log \dfrac{dw}{m^2 t}$
Correlated in pairs in $m(m-1)/2$ correlators, and outputs added	$m(m-1)\, wt \left(\dfrac{S}{N}\right)^2_{in}$	$5 \log \dfrac{dw}{m(m-1)t}$
Split into two halves and correlated	$\dfrac{m^2}{2} wt \left(\dfrac{S}{N}\right)^2_{in}$	$5 \log \dfrac{2dw}{m^2 t}$
Multiplied all together, m even	$\dfrac{m!m}{2} wt \left(\dfrac{S}{N}\right)^m_{in}$	$\dfrac{10}{m} \log \dfrac{2d}{m!m\, wt} + 10 \log w$

* Applies for a keyed signal. For an unkeyed signal, replace m^2 by $(m-1)^2$.

provide a record of past events. Yet the ear, as a detector and analyzer of tonal signals in noise backgrounds, is a remarkably compact and efficient device that still has a place in modern sonar. Surprisingly, its performance as a detector of sinusoidal signals approaches that of the optimum detector. In addition, it has other, and notably complex, characteristics that are still under intensive research.

Everything about the ear as a processor of acoustic signals appears to depend to some extent on everything else. Because of its complex behavior, its importance, and the ease of doing research on it, audition continues to be a popular subject for investigation; the number of papers devoted to it in the *Journal of the Acoustical Society of America* is comparable to, or exceeds, the number of papers dealing with the many and diverse aspects of underwater acoustics. Our concern will be restricted to but one aspect of the behavior of this wonderful sensory organ—its ability to detect sinusoidal signals in noise and reverberation backgrounds.

In the literature, the signal-to-noise ratio for auditory detection is frequently called the *recognition differential*. This term is defined as the "amount by which the signal level exceeds the noise level presented to the ear when there is a 50 percent probability of detection of the signal." Because there is no specification concerning false alarms, the term "recognition differential," as thus defined and as referred to in the older literature, is quantitatively almost meaningless.* In the current psychoacoustic literature, the use of recognition differential has all but disappeared, and has been replaced by a specification of the detection index d' ($=d^{1/2}$), equivalent to a pair of probabilities, needed for detection under stated conditions of signal and noise.

Physical model of audition Of the many component parts of the human hearing mechanism, one particular structure, called the *basilar membrane,* is of particular importance. This is a thin membrane lying coiled up inside a bony tube termed the *cochlea*—so called from its resemblance to a small snail shell. Embedded in the basilar membrane throughout its length are a great number of fine fibers, called the *auditory strings,* that initiate the complex sensory and nervous process resulting in the tonal sensation. According to the resonance theory of hearing, the auditory strings are analogous to the strings of a piano or harp; different portions of the basilar membrane and the corresponding embedded auditory strings resonate at different frequencies of the impressed sound. The ear is thus imagined to be like a comb filter consisting of a large number of narrow adjacent filter bands, called the *critical bands* of hearing. The bandwidth of the critical bands is the *critical bandwidth*.

The simple model of the ear for detection accordingly consists of a series of adjacent input filter bands corresponding to the various critical bands. These are followed, in the model, by a square-law detector and an integrator, or low-

---

* It appears likely that a false-alarm rate somewhere between 1 and 10 percent prevailed during many early laboratory measurements of recognition differential.

pass filter, to model the integration, or energy summation, time of the hearing process. Figure 12.13 shows the energy-detection model of the ear for the critical band centered at frequency f_n. The entire model would consist of a large number of such adjacent bandpass filters, each corresponding in the ear itself to a particular portion of the basilar membrane and its associated neural structure.

This simple model of the auditory detection process by no means accounts for all the phenomena of audition; moreover, its parameters depend on many factors, including the input level and the immediate past stimulus of the ear, in a way unfamiliar to electronic circuitry. Nevertheless the model appears to account (37, 38) for the principal phenomena of auditory detection and, as will be seen, yields reasonable agreement with observed detection thresholds.

The critical band concept was first proposed by Fletcher (39) to explain certain observations of the masking of tones by broadband noise. A variety of methods have been devised to measure the critical bandwidth. One method, first used by Fletcher and later by Schafer et al. (40), employs a tone in a narrow, adjustable band of noise. Observations are made of the bandwidth of the noise above which the detectability of the tone is constant; this bandwidth equals the critical band of the ear. Figure 12.14 is a compilation of measurements by this and other methods of the critical bandwidth at various frequencies. In the vicinity of 500 Hz, the critical bandwidth lies in the vicinity of 50 Hz. Measurements by some methods, such as by loudness summation, the detection of multiple tones, and the masking of noise by tones (the reverse of the method just mentioned), yield bandwidths appreciably larger than this, although the variation with frequency is the same. The contradictory findings concerning the critical bandwidth of the ear have been reviewed by Zwicker, Flottorp, and Stevens (41); DeBoer (42); and Swets, Green, and Tanner (43); and in a Soviet review paper (44). The dashed curve of Fig. 12.14 is an

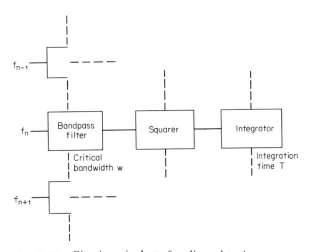

fig. 12.13 *Circuit equivalent of auditory detection.*

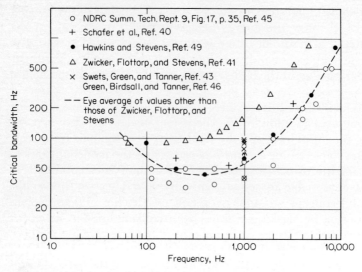

fig. 12.14 *Measured values of the critical bandwidth of the ear.*

estimated eye average of the smaller bandwidth values. More recent measurements by Patterson (57) at 500, 1,000, and 2,000 Hz gave a critical bandwidth of 0.13 times the center frequency, in rough agreement with the higher values of Zwicker, Flottorp, and Stevens.

The other parameter of the model—the integration time for audition—has received much less attention. Some measurements (45) indicate that the detectability of tone pulses is constant for durations greater than 1 second, indicating that the ear integrates energy up to, but not beyond, an interval of 1 second. More recent data (46) show an integration time of the order of 0.2 second. Apparently this parameter lies somewhere in the decade 0.1 to 1.0 second and is the time interval over which the auditory process, according to the physical model, effectively integrates acoustic energy.

Detection threshold for tones in broadband noise The detectability of tone pulses masked by broadband gaussian noise varies with the duration of the tone. Figure 12.15 shows the detection threshold of tonal pulses as compiled from measured values in the literature. The values plotted refer to a detection probability of 0.5 with a false-alarm rate of the order of 0.05; in all but the most recent experiments, the false-alarm rate was not stated. In the work of Green, Birdsall, and Tanner (46), the plotted thresholds correspond to a value of $d' = 1.8$ or a detection index $d = (d')^2 = 3.3$; which, according to the ROC curves, implies a false-alarm probability of 0.05 at the 50 percent level of detection probability.

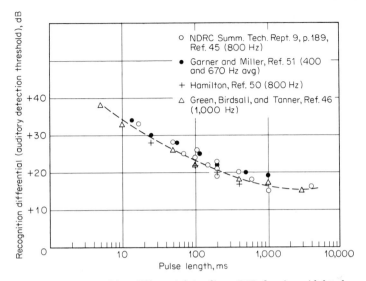

fig. 12.15 *Recognition differential (auditory DT) for sinusoidal pulses in broadband noise reduced to 1-Hz bands.*

Detectability of echoes in reverberation Against a background of reverberation, the detectability of tones (echoes) is remarkably different. In echo ranging with sinusoidal pulses, the governing parameter is the frequency difference between the echo and the reverberation background. This difference in frequency is the doppler shift caused by the motion of the sonar ship and target relative to the reverberation-producing scatterers. It amounts to 0.69 hertz per knot of relative velocity per kilohertz (Sec. 9.16).

Much attention was given in the World War II years to the detectability of recorded echoes of adjustable level in recorded reverberation backgrounds. Figure 12.16 shows smoothed curves of the aural echo-to-reverberation ratio required for the detection of echoes of different durations, as a function of

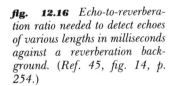

fig. 12.16 *Echo-to-reverberation ratio needed to detect echoes of various lengths in milliseconds against a reverberation background. (Ref. 45, fig. 14, p. 254.)*

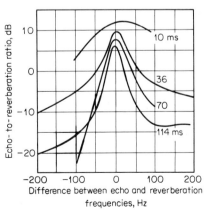

the frequency difference between echo and reverberation. In the experiments summarized by these curves, the echoes and reverberation obtained with a sonar operating near 25 kHz were heterodyned down to a frequency of about 800 Hz. Of particular note is the rapid improvement of detectability with frequency difference. Also noteworthy is the asymmetry of these curves; "down doppler" echoes having a frequency lower than the reverberation are more readily detected than "up doppler" echoes of the same frequency difference. This suggests an asymmetry in the response characteristic of the critical bands of the ear—an effect that has been observed in later experiments (47) on the masking of one tone by another tone of slightly different frequency.

Comparison with the model Using the model of Fig. 12.13 and with some knowledge of its parameters, it is of interest to compare the performance of the ear as a detector of tones in broadband noise backgrounds with the performance of the theoretical model. We select a detection index d equal to 4, corresponding to $p(D) = 0.5$ and $p(FA) = 0.03$ and assume a critical bandwidth w of 50 Hz (Fig. 12.14) and an integration time T of 0.5 second. Then using the expression

$$\text{DT} = 5 \log \frac{dw}{t} + \left| 5 \log \frac{T}{t} \right|$$

we may compute DT as a function of the pulse length t. The result is shown in Fig. 12.17, together with the average observed curve taken from Fig. 12.15. Although they are similar in shape, the computed curve lies lower than the observed curve, indicating that the ear falls somewhat short, by about 4 dB, of what would be expected from the model; the difference can be lessened by a more favorable choice of parameters (such as $w = 100$ Hz and $T = 1.0$ second), or by including a correction to the model for small bandwidth-time

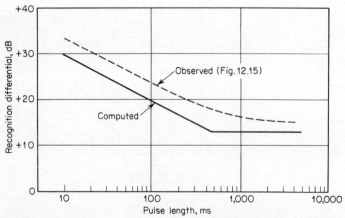

fig. 12.17 *Comparison of observed and computed recognition differentials.*

products (Fig. 12.11). In any case, it appears that the simple physical model provides a reasonable replica of the detection performance of the ear. In terms of the model, the ear is a remarkably efficient detector of tonal signals in gaussian noise backgrounds. It has been shown (48), by laboratory comparisons of a human and an electronic detector, that the ear approaches the performance of a theoretical energy detector having the same bandwidth and integration time; the human ROC's closely approach the theoretical ROC's (52). In short, the energy model appears to be a valid model for describing the performance of human observers in detecting weak signals in noise (53, 54).

REFERENCES

1. Urick, R. J., and P. L. Stocklin: A Simple Prediction Method for the Signal Detectability of Acoustic Systems, *U.S. Nav. Ord. Lab. Tech. Rep.* 61–164, 1961.
2. Helstrom, C. W.: "Statistical Theory of Signal Detection," Pergamon Press, New York, 1960.
3. Bennett, W. R.: Methods of Solving Noise Problems, *Proc. IRE*, **44:**609 (1956).
4. Woodward, P. M.: "Probability and Information Theory," McGraw-Hill Book Company, New York, 1953.
5. Laning, J. H., and R. H. Battin: "Random Processes in Automatic Control," McGraw-Hill Book Company, New York, 1956.
6. Davenport, W. B., and W. L. Root: "Random Signals and Noise," McGraw-Hill Book Company, New York, 1958.
7. Freeman, J. J.: "Principles of Noise," John Wiley & Sons, Inc., New York, 1958.
8. Bendat, J. S.: "Principles and Applications of Random Noise Theory," John Wiley & Sons, Inc., New York, 1958.
9. Schwartz, M.: "Information Transmission Modulation and Noise," McGraw-Hill Book Company, New York, 1959.
10. Lee, Y. W.: "Statistical Theory of Communication," John Wiley & Sons, Inc., New York, 1963.
11. Harman, W. W.: "Principles of the Statistical Theory of Communication," McGraw-Hill Book Company, New York, 1963.
12. Stewart, J. L., and W. B. Allen: Pseudorandom Signal Correlation Methods of Underwater Acoustic Research, *J. Acoust. Soc. Am.*, **35:**810(A) (1963).
13. Allen, W. B., and E. C. Westerfield: Digital Compressed-Time Correlators and Matched Filters for Active Sonar, *J. Acoust. Soc. Am.*, **36:**121 (1964).
14. Persons, C. E., M. K. Brandon, and R. M. Zarnowitz: Linear and Nonlinear Correlators for Pseudo Random Signal Detection, *U.S. Navy Electron. Lab. Rep.* 1219, 1964.
15. Allen, W. B.: Pseudorandom Signal Correlation in Sonar Systems, *U.S. Navy Electron. Lab. Rep.* 1299, 1965.
16. Becken, B. A.: Sonar, "Advances in Hydroscience," vol. 1, Academic Press Inc., New York, 1964.
17. Peterson, W. W., and T. G. Birdsall: The Theory of Signal Detectability, *Univ. Mich. Eng. Res. Inst. Rep.* 13, 1953.
18. Lindgren, B. W.: "Statistical Theory," pp. 145–150, The Macmillan Company, New York, 1962.
19. Peterson, W. W., T. G. Birdsall, and W. C. Fox: The Theory of Signal Detectability, *Trans. IRE*, **PGIT-4:**171 (1954).
20. Faran, J. J., and R. Hills: Application of Correlation Techniques to Acoustic Receiving Systems, *Harvard Univ. Acoust. Res. Lab. Tech. Memo* 28, 1952.
21. Lawson, J. L., and G. E. Uhlenbeck: "Threshold Signals," M.I.T. Radiation Laboratory Series, vol. 24, sec. 8.10, McGraw-Hill Book Company, New York, 1950.

22. Hodgkiss, W. S., and V. C. Anderson: Detection of Sinusoids in Ocean Acoustic Background Noise, *J. Acoust. Soc. Am.*, **67:**214 (1980).
23. Urick, R. J., and G. C. Gaunaurd: Detection of Fluctuating Sonar Targets, *U.S. Nav. Ord. Lab. Tech. Rep.* 72–47, 1972.
24. Del Santo, R. F., and T. G. Bell: Comparison of Predicted vs. Actual Submarine Sonar Detection Ranges, *U. S. Navy Underwater Sound Lab. Rep.* 514, 1962.
25. Korn, G. A., and T. M. Korn: "Mathematical Handbook for Scientists and Engineers," 2nd ed., Sec. 18.6-5, McGraw-Hill Book Company, New York, 1968.
26. Nuttall, A. H., and A. F. Magaraci: Signal to Noise Ratios Required for Short Term Narrow-band Detection of Gaussian Processes, *U.S. Nav. Underwater Syst. Center Tech. Rep.* 4417, 1972.
27. Nuttall, A. H., and D. W. Hyde: Operating Characteristics for Continuous Square-Law Detection in Gaussian Noise, *U.S. Nav. Underwater Syst. Center Tech. Rep.* 4233, 1972.
28. Nolte, L. W., and D. Jaarsma: More on the Detection of One of M Orthogonal Signals, *J. Acoust. Soc. Am.*, **41:**497 (1967).
29. Middleton, D.: "Introduction to Statistical Communication Theory," chaps. 18–23, McGraw-Hill Book Company, New York, 1960.
30. Faran, J. J., and R. Hills: Correlators for Signal Reception, *Harvard Univ. Acoust. Res. Lab. Tech. Memo* 27, 1952.
31. Weston, D. C.: Correlation Loss in Echo Ranging, *J. Acoust. Soc. Am.*, **37:**119 (1965).
32. Remley, W. R., D. Upton, and D. del Giorno: DIMUS Processing with Seven-Element Active-Sonar Arrays, *J. Acoust. Soc. Am.*, **41:**439 (1967).
33. Winder, A. A.: Underwater Sound, a Review: Part II—Sonar System Technology, *IEEE Trans. on Sonics and Ultrasonics*, **SU 22:**291 (1975).
34. Skolnick, M. I.: "Introduction to Radar Systems," McGraw-Hill Book Company, New York, 1962.
35. Trunk, G. V., and S. F. George: Detection of Targets in Non-Gaussian Sea Clutter, *IEEE Trans. on Aerospace and Electronic Systems*, **AES-6:**620 (1970).
36. Kroenert, J. T.: Discussion of Detection Threshold with Reverberation Limited Conditions, *J. Acoust. Soc. Am.*, **71:**507 (1982).
37. Sheeley, E. C., and R. C. Bilger: Temporal Integration as a Function of Frequency, *J. Acoust. Soc. Am.*, **36:**1850 (1964).
38. Jeffress, L. A., and A. D. Gaston: ROC Curves from an Ear Model, *J. Acoust. Soc. Am.*, **38:**928(A) (1965).
39. Fletcher, H.: Auditory Patterns, *Rev. Mod. Phys.*, **12:**47 (1940).
40. Schafer, T. H., R. Gales, C. A. Shewmaker, and P. O. Thompson: Frequency-Selectivity of the Ear as Determined by Masking Experiments, *J. Acoust. Soc. Am.*, **22:**490 (1950).
41. Zwicker, E., G. Flottorp, and S. S. Stevens: Critical Bandwidth in Loudness Summation, *J. Acoust. Soc. Am.*, **29:**548 (1957).
42. DeBoer, E.: Note on the Critical Bandwidth, *J. Acoust. Soc. Am.*, **34:**985 (1962).
43. Swets, J. A., D. M. Green, and W. P. Tanner: On the Width of the Critical Bands, *J. Acoust. Soc. Am.*, **34:**108 (1962). Also, Swets, J. A.: "Signal Detection and Recognition by Human Observers," chap. 23, John Wiley & Sons, Inc., New York, 1964.
44. Freidin, A. A.: Role of the Critical Bands in the Processing of Information by the Human Auditory System (Review), *Sov. Phys. Acoust.*, **14:**271 (1969).
45. Recognition of Underwater Sounds, Office of Scientific Research and Development, *Nat. Def. Res. Comm. Div. 6 Sum. Tech. Rep.* 9, 1946.
46. Green, D. M., T. G. Birdsall, and W. P. Tanner: Signal Detection as a Function of Signal Intensity and Duration, *J. Acoust. Soc. Am.*, **29:**523 (1957). Also, Swets, J. A.: "Signal Detection and Recognition by Human Observers," chap. 11, John Wiley & Sons, Inc., New York, 1964.
47. Greenwood, D. G.: Auditory Masking and the Critical Band, *J. Acoust. Soc. Am.*, **33:**484 (1961).

48. Sherwin, C. W., et al.: Detection of Signals in Noise: A Comparison between the Human Detector and an Electronic Detector, *J. Acoust. Soc. Am.*, **28:**617 (1956).
49. Hawkins, J. E., and S. S. Stevens: Masking of Pure Tones and of Speech by White Noise, *J. Acoust. Soc. Am.*, **22:**6 (1950).
50. Hamilton, P. M.: Noise Masked Thresholds as a Function of Tonal Duration and Masking Bandwidth, *J. Acoust. Soc. Am.*, **29:**506 (1957).
51. Garner, W. A., and G. A. Miller: Mashed Threshold of Pure Tones as a Function of Duration, *J. Exp. Psychol.*, **37:**293 (1947).
52. Mulligan, B. F., and others: Prediction of Monaural Detection, *J. Acoust. Soc. Am.*, **43:**481 (1968).
53. Henning, G. B.: A Model for Auditory Discrimination and Detection, *J. Acoust. Soc. Am.*, **42:**1325 (1967).
54. Jeffress, L. A.: Mathematical and Electrical Models for Auditory Detection, *J. Acoust. Soc. Am.*, **44:**187 (1968).
55. Hunt, F. V.; Signal Rate Processing for Transit Detection, *J. Acoust. Soc. Am.*, **51:**1164 (1972).
56. Tuteur, F. B.: Detection of a Moving Noise Source in a Non-stationary Noise Background, *J. Acoust. Soc. Am.*, **44:**912 (1968).
57. Patterson, R. D.: Auditory Filter Shapes Derived with Noise Stimuli, *J. Acoust. Soc. Am.*, **59:**640 (1976).

thirteen

design and prediction in sonar systems

13.1 Sonar Design

The various applications of the sonar equations fall into two general classes. One involves *sonar design*, where a sonar system is to be devised to accomplish a particular purpose. In a sonar design problem, a set of sonar parameters that will provide the desired performance must be found. This can usually be expressed in terms of *range*, through its counterpart, by some assumed propagation condition, the parameter *transmission loss*.

This selection of parameters in sonar design is beset with difficulties arising from constraints that are of economic, mechanical, or electrical origin. Sonar systems must often be primarily inexpensive, as in expendable units such as sonobuoys. Sometimes they must fit in a confined space, as in a torpedo, where the maximum size of the transducer to be used is dictated by dimensions over which the design engineer has no control. Sonar systems may also have to be designed to consume only a limited amount of electric power, as in a battery-powered underwater acoustic beacon, where a limitation is placed on the available acoustic power output and the pulse length. Generally speaking, one or more of the parameters related to the system itself, such as directivity index or source level, may be fixed or limited by practical considerations not under the designer's control. The final design is achieved by "trade offs" and compromises between performance and achievable values of the equipment parameters. It is reached by what amounts to repeated solutions of the sonar equations—by a trial-and-error process wherein successive adjustments of parameters and performance are made until a reasonably satisfactory compromise is reached. Complications arise when the desired performance involves two or more of the variables. For example, a certain search rate, or area searched for a target in a given time, may be desired; this is a function of both range and beam width. In such problems, a number of trial solutions of the sonar equations will be needed to give a "feel" for the best set of conditions.

Sometimes the fortunate design engineer has a free choice of the operating frequency, or the operating frequency band, of the sonar under design. Then the choice will be influenced by the optimum frequency appropriate to the desired maximum range of the sonar. This choice will be considered in a section to follow.

In an active-sonar design problem, the design will depend in part on whether the echoes occur in a background of noise or reverberation. In active-sonar systems, the range increases with acoustic power output until the echoes begin to occur in a reverberation background. When this happens, the range is said to be *reverberation-limited*. Beyond this value of output power, no increase of range is available, since both echo level and reverberation increase together with increasing power. It follows, as a precept in active-sonar design, that the acoustic output power should be increased until the *reverberation level is equal to the level of the noise background at the maximum useful range of the system*. Unfortunately, although this is a useful general rule, it cannot always be followed because of limitations imposed by the amount of available power or because of interaction effects and cavitation at the sonar projector.

13.2 Sonar Prediction

The other broad class of problems has to do with *performance prediction*. Here the sonar system is of fixed design—and, indeed, may already by in operational use—and it is desired to predict its performance under a variety of conditions. Alternatively, if field trials of a system have already been made, it may be necessary to account for the performance that has been achieved—a kind of "postdiction," in which a numerical explanation is required for the results obtained. This class of problems normally requires solving the appropriate form of the sonar equation for the parameter containing the range. The passive-sonar equation may be written

$$TL = SL - NL + DI - DT$$
$$= FM$$

where the sum of the parameters on the right is called (Table 2.2) the *figure of merit* FM for the particular target referred to in the parameter SL. Similarly, the active-sonar equation for a noise background may be written as

$$2(TL) = SL + TS - NL + DI - DT$$
$$= FM$$

where FM is the figure of merit for the target implied by the value used for the parameter TS. The prediction of range requires the conversion into range of the value of transmission loss that is equal to the figure of merit. The conversion demands a specification of the propagation conditions, such as layer depth and propagation path, under which the equipment will be, or has been, used. With reverberation backgrounds, the transmission loss is usually the same for both the target and reverberation, and the range occurs implicitly in the terms $10 \log A$ or $10 \log V$, representing in decibel units the reverberating area or volume, respectively.

13.3 The Optimum Sonar Frequency

Existence of an optimum frequency When range calculations at different frequencies are made for a sonar set of a particular design and for some specified propagation and target conditions, it is often found that the range has a maximum at some particular frequency. This frequency is the *optimum frequency* for the particular equipment and target characteristics being considered. At the optimum frequency, a minimum figure of merit is required to reach a given range. Hence, the optimum frequency is a function of the detection range as well as the specified set of medium, target, and equipment parameters. If the operating frequency is made much higher than optimum, the absorption of sound in the sea reduces the range; if the operating frequency is made much lower than optimum, a number of other parameters become unfavorable and act to reduce the range. Examples of such parameters are the directivity index, background noise, and detection threshold (through a necessarily smaller bandwidth at the lower frequencies), all of which conspire to reduce the system figure of merit at low frequencies.

Curve AA of Fig. 13.1 shows a range-versus-frequency plot for a hypothetical sonar. If, by some means, the figure of merit of this sonar is raised by an amount that is the same for all frequencies, the range-frequency curve is shifted to BB. Although the range is increased at all frequencies, the optimum frequency, at which the maximum range occurs, has become lower. The locus of the peak values of a series of such curves gives the best frequency to use to obtain a desired range. Its shape and position depend on the system figure of merit and on the transmission loss and, more importantly, on how both vary with frequency.

Illustrative example The determination of the optimum frequency can best be illustrated by an example. Consider a passive listening system that employs a line hydrophone 5 ft long. It is desired to find the optimum frequency for

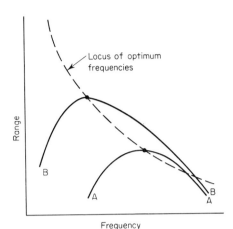

fig. 13.1 *Range as a function of frequency for two sonar systems of different figures of merit.*

the detection of a freighter traveling at a speed of 10 knots. The noise background is taken to be the ambient noise of the sea in sea state 3 (wind speed 11 to 16 knots), and the detection threshold is zero decibels. Let the transmission loss be determined by spherical spreading plus absorption according to the relationship TL = 20 log r + 0.01$f^2 r$ × 10^{-3}, where f is the frequency in kilohertz and r is the range in yards. Based on this expression, Fig. 13.2 shows curves of TL as a function of frequency for a number of different ranges. Superposed on the same plot is the line AA, equal to FM at different frequencies for the particular problem at hand, using appropriate values of the parameters.* At any frequency, the detection range is that for which TL = FM. This range has a maximum, for the curve AA, of 6,000 yd and occurs at 5 kHz. This is the optimum frequency for the assumed conditions. At this frequency, the slopes of the line AA and of an interpolated member of the family of TL curves are equal. At frequencies different from 5 kHz, the range is less, becoming reduced to 5,000 yd at both 2 and 10 kHz. If, by redesigning the system, the FM is increased by an amount that is constant with frequency, the line AA might be shifted to BB; the range will be increased to 19,500 yd and the optimum frequency lowered to 1.7 kHz. If the redesign is such as to change the slope of the FM curve, an altogether new optimum frequency will be obtained.

Analytic method When, as in the example just given, the transmission loss can be expressed as a particular function of range for the conditions of interest, the optimum frequency can be found analytically. In the equality TL = FM, the maximum (or minimum) range is obtained by differentiating both sides with respect to frequency and setting dr/df equal to zero. With the preceding expression for TL, we would have

$$\text{TL} = 20 \log r + 0.01 f^3 r \times 10^{-3} = \text{FM}$$

* SL: Table 10.2; NL: Fig. 7.5; DI: Fig. 3.6; DT = 0.

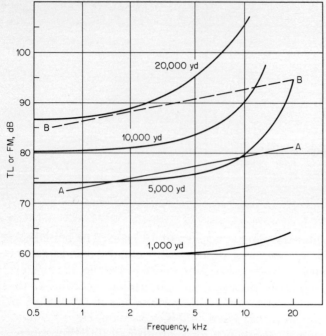

fig. 13.2 Curves of transmission loss and figure of merit as a function of frequency.

On differentiating and placing $dr/df = 0$, we obtain

$$0.02 f_0 r_0 \times 10^{-3} = \frac{d(\text{FM})}{df}$$

where f_0 = optimum frequency
r_0 = maximum range
$d(\text{FM})/df$ = rate of change of FM with frequency, dB/kHz

In terms of the more conventional unit of decibels per octave of frequency, we can write

$$\left.\frac{d(\text{FM})}{df}\right|_{\text{db/octave}} = \frac{f_0}{\sqrt{2}} \left.\frac{d(\text{FM})}{df}\right|_{\text{db/kHz}}$$

since the octave whose geometric mean frequency is f_0 is $f_0/\sqrt{2}$ Hz wide. It therefore follows that

$$\frac{0.02}{\sqrt{2}} f_0^2 r_0 = \frac{d(\text{FM})}{df}$$

and the optimum frequency becomes

$$f_0 = \left[\frac{70.7}{r_0} \frac{d(\text{FM})}{df}\right]^{1/2}$$

where $d(FM)/df$ is the rate of change of FM with frequency in units of decibels per octave, and f_0 and r_0 are in units of kilohertz and kiloyards, respectively. Since for a passive system

$$FM = SL - NL + DI - DT$$

it follows that

$$\frac{d(FM)}{df} = \frac{d(SL)}{df} - \frac{d(NL)}{df} + \frac{d(DI)}{df} - \frac{d(DT)}{df}$$

so that the quantity $d(FM)/df$ is the sum, with due regard for sign, of the rates of change with frequency of the sonar parameters of which it is composed. Considering the example given above (Fig. 13.2), $d(FM)/df$ would be found to be approximately equal to $-6 + 5 + 3 + 0 = +2$ dB/octave. On substituting in the above expression for f_0 and taking $r_0 = 6{,}000$ yd, f_0 becomes equal to 5.3 kHz. In echo ranging, where the two-way transmission loss is involved, the expression for f_0 becomes

$$f_0 = \left[\frac{35.4}{r_0} \frac{d(FM)}{df}\right]^{1/2}$$

The optimum frequency accordingly depends upon the frequency variation of all the sonar parameters and is especially sensitive to the frequency variation of the absorption coefficient. It is not sharply defined, but is the peak of a broad maximum extending over a frequency range of several octaves. The optimum frequency may be defined in terms other than range, as, for example, search rate or processing time, as discussed by Stewart, Westerfield, and Brandon (1). For reverberation backgrounds, the figure of merit is itself a function of range, and the optimum frequency is not as easily determined. Normally, an optimum frequency does not exist in reverberation-limited systems since the frequency-dependent absorption coefficient is ordinarily the same for the echo and for the reverberation background.

An extended discussion of the subject is given by Horton (2), who can be credited with having first recognized the existence of optimum frequencies in sonar applications. More recently, Stewart, Westerfield, and Brandon (3) have published curves of optimum frequency versus maximum range for active sonar detection using more recent expressions for attenuation as a function of frequency.

13.4 Applications of the Sonar Equations

Sonar Problem Solving

The following are some examples of how the sonar equations may be used to solve problems in a number of different applications of sonar. The examples given and the conditions assumed do not necessarily have any practical significance, but are selected more or less at random to illustrate how the equations are used in some specific problems concerning the many modern uses of sonar.

The approach to problem solving by means of the sonar equations is to select the equation appropriate to a particular problem and then to solve it for the unknown parameter in terms of the other parameters which are either specifiable or can be selected, with more or less uncertainty, from specified conditions on which they depend. Typical values for nearly all conditions of interest can be found in curves or tables given in earlier chapters.

In an actual design problem the usually straightforward computation should be accompanied by a plot of echo or signal level, together with the reverberation and noise masking levels, as a function of range. Such a plot will indicate most strikingly how the range, determined by intersection of the curves of signal and background, will vary with changes in level. This plot will lend confidence to the numerical computations. Once the range is determined, other quantities of perhaps greater significance, such as area searched per unit time, can be readily computed.

Active Submarine Detection

PROBLEM: An echo-ranging sonar mounted on a destroyer has a power output of 1,000 watts at a frequency of 8 kHz. Its DI is 20 dB and it uses a pulse length of 0.1 second, with a receiving bandwidth of 500 Hz. Find the range at which it can detect a beam-aspect submarine at a depth of 250 ft in a mixed layer 100 ft thick when the ship is traveling at a speed of 15 knots. Detection is required 50 percent of the time, using incoherent processing, with a probability of 0.01 percent of occurrence of a false alarm during the echo duration.

SOLUTION: The active-sonar equation, solved for TL, is

$$TL = \tfrac{1}{2}(SL + TS - NL + DI - DT)$$

SL is given by Fig. 4.4, using $DI_T = 20$ dB, as 221; by Table 9.3, TS = 25; by Fig. 11.11 and reducing from 25 kHz by assuming -6 dB/octave spectral slope, NL = $+53 + 20 \log (25/8) = +63$; DI = $+20$; by Fig. 12.6, $d = 15$ and DT = $5 \log (15 \times 500/0.1) = +24$. Therefore TL = $\tfrac{1}{2}(179) = 90$. Referring to Fig. 6.7b, for a layer depth of 100 ft, and assuming that the transmission is the same as that for a source depth of 50 ft, the range corresponding to this value of TL is 5,500 yd.

Passive Submarine Detection

PROBLEM: A submarine radiating a 500-Hz line component at a source level 160 dB crosses a convergence zone. Another submarine, located 30 miles away, listens with a nondirectional hydrophone. Assuming a noise background equivalent to that of the deep sea in sea state 3, how long an observation time will the second submarine need to detect the first if it uses incoherent (energy) processing in a receiver band 100 Hz wide and if a detection probability of 50 percent, with a 1 percent false-alarm probability, is satisfactory?

SOLUTION: When solved for the parameter of interest, the passive-sonar equation is

$$DT = SL - TL - NL + DI$$

SL is given as 160; TL is taken as being equal to spherical spreading to 30 miles plus a convergence gain of 10 dB, or TL = 20 log (30 × 2,000) − 10 = +86; by Fig. 7.5, NL = 66; DI = 0 for a nondirectional hydrophone. Therefore DT = +8. By the formula, DT = 5 log (dw/t), with w = 100 (given) and d = 6 (Fig. 12.7), we find an observation time of t = 15 seconds. The signal energy must therefore be integrated for this length of time in order for detection to occur at the required probability levels.

Minesweeping

PROBLEM: A minesweeper tows behind it, for the purpose of sweeping acoustic mines, a broadband sound source having a source spectrum level of 150 dB in a 1-Hz band. The mines to be swept are sensitive to noise in the band 100 to 300 Hz, and are suspected to be set to be actuated when the level of noise in this frequency band is 40 dB above the spectrum level of the ambient-noise background in coastal waters at a wind speed of 30 to 40 knots. If spherical spreading describes the transmission loss, at what range will the minesweeper sweep (actuate) these mines?

SOLUTION: Solving the passive equation for TL, we have

$$TL = SL - NL + DI - DT$$

Since the spectrum of the broadband source is 150 dB, the level in the sensitive frequency band of the mines is SL = 150 + 10 log 200 = 173; by Fig. 7.8, NL = +84; DI = 0 is implied by the nature of the problem; DT = 40 is given. Therefore, TL = 49. For spherical spreading, this corresponds to a swept range of 280 yd.

Depth Sounding

PROBLEM: A fathometer transducer is mounted on the keel of a destroyer and is pointed vertically downward. It has a DI of 15 dB with a source level of 200 dB at a frequency of 12 kHz. Assuming that reflection takes place at the sea bottom with a reflection loss of 20 dB, at what speed of the destroyer will the echo from the bottom in 15,000 ft of water be equal to the self-noise level of the ship in the 500-Hz receiving bandwidth of the fathometer receiver?

SOLUTION: Because reflection at the sea bottom has been postulated, the actual source can be replaced by an image source in the bottom at a range equal to twice the water depth. The transmission loss then will be

$$TL = 20 \log 2d + 2\alpha d \times 10^{-3} + 20$$

where d is the water depth in yards. With d = 5,000 yd and α taken at 1 dB/kyd, TL = 110 dB. The appropriate form of the sonar equation is

$$SL - TL = NL + 10 \log w - DI$$

where NL + 10 log w is the noise level in the bandwidth w of the receiver. Solving for the unknown parameter.

$$NL = SL - TL - 10 \log w + DI$$

With SL = 200 dB (given), 10 log w = 10 log 500 = 27 dB, DI = 15 dB, we find NL = 78 dB at 12 kHz. This would correspond at 25 kHz to a value of NL = 78 − 20 log (25/12) = 72 dB. Referring to Fig. 11.13, the ship speed at which the 25-kHz isotropic self-noise level is 72 dB is 25 knots.

Mine Hunting

PROBLEM: A mine of average aspect lies on a sand bottom. It is desired to detect the mine at a slant range of 100 yd by means of an active sonar located 20 yd from the bottom. If a pulse length of 10 ms is used, what horizontal beam width will be required if detection can be achieved at a detection threshold of zero decibels?

SOLUTION: The sonar equations for a reverberation background are

$$SL - 2TL + TS = RL + DT$$
$$RL = SL - 2TL + S_s + 10 \log A$$
$$A = \Phi r \frac{ct}{2}$$

Solving for A and eliminating common terms from the first two expressions, we obtain

$$10 \log A = TS - S_s - DT$$

By Table 9.3, we estimate $TS = -17$ dB; by Fig. 8.27 and estimating for a grazing angle equal to $\sin^{-1}(20/100) = 12°$, $S_s = -37$ dB; DT is given as zero. Therefore $10 \log A = 20$ dB and $A = 100$ yd². Solving the third equation for Φ, with $A = 100$, $r = 100$, and $ct/2 = 1{,}600 \times 0.01/2 = 8$ yd, we find $\Phi = 1/8$ rad $= 7.2°$. By Table 8.1, this would require a horizontal line transducer 11 wavelengths long.

Explosive Echo Ranging

PROBLEM: A 1-lb charge is used as a sound source for echo ranging on a submarine. Find the detection range of a bow-stern aspect submarine target in a background of deep-sea ambient noise in sea state 6. Detection is required 90 percent of the time with a 0.01 percent chance of a false alarm in the echo duration of 0.1 second. A nondirectional hydrophone with a 1-kHz bandwidth centered at 5 kHz is used for reception. Let the source and receiver depths be 50 ft in a mixed layer 100 ft thick, and let the target depth be 500 ft.

SOLUTION: For short transient sources, the source level is

$$SL = 10 \log E - 10 \log t_e$$

where E = source level in terms of energy density
t_e = echo duration

Solving the active sonar equation for TL, we obtain

$$TL = \tfrac{1}{2}(10 \log E - 10 \log t_e + TS - NL + DI - DT)$$

The quantities t_e and DI are given in the problem statement. Since the source is broadband, and using Fig. 4.19 at 5 kHz, $E = 180 + 10 \log 1{,}000 = 210$ dB in the 1-kHz receiver bandwidth; by Table 9.3, $TS = 10$ dB; by Fig. 7.5, $NL = 57$ dB; by formula, $DT = 5 \log (dw/t)$, using $d = 25$ (Fig. 12.7), $w = 1{,}000$ and $t = 0.1$, $DT = 27$. $DI = 0$ (given). With these values TL is found to be 73 dB. By Fig. 6.6c, the range is 2,600 yd at 2 kHz; by Fig. 6.7c, the range is 2,200 yd at 8 kHz; on interpolating for 5 kHz, the estimated range becomes 2,400 yd. However, it should be remarked that in this problem the range is likely to be reverberation-limited instead of noise-limited.

Torpedo Homing

PROBLEM: In an active homing torpedo, a detection range of 3,000 yd is required on an average-aspect submarine. A detection threshold of 30 dB is needed to cause the torpedo to "home" on its target. If the torpedo transducer is a plane-piston array restricted to a diameter of 15 in., how much acoustic power output is needed at an operating frequency of 40 kHz? The transmission loss is assumed to be adequately described by spherical spreading and absorption at a temperature of 60°F, and the self-noise is to be taken equal to the ambient noise of the deep sea in sea state 6.

SOLUTION: Solving the active-sonar equation for SL, we obtain

$$SL = 2TL - TS + NL - DI + DT$$

From Fig. 5.8, TL = 95 dB; by Table 9.3, TS = 15 dB; by Fig. 7.5, NL = 41 dB; by Fig. 3.6, DI = 30 dB; DT = 30, given. We therefore find SL = 216 dB, and by Fig. 4.4, with $DI_T = 30$, the required power output is found to be 30 watts.

Fish Finding

PROBLEM: A compact school of fish containing 1,000 members, each averaging 20 in. in length, lies 100 yd from a fishing boat equipped with a fish-finding sonar. What will be the level of the echo from this school of fish at a frequency of 60 kHz, assuming that the transducer has a beam pattern broad enough to contain the entire school? The sonar projector radiates 100 acoustic watts of power and is a circular plane array 10 in. in diameter.

SOLUTION: The echo level is the left-hand side of the active-sonar equation and is equal to SL − 2TL + TS. By Fig. 3.6, DI = 30 dB; by Fig. 4.4, SL = 221 dB; with spherical spreading and absorption, using $\alpha = 19$ dB/kyd (Fig. 5.5), TL = 42 dB; by Fig. 9.19, TS = −31 for a single fish 20 in. long; and for 1,000 fish, TS = −31 + 10 log 1,000 = −1 dB. The echo level becomes 136 dB re 1 μPa. If the transducer has a receiving sensitivity of −170 dB, the echo would appear as a voltage equal to 136 − 170 = −34 dB re 1 volt across the transducer terminals.

Communication

PROBLEM: In the sofar method of aviation rescue, a downed aviator drops a 4-lb explosive charge set to detonate on the axis of the deep sound channel. How far away can the detonation by heard by a nondirectional hydrophone, also located on the axis of the deep sound channel, at a location of moderate shipping in sea state 3? The receiving system uses a frequency band centered at 150 Hz and squares and integrates the received signals for an interval of 2 seconds—an interval estimated to be sufficiently long to accommodate all the energy of the signal. A signal-to-noise ratio of 10 dB is required for detection.

SOLUTION: Solving the passive equation for TL, we obtain TL = SL − NL + DI − DT. Recognizing the existence of severe signal distortion, we convert to energy-density and obtain TL = 10 log E_0 − (NL + 10 log t) + DI − DT, where E_0 is the source energy-density and t is the integration time. From Fig. 4.19, 10 log E_0 for a 4-lb charge at 150 Hz = 207 dB; by Fig. 7.5, NL = 68; DI = 0 dB, DT = 10 dB, and 10 log t = 3 dB are given in the problem statement. Therefore, TL = 126 dB. To convert to range, we write (Sec. 6.2) TL = 10 log r + 10 log r_0 + $\alpha r \times 10^{-3}$. Assume that $r_0 = 10,000$ yd. Using the formula (Sec. 5.3) $\alpha = 0.1 f^2/(1 + f^2)$, where f is in kilohertz, we find that $\alpha = 0.00225$ dB/kyd. Drawing a curve of TL against r, we read off, for TL = 126, the value of r = 8,000 kyd, or 4,000 miles.

An Echo Repeater

PROBLEM: It is desired to build an echo repeater which when suitably triggered will return a simulated echo to a range of 1,000 yd equal in level to the echo from a beam-aspect submarine at the same range. The echo that it must simulate is obtained with a sonar having a source level of 210 dB re 1 μPa. How much acoustic power should it radiate? How much electric power will be needed to drive it if its projector has an efficiency of 50 percent? How much power should it radiate at 100 yd? Assume spherical spreading plus absorption at the rate of 3 dB/kyd.

SOLUTION: The echo level is EL = SL − 2(TL) + TS = 210 − 2(20 log 1,000 + 3) + 25 = 109 dB re 1 μPa, where 25 is the target strength of the submarine (Table 9.3). The simulated echo level is SL' − TL = SL' − (20 log 1,000 + 3) = SL' − 63. Equating the two levels, we find SL' = 172. By the relation SL' = 171.5 + 10 log P + DI$_T$, we find 10 log P = ½; hence, P = 1 watt, if DI$_T$ = 0. At 50 percent efficiency, 2 electric watts will be needed to drive it. At 100 yd, the acoustic power rises to 220 watts! *Note:* A practical echo repeater would simulate much more than the level of the echo; its echoes would have a doppler shift and other realistic echo characteristics.

13.5 Concluding Remarks

A few words of caution must be said regarding the "pat" solutions of the problems just given. Everything depends upon the values of parameters assumed in their solution. These values are always accompanied by uncertainties arising from two sources: first, uncertainty that the conditions assumed are really those of actual interest and importance; and second, uncertainty that, under these conditions, the chosen values of the parameters are valid.

The first of these two sources of uncertainty involves the specification of the conditions, some natural, some of human origin, that the engineer feels will be representative of the environment and the target in and against which the system must operate. Here extreme cases will often need to be worked out in the hope that conditions beyond the selected limits will not be of practical significance. The second uncertainty arises from the presently crude state of underwater sound as a body of quantitative knowledge. Even when all the necessary nature and target conditions are specified, the associated acoustic parameter is likely to be uncertain by several decibels or more, simply because of insufficient quantitative information.

REFERENCES

1. Stewart, J. L., E. C. Westerfield, and M. K. Brandon: Optimum Frequencies for Sonar Detection, *J. Acoust. Soc. Am.*, **33**:1216 (1961).
2. Horton, J. W.: Fundamentals of Sonar, *U.S. Nav. Inst.*, art. 7C-3, 1957.
3. Stewart, J. L., E. C. Westerfield, and M. K. Brandon: Optimum Frequencies for Noise Limited Active Sonar Detection, *J. Acoust. Soc. Am.*, **70**:1336 (1981).

index

Absorption of sound in the sea, 103–110
  causes of, 104
  depth variation, 108
  frequency variation, 105
  measurement methods, 103
Absorption cross section, 253
Acoustic axis, 54
Active sonar, 1, 6
Adaptive beam forming, 64
Afternoon effect, 5, 118
Alberich (nonreflecting coating for submarines), 6, 321
Ambient noise, 202–236
  amplitude distribution, 223
  coherence, 231
  depth effect, 221
  directionality, 227
  ice-covered waters, 224
  at infrasonic frequencies, 206, 211
  intermittent sources, 216–221
  shallow-water, 211–215
  sources of, in deep water, 203–209, 216
  spatial coherence, 231
  spectra, 210–214
  variability of, 215
Ambient-noise levels:
  in coastal waters, 212
  in deep water, 209
  defined, 202
  in ice-covered waters, 224
  ship traffic in, 207
Anomalous depth effect, 337
A/N-SQS-26 sonar, 8, 369
A/N-SQS-35 sonar, 8
Arctic:
  ambient noise in ice-covered waters, 224
  transmission loss in, 168–172

Arctic (*Cont.*):
  under-ice reverberation, 280
Arctic propagation, 169–172
Array gain, 33–41
  defined, 34
  dependence on coherence, 35
  and directivity index, 42–43
Arrays, 54–60
  advantages of, 32
  multiplicative, 65–68
  steering of, 54
  synthetic aperture, 58
Asdic, 4
Attenuation, 100
  causes of, 102
  curves, 108–109
  loss due to, 100
Auditory detection, 396–403
AUTEC (Atlantic Undersea Test and Evaluation Center), 331

Backscattering cross section, 18, 239
Band level, 14, 330
Basic concepts, 11–14
Bathythermograph:
  airborne, 114
  description of, 114
  development of, 5
  expendable (XBT), 114
Beam forming, adoptive, 64
Beam patterns, 54–60
  of ambient noise, 229
  of line and plane arrays, 58
  nomogram for, 61
  of shaded arrays, 60

417

Beam width of arrays, 61
  for reverberation, 242
Binomial shading, 62
Biological sources:
  of ambient noise, 216–219
  of reverberation, 255
Bistatic sonar, 21, 239, 292
Blade-rate lines, model for noise, 348
Boric acid, 107
Bubble pulses, 87
Bubbles in the sea, 249–255
  attenuation by, 254
  cross sections of, 253
  damping of, 252
  resonance, 253
  sound velocity, 251
Butterfly pattern, 311

Cable strumming, 370
Calibration of transducers, 44–53
  comparison method, 46
  other methods, 45–46
  reciprocity method, 47–50
Caustics in the sea, 128, 164
Cavitation:
  on domes, 369
  noise generation by, 334–339
  of projectors, 76–80
  suppression by depth, 337
Cavitation index, 338
Cavitation threshold, 76
Channeling of sound:
  in Arctic, 169
  in deep sea, 147–168
  by mixed layer, 147–158
  in shallow water, 172–182
Chebyshev polynomials, 66
Clipping loss, 394
Coherence:
  of ambient noise, 231
  in array design, 35–41
  of isotropic noise, 38
  measurements of, 41
  of reverberation, 282
  of transmitted sound, 193
  of turbulent pressures, 363
CONGRATS (ray-trace program), 126
Continuous spectrum, 14, 329, 333
Convergence gain, 100, 164
Convergence zones, 159, 164
Correlation:
  of ambient noise, 231
  in flow noise, 363
  of reverberation, 282
  of signals and noise, 35–37

Correlation (*Cont.*):
  of temperature in the sea, 184
  of transmitted sound, 193
Correlation loss, 29
Correlators:
  in arrays, 65
  for detection, 384
Critical angle, 137
Critical bands, 398–400
Critical speed, 335
Cross section:
  absorption, 253
  extinction, 252–255
  for reverberation, 239
  scattering, 239
  for target strength, 294
Cutoff frequency:
  in the mixed layer, 151
  in shallow water, 175
Cylindrical spreading, 102

Decibel, 14
Deep scattering layer (DSL), 255–261
  characteristics of, 261
  effect on shadow zones, 136
  frequency effects, 257
  latitude variation, 261
  migration, 256
  organisms in, 255
  resonance effects, 257
Deep-sea paths and losses, transmission, 194–197
Deep sound channel, 159–168
  convergence zones in, 164
  ray paths in, 163
  signal characteristics of, 161
  transmission model, 161
Deep-water spectra of ambient noise, 209
Deflagration, 87
Depth sounding, 7
Design of sonars, 18, 407
Detection index, 382
Detection probability, 379
Detection threshold, 377–405
  auditory detection, 396–403
  computation example, 393
  concept, 380–383
  defined, 378
  effect of: averaging, 385
    duration and bandwidth, 392
    fluctuations, 386–389
    multiple signals, 390
    repeated signals, 385
    small bandwidth-time products, 389
  estimating, 390–392

Detection threshold (Cont.):
　formulas for, 384–385
　input signal-to-noise ratio, 383–385
　for reverberation, 395
DICANNE, 64
DIMUS (digital multibeam steering), 67
Directionality:
　of ambient noise, 227
　of arrays, 54–60
　of radiated noise, 338
　of reverberation, 244
Directivity index:
　defined, 42, 66
　limitations of, 42
　nomogram for, 43
　receiving, 42
　for reverberation, 244
　of simple transducers, 43
　transmitting, 71
Dolph-Chebyshev shading, 62
Dolphins, 218
Domes, sonar, 367–370
Doppler shift:
　of echoes, 322
　of reverberation, 284
DSL (see Deep scattering layer)

Echo duration, 27
Echo excess, 23
Echo level, 23, 220
Echoes:
　characteristics of, 322–324
　detection in reverberation, 362
　formation processes of, 317–320
ECR (Eyring-Christensen-Raitt) layer
　　(see Deep scattering layer)
"Eel" (listening device), 4
Efficiency of projectors, 76
Energy flux density:
　defined, 14
　of explosives, 91–95
　in reverberation, 248
　in sonar equations, 25–27
Estimator beam, 64
Explosions and ambient noise, 220
Explosive echo ranging, 10, 414
Explosives as sound sources, 86–97
　advantages of, 95
　energy spectra, 91
　pressure signature, 86–88
Extinction cross section, 253

False alarm probability, 379
Fathometer, 4

Figure of merit, 23, 407
Fish, target strength of, 315–317
Fish finding, 7, 315
Flexible line arrays, 371
Flow noise:
　discrimination against, 363
　as radiated noise, 341
　reduction of, 365
　as self-noise, 360–365
Fluctuation of transmitted sound:
　array effects, 41
　causes of, 183–188
　　deterministic vs. stochastic, 183
　effect on sonar equations, 29
　magnitude of, 185–193
　in shallow water, 188
　statistical model for, 188
　surface reflection as a source of, 129, 187
Frequency, optimum, 408–411

GHG (array listening equipment) sonar, 6
GRASS (ray-trace program), 126

History of sonar, 2–6
Homing torpedoes, 10
Hull drone, 333
Hydrodynamic noise:
　as flow noise, 359
　as radiated noise, 332, 341
Hydrophone, 1, 31

Ice edge noise, 227
Icebergs as noise sources, 226
Image interference, 131
Impedance, specific acoustic, 12
Intensity:
　defined, 13
　in transmission loss, 99
Interaction effects, 80
Internal sound channels, 169
Internal waves, 157
Intromission, angle of, 137
Ionic relaxation, 106

JP sonar, 5

Knudsen spectra, 210

Lambert's law, 278

Layered reverberation, 247
Leakage coefficient, 148, 153
Leonardo's air tube, 2
Level of a sound wave, 14
Limiting ray, 153
Line arrays, 55–60
  continuous, 58–60
  of equally spaced elements, 55–57
  nomogram for, 61
Line components in noise, 329
Lloyd mirror effect, 131

Machinery noise:
  as radiated noise, 332
  as self-noise, 357
Magnetostriction, 31, 73
Matched filter, 384
Micropascal, 13
Microstructure of the sea, 183–187
Military uses, 7–12
Mills Cross, 62
Mine-hunting sonar systems, 10
Mines:
  acoustic, 8
  pressure, 8
  target strength of, 314
Minesweeping, 8, 413
Mixed-layer channeling, 147–158
  depth variation, 149
  leakage out of, 153
  model for, 153
  occurrence of, 117, 158
  transmission loss in, 154–157
Models for propagation, 126
Monostatic sonar, 21, 292
Multipaths in the sea, 193
Multiplicative arrays, 65–68
  advantages of, 67–68
  types of, 65
MV device, 4

Near-field effects, 72
Noise (see Ambient noise; Flow noise; Radiated noise; Rain noise; Self-noise)
Noise level:
  of ambient noise, 202
  of self-noise, 354
  in sonar equations, 20
Noise-limited range, 24
Nonlinear effects, 80
  in ambient noise, 206
  in sonar, 80–84
Nonmilitary uses, 7
Normal-mode theory, 122, 174

Open-circuit response, 44
Optimum frequency, 408–411

Parametric sonar, 83–86
Passive sonar, 1, 6, 412
Performance figure, 23
Period meter, 284
Piezoelectricity, 31, 73
Porosity and bottom properties, 138
Postwar developments, 6
Prediction of performance, 16, 407
Processing gain, 379
Product theorem, 60–62
Projectors, 31
  beam patterns of, 72
  cavitation limitations, 76–80
  efficiency of, 76
  interaction effects, 80
  source level of, 73
Propagation theory, 120–125
Propeller, singing, 339
Propeller beats, 339
Propeller noise, 334–340
Pyroelectric effect, 204

QC sonar, 5

Radiated noise, 328–352
  acoustic power, total, 343
  levels of, 345–352
  measurement, methods of, 329–332
  noise spectra, 329
  source level, 328
  sources of: hydrodynamic noise, 341
    machinery noise, 332–334
    propeller noise, 334–340
    summary of, 341–343
Rain noise, 219
Ray diagrams:
  examples of, 135, 136, 148, 161
  ray theory in, 125
  transmission loss from, 128
Ray theory:
  compared with normal mode theory, 122
  for shallow water, 173
Ray tracing, 125–128
Rayleigh distribution, 282
Rayleigh parameter, 129
RBR (refracted bottom-reflected) rays, 164
Receiver-operating-characteristic (ROC) curves, 381–383
Reciprocity:
  calibration, 51–53

Reciprocity *(Cont.)*:
 parameter, 48–50
 principle of, 47
Recognition differential, 21, 398
Reference level conversion table, 15
Reflection:
 from sea bottom, 137–143
 from sea surface, 129
 from targets, 295–298, 317
Reflection loss:
 defined, 128
 of sea bottom, 137–143
 of sea surface, 128
Refracted bottom-reflected (RBR) rays, 164
Refracted surface-reflected (RSR) rays, 164
Reliable acoustic path, 195
Resonance:
 in the deep scattering layer, 255
 in noise generation, 333, 341
 of targets, 319
Response:
 open-circuit, 44
 receiving, 44
 transmitting-current, 44
Reverberating area, 245
Reverberating volume, 243
Reverberation, 237–290
 air bubbles, 249–255
 bottom, 271–280
 characteristics of, 281–285
 detection of echoes in, 395, 401
 equivalent plane-wave, 240
 frequency spread of, 282–285
 layered, 247
 prediction, 285
 in shallow water, 281
 for short transients, 248
 surface, 244–246, 262–271
 under-ice, 280
 volume, 240–244, 255–261
Reverberation level:
 equivalent plane-wave, defined, 240
 for short transients, 248
 as sonar parameter, 21
Reverberation-limited range, 24
ROC (receiver-operating-characteristic) curves, 381–383
RSR (refracted surface-reflected) rays, 164

Scattering cross section, 253
Scattering strength:
 compared with target strength, 246
 defined, 238
 integrated or columnar, 248
 for surface reverberation, 245

Scattering strength *(Cont.)*:
 and target strength, 246–247
 for volume reverberation, 239
Sea bottom, 136–143
 backscattering from, 271–280
 reflection by, 137–143, 271
 shadow cast by, 135
Sea surface, 128–136
 ambient noise from, 207
 fluctuation caused by, 130, 187
 reflection by, 128–129
 reverberation from, 244–246, 262–271
 scattering by a layer, 247
Seaquakes and volcanoes, 220
Sediments, acoustics of, 137–141
Seismic noise, 204
Self-noise, 354–376
 of cable-suspended hydrophones, 370
 domes, 367–370
 equivalent isotropic levels, 356
 flow noise, 360–365
  reduction of, 365–366
 levels, 372–375
 sources and paths of, 357–360
 of towed sonars, 371
Self-reciprocity, 52
Shading of arrays, 62
Shadow zone:
 of sea bottom, 136
 of sea surface, 135
Shallow water:
 ambient noise, 211–214
 fluctuation of transmission in, 188
 seasonal variability, 117, 179, 183
 seismic sensors in, 182
 transmission loss in, 172–182
 velocity profiles in, 120
Ship traffic in ambient noise, 207
Shock wave, 88
Similarity principle, 89
Slope enhancement effect, 192
Snell's law, 123
Sofar (sound fixing and ranging), 159
Sonar, origin of the word, 7
Sonar equations, 17–29, 372–377
 derivation of, 18–22
 examples of, 411–416
 limitations of, 29
 transient form, 25–28
 uses of, 17
Sonar parameters, 19
Sonobuoys, 10
SORAP (sonar overlay range prediction), 24
Sound channels:
 Arctic, 169
 internal, 169

Sound channels (*Cont.*):
  Mediterranean, 169
  mixed-layer, 147–158
  sofar, 159
Sound spectograph, 343
Sound wave, level of, 15
Source level, 71, 328
  of explosive sources, 86–97
  of projectors, 73
  for radiated noise, 328
  related to power output, 73–76
  for transients, 25–28
Spectograph, sound, 343
Spectrum level, 14
Specular reflection, 295–298, 317
Spheres:
  as reference targets 306–308
  reflection from, 292
  scattering by, 298–300
  target strength of, 293, 298
Spherical spreading, 100
Spreading loss, 100
  (*See also* Cylindrical spreading)
Square-law detector, 385
Steering of arrays, 54
Submarines:
  radiated noise levels of, 349
  self-noise levels of, 374
  target strength of, 308–314
Superdirectivity, 64
Surface duct, 147–158
  noise in, 223
Surface ships:
  radiated-noise levels of, 345–349
  self-noise levels of, 372
  target strength of, 314
Synthetic aperture arrays, 58

TAP (time-average-product) array, 65
Target strength, 291–327
  backscattering as a cause of, 294
  bistatic, 302
  defined, 291
  echo, 291–292
  echo characteristics, 322–324
  echo formation processes, 317–320
  of fish, 315–317
  measurement methods, 306–308
  of mines, 314
  reduction of, 320–322
  of simple forms, 273–278
  of small organisms, 317
  as sonar parameter, 20
  specular reflection, geometry of, 295–298
  of spheres, 293, 295–306

Target strength (*Cont.*):
  of submarines, 281–286, 308–314
  of surface ships, 314
  of torpedoes, 315
Thermal noise, 208
Thermocline, 117
Threshold:
  cavitation, 79
  detection, 380–382
Tides as noise sources, 203
Time stretching, 26
Torpedoes:
  homing, 10
  radiated-noise levels of, 351
  target strength of, 315
Towed sonars, self-noise of, 371
Transducer responses, 44
Transducers, 31
  calibration of (*see* Calibration of transducers)
Transduction, 2
Transition curves, 387
Transition range, 152
Transmission loss, 99
  absorption of sound in sea, 102–110
  arctic propagation, 169
  in bounded media, 102
  deep-sea paths and losses, 194–197
  deep sound channel, 159–163
  defined, 99
  fluctuation, 182–193
  in free field, 100
  internal sound channels, 169
  mixed-layer sound channel, 147–158
  propagation theory, 120–128
  ray tracing, 125–128
  sea bottom sound, 141–143
  sea surface sound, 128–129
  in shallow water, 177–182
  in sonar equations, 20–21
  spreading laws, 100–102
Transmitted sound (*see* Fluctuation of transmitted sound)
Triboelectric noise, 370
Twenty-cycle pulses, 218

Underwater telephone, 10
Units, 14–15
UQC-1 equipment, 10

Variability:
  of ambient noise, 215
  of echoes, 311
  of transmission, 182–193

Velocimeter, 115
Velocity profile:
  of the deep sea, 116–120
  in different oceans, 120
  in shallow water, 120
Velocity of sound, 111–116
  gradients of, 125
  measurement methods, 111–116
  in sea, 116–120
  variation with pressure, temperature, and depth, 114
Volume reverberation:
  causes of, 255
  deep scattering layer in, 255–261
  theory of, 240–244

Wake strength, 262
Wakes, 261
Wave equation, 121
Wave theory:
  for fluctuation, 183–186
  and ray theory, 122
  for shallow water transmission, 174
Waves:
  scattering by, 236–243
  as sources of noise, 203, 207
Wind noise:
  in deep water, 207
  in shallow water, 213

XBT (expendable bathythermograph), 114

About the Author

ROBERT J. URICK, *a consultant in underwater sound, is Adjunct Professor of Mechanical Engineering at the Catholic University of America, Washington, D.C. He was senior research physicist at the U.S. Navy Surface Weapons Center at Silver Spring, Maryland. Over the years, he has contributed to nearly all phases of underwater sound research and has published more than 200 reports and papers on the subject.*